Nucleation of Minerals: Precursors, Intermediates and Their Use in Materials Chemistry

Nucleation of Minerals: Precursors, Intermediates and Their Use in Materials Chemistry

Special Issue Editor

Denis Gebauer

MDPI • Basel • Beijing • Wuhan • Barcelona • Belgrade

MDPI

Special Issue Editor
Denis Gebauer
University of Konstanz
Germany

Editorial Office
MDPI
St. Alban-Anlage 66
Basel, Switzerland

This is a reprint of articles from the Special Issue published online in the open access journal *Minerals* (ISSN 2075-163X) from 2016 to 2018 (available at: http://www.mdpi.com/journal/minerals/special_issues/nucleation_minerals)

For citation purposes, cite each article independently as indicated on the article page online and as indicated below:

LastName, A.A.; LastName, B.B.; LastName, C.C. Article Title. *Journal Name* **Year**, *Article Number*, Page Range.

ISBN 978-3-03897-035-4 (Pbk)
ISBN 978-3-03897-036-1 (PDF)

Cover image courtesy of Julian Opel.

Contents

About the Special Issue Editor

Denis Gebauer, PD Dr., group leader at the University of Konstanz (Germany) since 2011. After completing his Ph.D. thesis at the Max-Planck-Institute of Colloids and Interfaces (Potsdam-Golm, Germany) in July 2008, he stayed at Stockholm University (Department of Materials and Environmental Chemistry, Arrhenius Laboratory, Stockholm, Sweden) as a postdoctoral researcher for two years. His research interest is currently focused on non-classical concepts of nucleation and crystallization, as well as biomineralization and materials chemistry in general.

Preface to "Nucleation of Minerals: Precursors, Intermediates and Their Use in Materials Chemistry"

This Special Issue features 18 articles, providing a cross-section of the current research activities in the field of mineral nucleation and growth. In one editorial, one commentary, two review articles, and 14 original research papers, various authors prove the vibrant and topical nature of this field. We hope that this Special Issue will serve as a resource and inspiration for future studies, and that this selection of papers will be useful for scientists and researchers of all career levels, who work on the exploration of mineral nucleation and growth mechanisms.

Denis Gebauer
Special Issue Editor

Editorial

Editorial for Special Issue "Nucleation of Minerals: Precursors, Intermediates and Their Use in Materials Chemistry"

Denis Gebauer

Department of Chemistry, Physical Chemistry, University of Konstanz, 78457 Konstanz, Germany;
denis.gebauer@uni-konstanz.de; Tel.: +49-(0)7531-88-2169

Received: 31 May 2018; Accepted: 2 June 2018; Published: 4 June 2018

Nucleation is the key event in mineralization, but a general molecular understanding of phase separation mechanisms is still missing, despite more than 100 years of research in this field [1]. In the recent years, many studies have highlighted the occurrence of precursors and intermediates, which seem to challenge the assumptions underlying classical theories of nucleation and growth. This is especially true for the field of biomineralization, where bio-inspired strategies take advantage of the precursors' and intermediates' special properties for the generation of advanced materials. All of this has led to the development of "non-classical" frameworks, which, however, often lack quantitative expressions for the evaluation and prediction of phase separation, growth and ripening processes, and are under considerable debate. It is, thus, evident that there is a crucial need for research into the early stages of mineral nucleation and growth, designed for the testing, refinement and expansion of the different existing notions. This special issue of *Minerals* aimed to bring together corresponding studies from all these areas, dealing with precursors and intermediates in mineralization with the hope that it may contribute to the achievement of a better understanding of nucleation precursors and intermediates, and their target-oriented use in materials chemistry.

In his commentary, Evans [2] summarises different existing nucleation theories and discusses them from the point of view of biomineralization. The focus lies on proteins and their role in mineral precursor formation, stabilisation, and assembly into crystalline polymorphs. It is stressed that a limitation of the advancement of the understanding of protein-controlled mineralization processes [3] is, at least, partly due to variations in techniques, methodologies and the lack of standardisation in mineral assay experimentation. Evans argues that the protein community should adopt standardized nucleation assays [4,5], allowing for cross-comparisons and kinetic observations. Burgos-Cara et al. [6] use such an experimental approach for studying the effects of background ionic species on the formation and stability of $CaCO_3$ pre-nucleation species in aqueous solutions. They find that the effective critical supersaturation in the presence of background ions with a decreasing ionic radius becomes systematically higher, and propose that the stabilisation of hydration water molecules impedes dehydration processes, which are essential steps during mineral precipitation, according to the notions of the so-called pre-nucleation cluster pathway [1].

Several papers of this *Minerals* special issue address further methodological aspects associated with research into mineral nucleation and growth. The contribution of Kröger and Verch [7] deals with studies employing liquid cell transmission electron microscopy (LCTEM). 2D finite element simulations highlight that the confinement, which occurs in typical LCTEM cells and can significantly reduce the concentration of available ions, can explain the necessity to substantially increase the supersaturation in LCTEM cells in order to induce precipitation. Zeng et al. [8] focus on another important microscopic technique for studying nucleation and self-assembly, in situ atomic force microscopy (AFM). They review the latest contributions in this field and also address the theoretical background of AFM. Kuwahara et al. [9] use AFM for exploring the growth behaviour and kinetics of the barite (001) surface in supersaturated solutions, predicting a critical supersaturation, at which a 2D nucleation growth mechanism becomes important, significantly altering the crystal morphology. Harris and Wolf [10] focus on an experimental methodological issue, demonstrating that the desiccator size in commonly used vapor diffusion-based crystallisation assays can alter the crystallisation mechanism. This shows that a careful experimental design is required in order to identify and explore additive effects when this method is applied. Gebauer et al. [11] present a thorough investigation of a urinary stone of a guinea pig. The suggested role of amorphous calcium carbonate in pathological mineralization highlights that future studies of such stones should not be based on analytical techniques that are sensitive only to crystalline $CaCO_3$. Liquid [12] and amorphous mineral precursors can, in principle, be used to fill cavities, which may occur in historical artefacts or dental lesions. Gruber et al. [13] present a transparent, inexpensive, and reusable test system for the investigation of infiltration and crystallization processes by using a micro-comb test system. Ossorio et al. [14] perform synchrotron-based small- and wide-angle X-ray scattering to examine the precipitation of gypsum from solution, which occur via primary particles that aggregate and transform/re-organize towards the final precipitate, with and without the addition of Mg^{2+} and citrate. Self-assembly processes of alkaline earth carbonates in the presence of silica, on the other hand, produce a unique class of composite materials with complex morphologies, as studied by Opel et al. [15] (also see the cover page of this *Minerals* special issue). Trumpet- and coral-like structures form especially at an elevated temperature and in the presence of additional ions.

Several papers of this *Minerals* special issue deal with calcium phosphate—the most important biomineral for the human race. Pastero et al. [16] provide a review of the genetic mechanisms of apatite, paying close attention to the structural complexity of hydroxyapatite, the richness of its surfaces and their role in interactions with precursor phases regarding growth kinetics and morphology. Ibsen et al. [17] study the impact of pyrophosphate on apatite formation by synchrotron-based in situ X-ray diffraction, revealing a strong inhibition of apatite nucleation and growth. Ross et al. [18] investigate the possibility of precipitating carbonate apatite from municipal wastewater treatment plants, where a suggestive amorphous precursor transforms without changing morphology.

The transformation of aragonite into calcite by solid-state transformation is explored by Kezuka et al. [19], yielding fascinating, single-crystalline calcite needle-like particles with zigzag surface structures. Jones [20] studies the formation of jarosite–alunite solid solutions, finding a new spherical morphology of pure alunite. Notably, a distinct nucleation behaviour for jarosite and Fe-containing alunite is found, as the latter nucleates continuously rather than in a single event. Ochiai and Utsunomiya [21] investigate the crystal chemical properties of hydrous rare-earth phosphates, forming at an ambient temperature with fractions of an amorphous component with an increasing ionic radius. Finally, Bacsik et al. [22] demonstrate that amine–CO_2 chemistry, which is important for carbon dioxide capture, can be used to prepare amorphous calcium carbonate.

We hope that this special issue will contribute to a better understanding of nucleation and growth phenomena, and will serve as a resource and inspiration for future studies in this vibrant and topical field of research.

Acknowledgments: D.G. is a Research Fellow of the Zukunftskolleg of the University of Konstanz. We thank Julian Opel for providing the image for the cover page of this special issue.

Conflicts of Interest: The author declares no conflicts of interest.

References

1. Gebauer, D.; Kellermeier, M.; Gale, J.D.; Bergström, L.; Cölfen, H. Pre-nucleation clusters as solute precursors in crystallisation. *Chem. Soc. Rev.* **2014**, *43*, 2348–2371. [CrossRef] [PubMed]
2. Evans, J. Polymorphs, Proteins, and Nucleation Theory: A Critical Analysis. *Minerals* **2017**, *7*, 62. [CrossRef]
3. Gebauer, D. How Can Additives Control the Early Stages of Mineralisation? *Minerals* **2018**, *8*, 179. [CrossRef]

4. Habraken, W.J.E.M. The Integration of Ion Potentiometric Measurements with Chemical, Structural, and Morphological Analysis to Follow Mineralization Reactions in Solution. *Methods Enzymol.* **2013**, *532*, 25–44. [PubMed]

5. Kellermeier, M.; Cölfen, H.; Gebauer, D. Investigating the Early Stages of Mineral Precipitation by Potentiometric Titration and Analytical Ultracentrifugation. *Methods Enzymol.* **2013**, *532*, 45–69. [PubMed]

6. Burgos-Cara, A.; Putnis, C.; Rodriguez-Navarro, C.; Ruiz-Agudo, E. Hydration Effects on the Stability of Calcium Carbonate Pre-Nucleation Species. *Minerals* **2017**, *7*, 126. [CrossRef]

7. Kröger, R. Andreas Verch Liquid Cell Transmission Electron Microscopy and the Impact of Confinement on the Precipitation from Supersaturated Solutions. *Minerals* **2018**, *8*, 21. [CrossRef]

8. Zeng, C.; Vitale-Sullivan, C.; Ma, X. In Situ Atomic Force Microscopy Studies on Nucleation and Self-Assembly of Biogenic and Bio-Inspired Materials. *Minerals* **2017**, *7*, 158. [CrossRef]

9. Kuwahara, Y.; Liu, W.; Makio, M.; Otsuka, K. In Situ AFM Study of Crystal Growth on a Barite (001) Surface in BaSO4 Solutions at 30 °C. *Minerals* **2016**, *6*, 117. [CrossRef]

10. Harris, J.; Wolf, S. Desiccator Volume: A Vital Yet Ignored Parameter in $CaCO_3$ Crystallization by the Ammonium Carbonate Diffusion Method. *Minerals* **2017**, *7*, 122. [CrossRef]

11. Gebauer, D.; Jansson, K.; Oliveberg, M.; Hedin, N. Indications that Amorphous Calcium Carbonates Occur in Pathological Mineralisation—A Urinary Stone from a Guinea Pig. *Minerals* **2018**, *8*, 84. [CrossRef]

12. Gower, L.B. Biomimetic model systems for investigating the amorphous precursor pathway and its role in biomineralization. *Chem. Rev.* **2008**, *108*, 4551–4627. [CrossRef] [PubMed]

13. Gruber, D.; Wolf, S.; Hoyt, A.-L.; Konsek, J.; Cölfen, H. A Micro-Comb Test System for In Situ Investigation of Infiltration and Crystallization Processes. *Minerals* **2017**, *7*, 187. [CrossRef]

14. Ossorio, M.; Stawski, T.; Rodríguez-Blanco, J.; Sleutel, M.; García-Ruiz, J.; Benning, L.; Van Driessche, A. Physicochemical and Additive Controls on the Multistep Precipitation Pathway of Gypsum. *Minerals* **2017**, *7*, 140. [CrossRef]

15. Opel, J.; Kellermeier, M.; Sickinger, A.; Morales, J.; Cölfen, H.; García-Ruiz, J.M. Structural Transition of Inorganic Silica–Carbonate Composites Towards Curved Lifelike Morphologies. *Minerals* **2018**, *8*, 75. [CrossRef]

16. Pastero, L.; Bruno, M.; Aquilano, D. About the Genetic Mechanisms of Apatites: A Survey on the Methodological Approaches. *Minerals* **2017**, *7*, 139. [CrossRef]

17. Ibsen, C.J.S.; Birkedal, H. Pyrophosphate-Inhibition of Apatite Formation Studied by In Situ X-ray Diffraction. *Minerals* **2018**, *8*, 65. [CrossRef]

18. Ross, J.; Gao, L.; Meouch, O.; Anthony, E.; Sutarwala, D.; Mamo, H.; Omelon, S. Carbonate Apatite Precipitation from Synthetic Municipal Wastewater. *Minerals* **2017**, *7*, 129. [CrossRef]

19. Kezuka, Y.; Kawai, K.; Eguchi, K.; Tajika, M. Fabrication of Single-Crystalline Calcite Needle-Like Particles Using the Aragonite–Calcite Phase Transition. *Minerals* **2017**, *7*, 133. [CrossRef]

20. Jones, F. Crystallization of Jarosite with Variable Al^{3+} Content: The Transition to Alunite. *Minerals* **2017**, *7*, 90. [CrossRef]

21. Ochiai, A.; Utsunomiya, S. Crystal Chemistry and Stability of Hydrated Rare-Earth Phosphates Formed at Room Temperature. *Minerals* **2017**, *7*, 84. [CrossRef]

22. Bacsik, Z.; Zhang, P.; Hedin, N. Ammonium-Carbamate-Rich Organogels for the Preparation of Amorphous Calcium Carbonates. *Minerals* **2017**, *7*, 110. [CrossRef]

minerals

MDPI

Commentary

Polymorphs, Proteins, and Nucleation Theory: A Critical Analysis

John Spencer Evans

Department of Basic Sciences, Skeletal Biology and Craniofacial Medicine, New York University, 345 E. 24th Street, New York, NY 10010, USA; jse1@nyu.edu; Tel.: +1-212-998-9605

Academic Editor: Denis Gebauer
Received: 1 April 2017; Accepted: 18 April 2017; Published: 21 April 2017

Abstract: Over the last eight years new theories regarding nucleation, crystal growth, and polymorphism have emerged. Many of these theories were developed in response to observations in nature, where classical nucleation theory failed to account for amorphous mineral precursors, phases, and particle assembly processes that are responsible for the formation of invertebrate mineralized skeletal elements, such as the mollusk shell nacre layer (aragonite polymorph) and the sea urchin spicule (calcite polymorph). Here, we summarize these existing nucleation theories and place them within the context of what we know about biomineralization proteins, which are likely participants in the management of mineral precursor formation, stabilization, and assembly into polymorphs. With few exceptions, much of the protein literature confirms that polymorph-specific proteins, such as those from mollusk shell nacre aragonite, can promote polymorph formation. However, past studies fail to provide important mechanistic insights into this process, owing to variations in techniques, methodologies, and the lack of standardization in mineral assay experimentation. We propose that the way forward past this roadblock is for the protein community to adopt standardized nucleation assays and approaches that are compatible with current and emerging nucleation precursor studies. This will allow cross-comparisons, kinetic observations, and hopefully provide the information that will explain how proteins manage polymorph formation and stabilization.

Keywords: polymorphs; sea urchin; mollusk; calcite; vaterite; aragonite; biomineralization; proteins; classical nucleation theory; crystallization by particle attachment; non-classical nucleation

1. Introduction

In nature, invertebrate organisms primarily utilize calcium carbonates as the building materials for extracellular skeletal elements such as the mollusk shell [1–4] sea urchin spicules [5–11] and corals [12–14]. Biological calcium carbonates can exist in an amorphous state (amorphous calcium carbonate, ACC) [9–11,15–19] and as three different crystalline polymorphs: calcite, aragonite, and vaterite, where calcite is the more stable form and aragonite and vaterite are metastable relative to calcite [15,16]. Polymorphism, or the ability of a solid to exist in more than one lattice structure [15,16,20], is a feature of both naturally-occurring and artificial compounds [15–24]. Thus, the study of polymorphism in invertebrate skeletal elements can provide important insights into polymorph formation and stabilization over a wide range of conditions.

The attraction of biological calcium carbonate polymorphism to the scientific community is that this process occurs largely under ambient conditions [1–19] and thus represents a "game changer" for materials science and chemistry communities who have utilized non-ambient and sometimes extreme conditions to generate a given polymorph. What is often overlooked in the discussion of biological calcium carbonate polymorphism is the role and identity of extrinsic agents in the formation and stabilization processes. For example, the literature shows that extrinsic agents such as alcohols [21], gels [22,23], and Mg(II) ions [24] can foster metastable aragonite formation instead of calcite in vitro. Thus, we might assume that in biological organisms, polymorph stabilizing or promoting agents may exist as well. Some examples would include Mg(II), which is present in a 5:1 ratio relative to Ca(II) in seawater [24], or, small molecule metabolic end products, such as phosphate, which have been shown to stable ACC [25].

We must, however, also acknowledge that invertebrate skeletal elements are composite structures consisting of the mineral phase(s) and macromolecular components [1–14]. Here, the mineral phase nucleates and matures within this matrix. Therefore the potential role of matrix macromolecules such as proteins looms large in polymorphism. If proteins do play a role in calcium carbonate polymorph selection and stabilization, then this process must fall under genomic and proteomic guidance [5–8,13,14,26–28]. Hence, polymorphism could be controlled by the temporal, spatial, and quantitative aspects of protein components within the extracellular space. If true, then the scientific community could avail itself of this tremendous molecular variation to guide a given polymorph to form, possibly with higher efficiency and purity at the endpoint that present capabilities do not allow.

The intent of this review is to "jumpstart" the discussion of protein-mediated polymorphism as it pertains to invertebrate calcium carbonate skeletal elements alone in the absence of additives such as Mg(II). In other words, do proteins possess the inherent capabilities to direct specific polymorph formation without the assistance of other agents? Admittedly, scientific progress and understanding in this area have been slow, primarily due to two factors: first, our understanding of the nucleation process is still undergoing evolution and testing, and within the last few years alternatives to classical nucleation theory have emerged [15,16,29–33], which will be discussed in the next section. Thus, the challenge is to figure out where in these schemes aragonite, calcite, and vaterite-specific proteins might avail themselves to manipulate and regulate the nucleation and polymorph selection process. Second, and perhaps most importantly, genomic and proteomic studies of biomineralization proteins have uncovered a vast repertory of proteins [6–8,13,14,26–28] that could potentially manage many aspects of proposed nucleation processes, including polymorphism. However, although many protein studies have confirmed that polymorphs do form when proteins are present [34–53], these studies have not provided information that is needed for detailing the mechanisms of polymorph formation, or, have utilized a variety of testing methods that prevent cross-comparisons of datasets necessary for mechanism building [34–53]. Here, we believe that the challenge for the protein community will be to adopt consistent mineralization assay standards and approaches that are compatible with current and emergent nucleation precursor studies [15,16,29–33]. These studies will allow time-dependent, careful monitoring of protein effects on the early events in nucleation and polymorph selection processes that can yield information on mechanisms. Only then will we be able to determine what nucleation schemes are at work in nature, how proteins foster polymorph formation and stabilization, and how to apply this information to materials science.

2. The Current State of Knowledge Regarding Nucleation and Polymorph Formation

In terms of classical nucleation theory, polymorph formation is a process dictated by thermodynamic and kinetics factors, where monomer-by monomer addition to an isolated ordered cluster results in a crystal with a given lattice structure (Figure 1) [15,16,29,54]. In the classical scheme polymorph selection is thus determined by the correct combination of polymorph-specific thermodynamic (e.g., temperature, pressure, volume) and kinetic (i.e., timeline or rate) factors which lower the nucleation energy barrier to crystal formation [54]. Hypothetically, this process could be managed by agents or additives in many ways to lower the energy barrier for nucleation and generate the appropriately structured cluster that would subsequently grow into a specific polymorph via monomer-by-monomer addition [15,16,29,54]. In laboratory practice, however, there is sometimes a lack of efficiency in generating a desired polymorph by simply manipulating these factors, and this leads to the formation of undesired polymorphs as impurities [20].

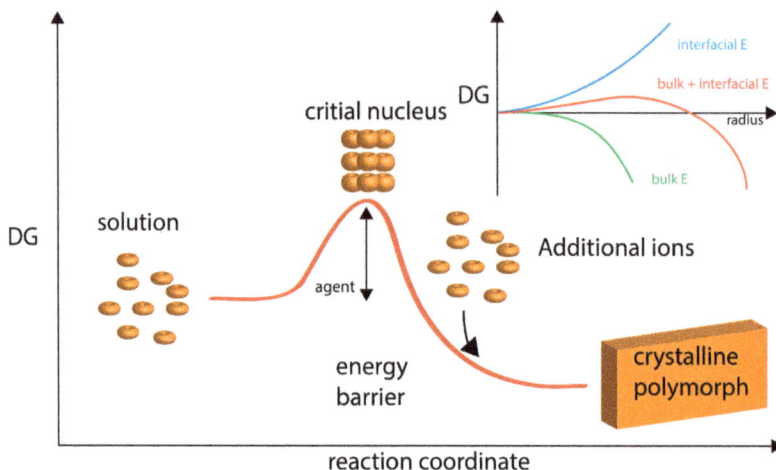

Figure 1. Graphical representation of classical nucleation theory and polymorph formation. Here, ions attain a concentration level (supersaturation) which fosters the formation of a critical nucleus. This nucleus then continues to growth via ion apposition at interfaces. "Energy barrier" refers to the energy of nucleation. "Agent" refers to additives, such as proteins, that would lower the nucleation energy barrier and foster polymorph formation and stabilization. The inset graph describes the formation of nuclei in supersaturated homogeneous solution, which is governed by the balance between the bulk and surface energy of the new phase. The point at which the bulk contribution compensates for the energetic costs arising from the interfacial surface is referred to as the critical size. Thermodynamically, the critical size reflects a metastable state since any infinitesimal change towards either smaller or larger radii will render the system unstable and lead to nucleus dissolution or unlimited growth, respectively. Adapted from references [15,16,54].

As usual, in nature, things are not always so straightforward. One of the most dramatic observations is that prior to the crystallization event, organisms first form an amorphous precursor, ACC, which then transforms into a given crystalline calcium carbonate polymorph [9–11,15–19].

Here, the nucleation of ACC takes place at concentrations lower than those necessitated by classical theory and the monomer addition process [15,16,29,54]. This is compounded by another important observation that is clearly inconsistent with classical nucleation theory [54]: the formation of mineralized tissues, such as mollusk shell nacre [1–4] or sea urchin spicules [5–11] occurs via the assembly of individual mineral nanoparticles into mesoscale structures, rather than by monomer-by-monomer addition. Thus, we must either look outside or amend classical nucleation theory to describe applicable mechanisms that explain biological ACC formation and subsequent crystal growth. In doing so, we stand a good chance of developing new laboratory processes for practical polymorph selection that are more efficient.

Owing to the complexity of the extracellular matrix (ECM) it has been difficult to establish in situ ACC and polymorph formation mechanisms in organisms. As a result, in vitro experiments [16,17,25,29–33] and theoretical modeling [16,55,56] have provided an alternative means of understanding ACC and calcium carbonate crystal formation, and it now appears that nucleation may be explained and understood by alternative nucleation theories. The first of these is the non-classical or pre-nucleation cluster theory, where ACC formation results from the assembly of 1–3 nm ionic Ca^{2+}-CO_3^{2-} clusters, known as pre-nucleation clusters, during the early stages of the nucleation process (Figure 2) [29–33]. What is intriguing about this theory is that the prenucleation cluster formation and assembly processes have been shown to be very sensitive to pH and additives [29–32]. Moreover, it has been proposed that ACC can adopt different proto-structures, i.e., short-range structures that correlate to a given crystalline polymorphs [33]. Thus, mineral cluster proto-structure may influence polymorph selection during the ACC-to-crystal transformation process either via a dissolution-re-precipitation Ostwald mechanism [54] or solid-state transformation-like pathways [15,16,31]. If true, then the selection of a given polymorph could be determined at early stages of mineral cluster formation by the stabilization of a given proto-structure of ACC [33].

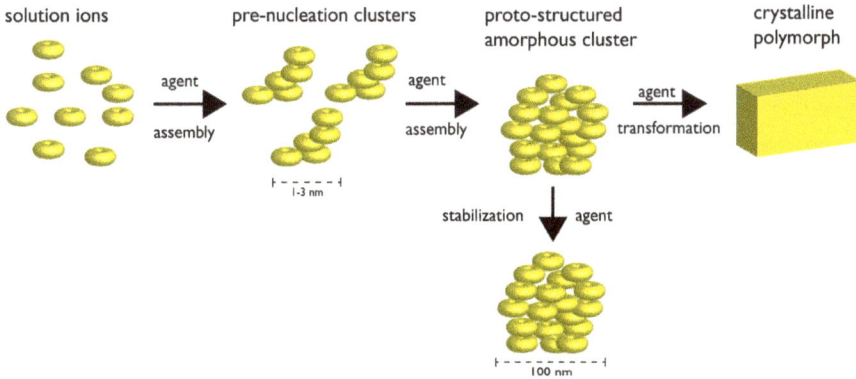

Figure 2. Non-classical nucleation scheme involving pre-nucleation clusters. The formation of pre-nucleation clusters (PNCs) from constituent ions represents the first step towards mineral formation. The PNCs are linear, chain-like ionic polymers (1–3 nm in diameter) that undergo further assembly into amorphous clusters (AC) (~102 nm in diameter). The AC particle can be stabilized indefinitely by agents, or, can proceed through a transformation mechanism (agent-controlled) and form crystalline solids. In the case where different lattice structures of the same given crystalline compound exist (i.e., polymorph), the choice of polymorph may be dictated at the PNC stage and/or by agents controlling the transformation process. Note that each stage of this non-classical process may be under the control of agent(s) (e.g., proteins, polymers, ions, etc.). Note also that certain steps of the non-classical pathway may be thermodynamically reversible. Adapted from references [27–31].

The second, alternative view on nucleation processes is termed crystallization by particle attachment (CPA) [15]. Here, a crystalline or amorphous mineral forms via the addition of higher-order species such as multi-ion complexes, liquid phases, droplets, gels, oligomers, and fully formed nanoparticles (Figure 3) [15]. In addition to a classical monomer addition pathway [54], in CPA there exist multiple particle-based pathways that eventually lead to the final crystal, and, multiple growth mechanisms can occur simultaneously within a single crystallizing system. The process of particle attachment or assembly can be influenced by several factors, including interfacial or surface free energies. It is argued that the surface free energy is an important determinant in polymorph formation pathways for two reasons: (1) This term affects the magnitude of the free energy barrier for nucleating a given polymorph; and (2) The surface free energy impacts particle size, and thus to form a given polymorph one needs to first nucleate particles with a relatively uniform size distribution within the size range in which that polymorph is stable, then assemble these particles to form the final polymorph [15]. Furthermore, surface free energy can be tuned by extrinsic factors such as: (1) the presence of surfaces, which can reduce interfacial free energy; (2) volume confinement, which can stabilize phases; and (3) the presence of agents, such as small molecules or polymers, which can affect the kinetics of particle formation, particle assembly, and phase stabilities [15]. Thus, if we consider the particle attachment process to be representative of polymorph nucleation, then control over polymorph expression would minimally require the manipulation of both particle formation and the subsequent assembly process.

Pathway/Particle Categories

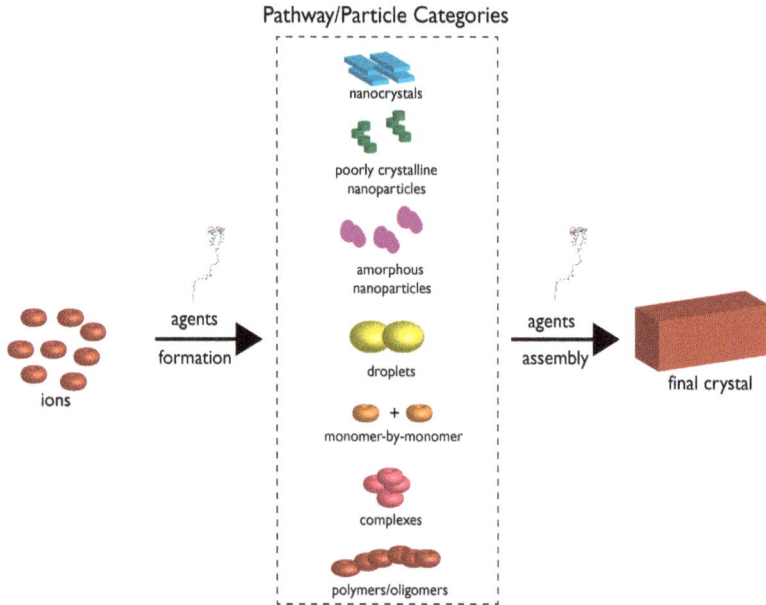

Figure 3. Crystallization by particle attachment (CPA) nucleation scheme. Constituent ions can form several types of particles (described in box region) which can then further assemble into the final crystal polymorph. Thus, the multiple pathways involve ion assembly into a particle type and then further assembly of a given particle type into the crystal. Note that this theory still allows for the classical nucleation scheme which involves monomer-by-monomer addition [54]. Here, formation and assembly pathways are subject to regulation and control by agents or additives. Above the word "agents" we have placed a protein structure to indicate that biomineralization proteins could conceivably perform these proposed assembly tasks. Size of components is not drawn to scale. Adapted from reference [15].

3. How Biomineralization Proteomes Might Fit into Recently Proposed Nucleation Schemes

On the assumption that nature does not employ a purely classical nucleation scheme [54] to form calcium carbonate polymorphs, our next task is to take the alternative views on the nucleation process [15,29–33] and determine what role, if any, proteins might play in the polymorph formation process. From the foregoing, it is clear that non-classical [29–33] and CPA [15] pathways to polymorph formation involve multiple steps and processes. Thus, the most efficient path to form calcite, aragonite, or vaterite would depend upon the action of agent(s) that can manipulate key steps along a specific pathway (Figures 2 and 3). This is where biomineralization proteins may be involved. Organisms can regionally and temporally express numerous proteins that cooperatively and synergistically handle tasks within chemical scenarios. As recent studies have shown, invertebrates such as the pearl oyster and sea urchin spicule genomes encode several hundred mineral-associated protein genes apiece [5–8,26–28]. Granted, we would not expect every member of these proteomes to be directly responsible for nucleation and crystal growth, but two possibilities emerge: (1) It is plausible that specialized proteins may be assigned to cover multiple nucleation and crystal growth pathways within a single skeletal element; (2) the formation of a given polymorph is not entrusted to a single protein but spread over multiple proteins, thus ensuring organism survival if one or more proteins experienced a debilitating sequence mutation.

Regarding the non-classical pre-nucleation cluster scheme, it has been proposed that there are several of steps that could be potentially regulated by additives such as proteins, such that a given polymorph crystal is formed: (1) the kinetics of pre-nucleation cluster formation; (2) the assembly and stabilization of PNCs into proto-structured ACC; (3) the stabilization and transformation of ACC to a specific polymorph [29–33]. Regarding the CPA scheme, we are confronted with a myriad of possibilities: there may exist multiple assembly pathways to crystal polymorph formation, multiple particle species available for assembly, and multiple growth mechanisms running concurrently [15]. Here, it has been suggested that proteins could lower activation energy barriers to particle assembly, act as stabilizers of liquid-like mineral phases, provide volume confinement for nucleation, simultaneously organize and regulate multiple mechanisms of mineral growth, and so on [15]. In short, each nucleation theory invokes ample control points that could be manipulated by proteins to direct specific polymorph formation.

4. Protein-Polymorph Formation and Stabilization—What Do We Currently Know?

Although hypothetical roles and scenarios can be postulated for protein-mediated polymorph formation and stabilization within calcium carbonate nucleation schemes, the more relevant issue is this: what have in vitro and in situ studies revealed about protein participation in the formation of either metastable aragonite or vaterite?

For aragonite, what we must draw from are studies conducted with mollusk shell nacre-associated protein sequences [34–51]. For in vitro-based studies, several different carbonate-based assays have been used to achieve supersaturation conditions necessary for calcium carbonate crystal growth, with no control over pH, and the duration of these nucleation experiments were variable (i.e., from several hours up to 7 days) [36,38,39,41–44,46–51]. Further, some studies utilized Mg(II) ions in the assay mixture to mimic seawater conditions and promote and stabilize aragonite [36,44]. For in situ studies, measurements of aragonite formation were conducted within shells over defined periods of time when certain proteins are expressed or introduced into the shell environment [34,35,37,40,45]. In both sets of studies aragonite formation was confirmed in the presence of multiple nacre proteins [34–51]. In addition, some nacre proteins exhibited the ability to inhibit calcite growth [41,46–48] or induce metastable vaterite formation as well [48,50,51].

In comparison to the aragonite studies, there have been fewer published studies that examine the role of proteins in metastable vaterite formation [48,52,53], most likely because of fewer organisms utilizing this polymorph over calcite or aragonite [52,53,57]. These studies were in vitro in nature and utilized corresponding calcium carbonate mineralization assays as described for aragonite studies [48,52,53]. Again, as per the aragonite studies, these studies confirmed the formation of metastable vaterite crystals in the presence of the proteins under investigation [48,52,53].

But what do these protein-polymorph studies tell us about the mechanism(s) by which proteins control polymorph formation on their own? Unfortunately, very little. The major reason for this is the lack of standardization: even though conditions for studying non-classical nucleation schemes were published nearly 8 years ago [30] there are few examples in the protein literature that utilize these methods of generating ion supersaturation and subsequent nucleation. Further, many studies relied on different ion concentration levels, control conditions, time periods, temperatures, pH values, and most importantly, protein quantities, in their assay experiments [34–53]. Typically, in most cases protein quantities have been reported as mass/volume values [34–45,52,53] rather than mole values [46–51], and this lack of standardization prevents cross-comparisons between protein studies. Thus, with all the published experimental variations, it is almost impossible to quantitatively compare the published nucleation results obtained with protein "A" from one study with protein "B" of another study. As a result, we cannot effectively piece together the information required to identify a polymorph formation mechanism.

The other important issue regarding past protein—polymorph studies is the intent: These studies focused exclusively on a "yes or no" outcome—i.e., did aragonite or vaterite form [34–53]?—rather than monitor the stepwise or temporal process of polymorph formation that is a prerequisite for mechanism development. These studies confirm that certain calcium carbonate polymorphs do form when proteins are present [34–53], but tell us nothing about how this process occurs, particularly within non-classical nucleation [29–33] or CPA [15] schemes. This leaves us at an impasse regarding the verification of protein-mediated polymorph formation.

5. Moving towards More Informative Studies

If we are to understand how proteins might control the polymorph selection and stabilization processes, then we need to move towards away from random, non-complementary studies [34–53] and towards standardized assay systems and methodologies [16,29–33] that place biomineralization proteins within the context of non-classical nucleation and CPA schemes. As an example, recent in vitro studies involving five recombinant mollusk shell nacre-associated proteins [58–66] and ACC-stabilizing sea urchin spicule matrix proteins [67,68] have utilized mineralization assay conditions identical to those employed in non-classical nucleation studies [29–33]. What links these different biomineralization proteins together are two common molecular themes [69,70]: intrinsic disorder, or absence of folding protein structure, and amyloid-like aggregation-prone domains, which promote protein-protein association leading to the formation of protein phases or hydrogels [58–68]. These reported studies utilized mole protein quantities, Ca(II) potentiometric methods, and parallel mineralization assay systems to study the early and later events in calcite-based calcium carbonate nucleation and monitor the formation and stabilization of PNCs and ACC in a time-dependent fashion. What was discovered was very informative: these proteins are distinguishable in terms of what mineral species or steps in the non-classical scheme they affect (Figure 4) [58–67]. Further, using mineralization assays which overlap with the time periods of the potentiometric titrations and utilize similar solution and supersaturation conditions [58–67] these studies demonstrated that these proteins form hydrogels that can capture, assemble, and organize mineral nanoparticles (Figure 5) [62,65] consistent with CPA theory [15]. The major conclusions one can draw from these studies are the following: (1) nacre [58–66] and sea urchin spicule matrix proteins [67,68] perform seminal tasks that would be critical for eventual polymorph selection and stabilization; (2) the functionalities of these six nacre proteins [58–65] are consistent with either non-classical [29–33] or CPA [16] theories [15]. Note that since these protein assay studies were calcite-based, there were no opportunities to study metastable aragonite or vaterite formation and stabilization. However, adjustment of in vitro assay and potentiometric conditions should allow these types of studies to move forward, and thus polymorph-specific studies under controlled conditions with defined protein content will be accomplished.

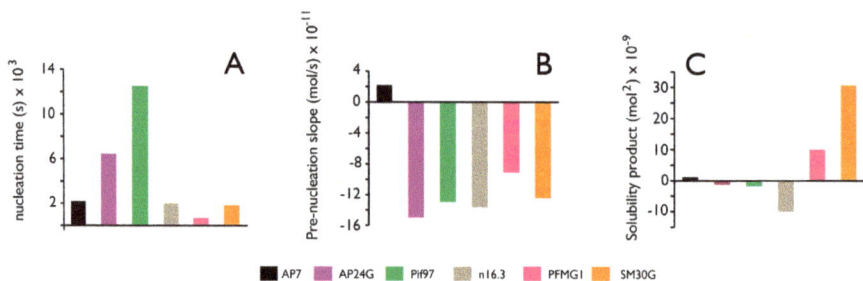

Figure 4. Comparative Ca(II) potentiometric datasets obtained from nacre (AP7, AP24G, Pif97, n16.3, PFMG1) and sea urchin spicule matrix (SM30G) proteins. All datasets are normalized relative to protein-deficient controls and were conducted at pH 9.0. Quantities of protein tested ranged from 500 nM to 1 μM. (**A**) Nucleation time refers to the time required for PNC formation; (**B**) "Prenucleation slope" refers to the slope of the linear prenucleation regime in each titration curve. Here, negative value indicates PNC stabilization; positive value indicates PNC destabilization; (**C**) Solubility product after nucleation. Negative values indicate the presence of less soluble, possibly crystalline phase; positive values reflect transient stabilization of ACC. Data takes from references [58–67].

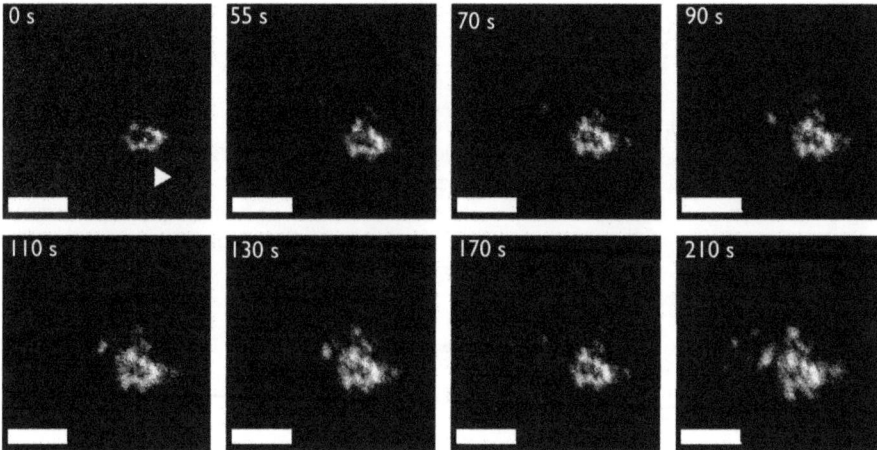

Figure 5. An example of protein-mediated mineral nanoparticle attachment: Scanning Transmission Electron Microscopy (STEM) flowcell video stills of mineral nanodeposits forming in 10 mM calcium carbonate solutions in the presence of 13.3 μM Haliotis rufescens abalone shell nacre protein AP7. AP7 is a hydrogelator that forms aggregating gels that contain mineral nanoparticles. The sequence of images starts 45 min after injection of the calcium solution into the protein-containing flow cell (first frame, denoted as 0 s) when mineral nanodeposits are first detected (white arrow, first frame) and nucleation evolves slowly (i.e., 210 s). As one can see, mineral particles are assembling from solution, presumably directed by the protein hydrogel. Note the formation of ring-like nanoclusters as time evolves; this would be atypical for a protein-free assay. By comparison, in the protein-deficient experiments mineralization commences 20 min after injection of the calcium solution and evolves much more rapidly with no evidence of organized particle attachment (data not shown). Scale bars = 100 nm. Taken from reference [62].

Although in vitro studies provide basic insights into polymorph formation, there remains the task of verifying that these processes occur in situ within mineralized tissues where polymorphs are formed. Although it is quite feasible to dissect time- and location-dependent nucleation events in parallel with protein expression, transport, and appearance [35,40,42,43,71–73] it is not yet feasible to correctly monitor, interpret, and integrate PNC, ACC, phase formation, particle assembly and organization simultaneously within the context of protein-specific events in situ. These limitations are primarily technological in scope, meaning that the methods and techniques to obtain this critical information are currently unavailable or do not possess sufficient sensitivity or spatio-temporal resolution to achieve the desired result. Once again, these issues should be resolved in time.

6. The Next Steps

We are at a crossroads in biomineralization and nucleation theory. Earlier models [54] which described classical nucleation of polymorphs have undergone a sophisticated evolution towards a multicomponent, multiprocess—based set of theories [15,16,29–33]. Although such theories might seem more complex at the outset, they are quite simple from the biological perspective: they allow for multiple pathways and approaches to achieve a desired mineral phase result specific for a given skeleton and organism. Again, this is where biology shines, for each organism could evolve its own version of nucleation pathway manipulation using specialized proteins [26–28] yet still obey the same thermodynamic laws and limitations [15,16,29–33]. As we are starting to see, a common theme that emerges from proteome studies is that these proteins share common traits, such as intrinsic disorder or unfolded structure [69] and aggregation propensity [69] that leads to protein phase or hydrogel formation that provides volume confinement and a pathway to particle assembly and organization [58–67]. Yet, the sequences of these proteins are unlike [58–67] and so genetic variation is permissible and even welcomed in biological organisms so long as specific nucleation goals are being met.

Thus, protein-mediated nucleation and polymorph selection and stabilization could be tweaked for an aragonite, vaterite, or calcite-based organism using a common blueprint—protein disorder and aggregation propensity [69]—that is modified to influence nucleation along a specific pathway, and, to manipulate certain steps within a given pathway. It is hoped that future research, involving standardized conditions and approaches, will firmly establish protein-based mechanisms of polymorph selection and stabilization in nature, and thus facilitate the transfer of this information to materials science and chemistry.

Acknowledgments: Themes concerning nacre proteins were supported by the U.S. Department of Energy, Office of Basic Energy Sciences, Division of Materials Sciences and Engineering under Award DE-FG02-03ER46099. Themes concerning sea urchin proteins were supported by the U.S. Army Research Laboratory and the U.S. Army Research Office under grant number W911NF-16-1-0262. This paper represents contribution number 87 from the Laboratory for Chemical Physics, New York University.

Conflicts of Interest: The author declares no conflict of interest.

References

1. Wegst, U.G.K.; Bai, H.; Saiz, E.; Tomsia, A.P.; Ritchie, R.O. Bioinspired materials. *Nat. Mater.* **2015**, *14*, 23–36. [CrossRef] [PubMed]

2. Zhang, G.; Li, X. Uncovering aragonite nanoparticle self-assembly in nacre—A natural armor. *Cryst. Growth Des.* **2012**, *12*, 4306–4310. [CrossRef]

3. Li, X.; Chang, W.C.; Chao, Y.J.; Wang, R.; Chang, M. Nanoscale structural and mechanical characterization of a natural nanocomposite material: The shell of red abalone. *NanoLetters* **2004**, *4*, 613–617. [CrossRef]

4. Sun, J.; Bhushan, B. Hierarchical structure and mechanical properties of nacre: A review. *RSC Adv.* **2012**, *2*, 7617–7632. [CrossRef]

5. Berman, A.; Addadi, L.; Kvick, A.; Leiserowitz, L.; Nelson, M.; Weiner, S. Intercalation of sea urchin proteins in calcite: Study of a crystalline composite material. *Science* **1990**, *250*, 664–667. [CrossRef] [PubMed]

6. Sea Urchin Genome Sequencing Consortium. The genome of the sea urchin *Strongylocentrotus purpuratus*. *Science* **2006**, *314*, 941–952.

7. Mann, K.; Poustka, A.J.; Mann, M. The sea urchin (*Strongylocentrotus purpuratus*) test and spine proteomes. *Proteome Sci.* **2008**, *6*, 1–10. [CrossRef] [PubMed]

8. Mann, K.; Wilt, F.H.; Poustka, A.J. Proteomic analysis of sea urchin (*Strongylocentrotus purpuratus*) spicule matrix. *Proteome Sci.* **2010**, *8*, 1–12. [CrossRef] [PubMed]

9. Tester, C.C.; Wu, C.H.; Krejci, M.R.; Mueller, L.; Park, A.; Lai, B.; Chen, S.; Sun, C.; Joester, D. Time-resolved evolution of short- and long-range order during the transformation of ACC to calcite in the sea urchin embryo. *Adv. Funct. Mater.* **2013**, *23*, 4185–4194. [CrossRef]

10. Politi, Y.; Metzler, R.A.; Abrecht, M.; Gilbert, B.; Wilt, F.H.; Sagi, I.; Addadi, L.; Weiner, S.; Gilbert, P. Transformation mechanism of amorphous calcium carbonate into calcite in the sea urchin larval spicule. *Proc. Natl. Acad. Sci. USA* **2008**, *105*, 17362–17366. [CrossRef] [PubMed]

11. Gong, Y.U.T.; Killian, C.E.; Olson, I.C.; Appathurai, N.P.; Amasino, A.L.; Martin, M.C.; Holt, L.J.; Wilt, F.H.; Gilbert, P. Phase transitions in biogenic amorphous calcium carbonate. *Proc. Natl. Acad. Sci. USA* **2012**, *109*, 6088–6093. [CrossRef] [PubMed]

12. Drake, J.L.; Mass, T.; Haramaty, L.; Zelzion, E.; Bhattacharya, D.; Falkowski, P.G. Proteomic analysis of skeletal organic matrix from the stony coral Stylophora pistillata. *Proc. Natl. Acad. Sci. USA* **2013**, *110*, 3788–3793. [CrossRef] [PubMed]

13. Moya, A.; Luisman, L.; Ball, E.E.; Hayward, D.C.; Grasso, L.C.; Chua, C.M.; Woo, H.N.; Gattuso, J.P.; Forêt, S.; Miller, D.J. Whole transcriptome analysis of the coral Acropora millepora reveals complex responses to CO_2-driven acidification during the initiation of calcification. *Mol. Ecol.* **2012**, *21*, 2440–2454. [CrossRef] [PubMed]

14. Goffredo, S.; Vergni, P.; Reggi, M.; Caroselli, E.; Sparla, F.; Levy, O.; Dubinsky, Z.; Falini, G. The skeletal organic matrix from Mediterranean coral Balanophyllia europaea influences calcium carbonate precipitation. *PLoS ONE* **2011**, *6*, e22338. [CrossRef] [PubMed]

15. De Yoreo, J.J.; Gilbert, P.U.P.A.; Sommerdijk, N.A.J.M.; Penn, R.L.; Whitelam, S.; Joester, D.; Zhang, H.; Rimer, J.D.; Navrotsky, A.; Banfield, J.F.; et al. Crystallization by particle attachment in synthetic, biogenic, and geologic environments. *Science* **2015**, *349*, 498–510. [CrossRef] [PubMed]

16. Wallace, A.F.; Hedges, L.O.; Fernandez-Martinez, A.; Raiteri, P.; Gale, J.D.; Waychunas, G.A.; Whitelam, S.; Banfield, J.F.; De Yoreo, J.J. Liquid-liquid separation in supersaturated $CaCO_3$ solutions. *Science* **2013**, *341*, 885–889. [CrossRef] [PubMed]

17. Lee, K.; Wagermaier, W.; Masic, A.; Kommareddy, K.P.; Bennet, M.; Manjubala, I.; Lee, S.W.; Park, S.B.; Cölfen, H.; Fratzl, P. Self-assembly of amorphous calcium carbonate microlens arrays. *Nat. Commun.* **2012**, *3*, 725–727. [CrossRef] [PubMed]

18. Politi, Y.; Arad, T.; Klein, E.; Weiner, S.; Addadi, L. Sea urchin spine calcite forms via a transient amorphous calcium carbonate phase. *Science* **2004**, *306*, 1161–1164. [CrossRef] [PubMed]

19. Weiss, I.M.; Tuross, N.; Addadi, L.; Weiner, S. Mollusc larval shell formation: Amorphous calcium carbonate is a precursor phase for aragonite. *J. Exp. Zool.* **2002**, *293*, 478–491. [CrossRef] [PubMed]

20. Bernstein, J. Polymorphism—A perspective. *Cryst. Growth Des.* **2011**, *11*, 632–650. [CrossRef]

21. Sand, K.K.; Rodriguez-Blanco, M.; Makovicky, E.; Benning, L.G.; Stipp, S.L.S. Crystallization of $CaCO_3$ in water-alcohol mixtures: Spherulitic growth, polymorph stabilization, and morphology change. *Cryst. Growth. Des.* **2012**, *12*, 842–853. [CrossRef]

22. Olderoy, M.O.; Xie, M.; Strand, B.L.; Draget, K.I.; Sikorski, P.; Andreassen, J.P. Polymorph switching in the calcium carbonate system by well-defined alginate oligomers. *Cryst. Growth Des.* **2011**, *11*, 520–529. [CrossRef]

23. Xiao, J.; Zhu, Y.; Liu, Y.; Liu, H.; Xu, F.; Wang, L. Vaterite selection by chitosan gel: An example of polymorph selection by morphology of biomacromolecules. *Cryst. Growth Des.* **2008**, *8*, 2887–2891. [CrossRef]

24. Sun, W.; Jayaraman, S.; Chen, W.; Persson, K.A.; Ceder, G. Nucleation of metastable aragonite $CaCO_3$ in seawater. *Proc. Natl. Acad. Sci. USA* **2015**, *112*, 3199–3204. [CrossRef] [PubMed]

25. Kababya, S.; Gal., A.; Kahil, K.; Weiner, S.; Addadi, L.; Schmidt, A. Phosphate-water interplay tunes amorphous calcium carbonate metastability: Spontaneous phase separation and crystallization versus stabilization viewed by solid-state NMR. *J. Am. Chem. Soc.* **2015**, *137*, 990–998. [CrossRef] [PubMed]

26. Immel, F.; Gaspard, D.; Marie, A.; Guichard, N.; Cusack, M.; Marin, F. Shell proteome of rhynchonelliform brachiopods. *J. Struct. Biol.* **2015**, *190*, 360–384. [CrossRef] [PubMed]

27. Zhang, G.; Fang, X.D.; Guo, X.M.; Li, L.; Luo, R.B.; Xu, F.; Yang, P.C.; Zhang, L.L.; Wang, X.T.; Qi, H.G.; et al. The oyster genome reveals stress adaptation and complexity of shell formation. *Nature* **2012**, *490*, 49–54. [CrossRef] [PubMed]

28. Jackson, D.J.; McDougall, C.; Woodcroft, B.; Moase, P.; Rose, R.A.; Kube, M.; Reinhart, R.; Rokhsar, D.S.; Montagnani, C.; Joube, C.; et al. Parallel evolution of nacre building gene sets in mollusks. *Mol. Biol. Evol.* **2010**, *27*, 591–608. [CrossRef] [PubMed]

29. Gebauer, D.; Kellermeier, M.; Gale, J.D.; Bergstrom, L.; Cölfen, H. Pre-nucleation clusters as solute precursors in crystallization. *Chem. Soc. Rev.* **2014**, *43*, 2348–2371. [CrossRef] [PubMed]

30. Gebauer, D.; Volkel, A.; Cölfen, H. Stable prenucleation of calcium carbonate clusters. *Science* **2008**, *322*, 1819–1822. [CrossRef] [PubMed]

31. Gebauer, D.; Cölfen, H. Prenucleation clusters and non-classical nucleation. *Nano Today* **2011**, *6*, 564–584. [CrossRef]

32. Demichelis, R.; Raiteri, P.; Gale, J.D.; Quigley, D.; Gebauer, D. Stable prenucleation mineral clusters are liquid-like ionic polymers. *Nat. Commun.* **2011**, *2*, 1–12. [CrossRef] [PubMed]

33. Gebauer, D.; Gunawidjaja, P.N.; Ko, J.Y.P.; Bascik, Z.; Aziz, B.; Liu, L.; Hu, Y.; Bergstrom, L.; Tai, C.W.; Sham, T.K.; et al. Proto-calcite and proto-vaterite in amorphous calcium carbonates. *Angew. Chem. Int. Ed.* **2010**, *49*, 8889–8891. [CrossRef] [PubMed]

34. Belcher, A.M.; Wu, X.H.; Christensen, R.J.; Hansma, P.K.; Stucky, G.D.; Morse, D.E. Control of crystal phase switching and orientation by soluble mollusk shell proteins. *Nature* **1996**, *381*, 56–58. [CrossRef]

35. Fritz, M.; Belcher, A.M.; Radmacher, M.; Walters, D.A.; Hansma, P.K.; Stucky, G.D.; Morse, D.E.; Mann, S. Flat pearls from biofabrication of organized composites on inorganic substrates. *Nature* **1994**, *371*, 49–51. [CrossRef]

36. Su, J.; Zhu, F.; Zhang, G.; Wang, H.; Xie, L.; Zhang, R. Transformation of amorphous calcium carbonate nanoparticles into aragonite controlled by ACCBP. *CrystEngComm* **2016**, *18*, 2125–2134. [CrossRef]

37. Thompson, J.B.; Paloczi, G.T.; Kindt, J.H.; Michenfelder, M.; Smith, B.L.; Stucky, G.; Morse, D.E.; Hansma, P.K. Direct observation of the transition from calcite to aragonite growth as induced by abalone shell proteins. *Biophys. J.* **2000**, *79*, 3307–3312. [CrossRef]

38. Seto, J.; Picker, A.; Evans, J.S.; Cölfen, H. A nacre protein sequence organizes the mineralization space for polymorph formation. *Cryst. Growth Des.* **2014**, *14*, 1501–1505. [CrossRef]

39. Suzuki, M.; Saruwatari, K.; Kogure, T.; Yamamoto, Y.; Nishimura, T.; Kato, T.; Nagasawa, H. An acidic matrix protein, Pif, is a key macromolecule for nacre formation. *Science* **2009**, *325*, 1388–1390. [CrossRef] [PubMed]

40. Xiang, L.; Su, J.; Zheng, J.G.; Liang, J.; Zhang, G.; Wang, H.; Xie, L.; Zhang, R. Patterns of expression in the matrix proteins responsible for nucleation and growth of aragonite crystals in flat pearls of Pinctada fucata. *PLoS ONE* **2013**, *8*, e66564. [CrossRef]

41. Yan, Z.; Jing, G.; Gong, N.; Li, C.; Zhou, Y.; Xie, L.; Zhang, R. N40, a novel nonacidic matrix protein from pearl oyster nacre, facilitates nucleation of aragonite in vitro. *Biomacromolecules* **2007**, *8*, 3597–3601. [CrossRef] [PubMed]

42. Jiao, Y.; Wang, H.; Du, X.; Zhao, X.; Wang, Q.; Huang, R.; Deng, R. Dermatopontin, a shell matrix protein gene from pearl oyster Pinctada martensii, participates in nacre formation. *Biochem. Biophys. Res. Commun.* **2012**, *425*, 679–693. [CrossRef] [PubMed]

43. Ma, Z.; Huang, J.; Sun, J.; Wang, G.; Li, C.; Xie, L.; Zhang, R. A novel extrapallial fluid protein controls the morphology of nacre lamellae in the pearl oyster, *Pinctada fucata*. *J. Biol. Chem.* **2007**, *282*, 23253–23263. [CrossRef] [PubMed]

44. Samata, T.; Hayashi, N.; Kono, M.; Hagesawa, K.; Horita, C.; Akera, S. A new matrix protein family related to the nacreous layer formation of *Pinctada fucata*. *FEBS Lett.* **1999**, *462*, 225–229. [CrossRef]

45. Falini, G.; Albeck, S.; Weiner, S.; Addadi, L. Control of aragonite or calcite polymorphism by mollusk shell macromolecules. *Science* **1996**, *271*, 67–69. [CrossRef]

46. Amos, F.F.; Evans, J.S. AP7, a partially disordered pseudo C-RING protein, is capable of forming stabilized aragonite in vitro. *Biochemistry* **2009**, *48*, 1332–1339. [CrossRef] [PubMed]

47. Ponce, C.B; Evans, J.S. Polymorph crystal selection by n16, an intrinsically disordered nacre framework protein. *Cryst. Growth Des.* **2011**, *11*, 4690–4696. [CrossRef]

48. Amos, F.F.; Ponce, C.B.; Evans, J.S. Formation of framework nacre polypeptide supramolecular assemblies that nucleate polymorphs. *Biomacromolecules* **2011**, *12*, 1883–1890. [CrossRef] [PubMed]

49. Metzler, R.A.; Evans, J.S.; Kilian, C.E.; Zhou, D.; Churchill, T.H.; Appathurai, P.N.; Coppersmith, S.N.; Gilbert, P.U.P.A. Lamellar self-assembly and aragonite polymorph selection by a single intrinsically disordered protein fragment. *J. Am. Chem. Soc.* **2010**, *132*, 6329–6334. [CrossRef] [PubMed]

50. Keene, E.C.; Evans, J.S.; Estroff, L.A. Matrix interactions in biomineralization: Aragonite nucleation by an intrinsically disordered nacre polypeptide, n16N, associated with a β-chitin substrate. *Cryst. Growth Des.* **2010**, *10*, 1383–1389. [CrossRef]

51. Amos, F.F.; Destine, E.; Ponce, C.B.; Evans, J.S. The N- and C-terminal regions of the pearl-associated EF Hand protein, PFMG1, promote the formation of the aragonite polymorph in vitro. *Cryst. Growth Des.* **2010**, *10*, 4211–4216. [CrossRef]

52. Natoli, A.; Wiens, M.; Schroder, H.C.; Stifanic, M.; Batel, R.; Soldati, A.L.; Jacob, D.E.; Muller, W.E.G. Bio-vaterite formation by glycoproteins from freshwater pearls. *Micron* **2010**, *41*, 359–366. [CrossRef] [PubMed]

53. Wang, Y.Y.; Yao, Q.Z.; Zhou, G.T.; Sheng, Y.M. Formation of vaterite mesocrystals in biomineral-like structures and implication for biomineralization. *Cryst. Growth Des.* **2015**, *15*, 1714–1725. [CrossRef]

54. Kalikmanov, V.I. *Nucleation Theory in Lecture Notes in Physics*; Springer Science+Business Media: Dordrecht, The Netherlands, 2013; Volume 860, pp. 17–41.

55. Bano, A.M.; Rodger, P.M.; Quigley, D. New insight into the stability of $CaCO_3$ surfaces and nanoparticles via molecular simulation. *Langmuir* **2014**, *30*, 7513–7521. [CrossRef] [PubMed]

56. Singer, J.W.; Yazaydin, A.O.; Kirkpatrick, R.J.; Bowers, G.M. Structure and transformation of amorphous calcium carbonate: A solid-state 43Ca NMR and computational molecular dynamics investigation. *Chem. Mater.* **2012**, *24*, 1828–1836. [CrossRef]

57. Ren, D.; Albert, O.; Sun, M.; Muller, W.E.G.; Feng, Q. Primary cell culture of fresh water *Hyriopsis comingii* mantle/pearl sac tissues and its effect on calcium carbonate mineralization. *Cryst. Growth Des.* **2014**, *14*, 1149–1157. [CrossRef]

58. Chang, E.P.; Roncal-Herrero, T.; Morgan, T.; Dunn, K.E.; Rao, A.; Kunitake, J.A.M.R.; Lui, S.; Bilton, M.; Estroff, L.A.; Kröger, R.; et al. Synergistic biomineralization phenomena created by a nacre protein model system. *Biochemistry* **2016**, *55*, 2401–2410. [CrossRef] [PubMed]

59. Chang, E.P.; Perovic, I.; Rao, A.; Cölfen, H.; Evans, J.S. Insect cell glycosylation and its impact on the functionality of a recombinant intracrystalline nacre protein, AP24. *Biochemistry* **2016**, *55*, 1024–1035. [CrossRef] [PubMed]

60. Chang, E.P.; Evans, J.S. Pif97, a von Willebrand and Peritrophin biomineralization protein, organizes mineral nanoparticles and creates intracrystalline nanochambers. *Biochemistry* **2015**, *54*, 5348–5355. [CrossRef] [PubMed]

61. Chang, E.P.; Williamson, G.; Evans, J.S. Focused ion beam tomography reveals the presence of micro-, meso-, and macroporous intracrystalline regions introduced into calcite crystals by the gastropod nacre protein AP7. *Cryst. Growth Des.* **2015**, *15*, 1577–1582. [CrossRef]

62. Perovic, I.; Chang, E.P.; Verch, A.; Rao, A.; Cölfen, H.; Kröger, R.; Evans, J.S. An oligomeric C-RING nacre protein influences pre-nucleation events and organizes mineral nanoparticles. *Biochemistry* **2014**, *53*, 7259–7268. [CrossRef] [PubMed]

63. Chang, E.P.; Russ, J.A.; Verch, A.; Kröger, R.; Estroff, L.A.; Evans, J.S. Engineering of crystal surfaces and subsurfaces by an intracrystalline biomineralization protein. *Biochemistry* **2014**, *53*, 4317–4319. [CrossRef] [PubMed]

64. Chang, E.P.; Russ, J.A.; Verch, A.; Kröger, R.; Estroff, L.A.; Evans, J.S. Engineering of crystal surfaces and subsurfaces by framework biomineralization protein phases. *CrystEngComm* **2014**, *16*, 7406–7409. [CrossRef]

65. Perovic, I.; Chang, E.P.; Lui, M.; Rao, A.; Cölfen, H.; Evans, J.S. A framework nacre protein, n16.3, self-assembles to form protein oligomers that participate in the post-nucleation spatial organization of mineral deposits. *Biochemistry* **2014**, *53*, 2739–2748. [CrossRef] [PubMed]

66. Perovic, I.; Mandal, T.; Evans, J.S. A pseudo EF-hand pearl protein self-assembles to form protein complexes that amplify mineralization. *Biochemistry* **2013**, *52*, 5696–5703. [CrossRef] [PubMed]

67. Jain, G.; Pendola, M.; Rao, A.; Cölfen, H.; Evans, J.S. A model sea urchin spicule matrix protein self-associates to form mineral-modifying hydrogels. *Biochemistry* **2016**, *55*, 4410–4421. [CrossRef] [PubMed]

68. Rao, A.; Seto, J.; Berg, J.K.; Kreft, S.G.; Scheffner, M.; Cölfen, H. Roles of larval sea urchin spicule SM50 domains in organic matrix self-assembly and calcium carbonate mineralization. *J. Struct. Biol.* **2013**, *183*, 205–215. [CrossRef] [PubMed]

69. Evans, J.S. Identification of intrinsically disordered and aggregation—Promoting sequences within the aragonite-associated nacre proteome. *Bioinformatics* **2012**, *28*, 3182–3185. [CrossRef] [PubMed]

70. Evans, J.S. "Liquid-like" biomineralization protein assemblies: A key to the regulation of non-classical nucleation. *CrystEngComm* **2013**, *15*, 8388–8394. [CrossRef]

71. Yan, F.; Jiao, Y.; Deng, Y.; Du, X.; Huang, R.; Wang, Q.; Chen, W. Tissue inhibitor of metalloprotease gene from pearl oyster *Pinctada martensii* participates in nacre formation. *Biochem. Biophys. Res. Commun.* **2014**, *450*, 300–305. [CrossRef] [PubMed]

72. Liu, H.L.; Liu, S.F.; Ge, Y.J.; Liu, J.; Wang, X.Y.; Xie, L.P.; Zhang, R.; Wang, Z. Indentification and characterization of a biomineralization related gene PFMG1 highly expresses in the mantle of *Pinctada fucata*. *Biochemistry* **2007**, *46*, 844–851. [CrossRef] [PubMed]

73. Yan, Z.; Fang, Z.; Ma, Z.; Deng, J.; Li, S.; Xie, L.; Zhang, R. Biomineralization: Functions of calmodulin-like protein in the shell formation of pearl oyster. *Biochim. Biophys. Acta* **2007**, *1770*, 1338–1344. [CrossRef] [PubMed]

![minerals logo] *minerals*

MDPI

Article

Hydration Effects on the Stability of Calcium Carbonate Pre-Nucleation Species

Alejandro Burgos-Cara [1], Christine V. Putnis [2,3], Carlos Rodriguez-Navarro [1] and Encarnacion Ruiz-Agudo [1,*]

1   Mineralogy and Petrology Department, University of Granada, 18071 Granada, Spain; aburgoscara@ugr.es (A.B.-C.); carlosrn@ugr.es (C.R.-N.)
2   Institut für Mineralogie, University of Münster, 48149 Münster, Germany; putnisc@uni-muenster.de
3   Department of Chemistry, Curtin University, Perth 6845, Australia
*   Correspondence: encaruiz@ugr.es; Tel.: +34-958-240-473

Received: 1 June 2017; Accepted: 14 July 2017; Published: 20 July 2017

Abstract: Recent experimental evidence and computer modeling have shown that the crystallization of a range of minerals does not necessarily follow classical models and theories. In several systems, liquid precursors, stable pre-nucleation clusters and amorphous phases precede the nucleation and growth of stable mineral phases. However, little is known on the effect of background ionic species on the formation and stability of pre-nucleation species formed in aqueous solutions. Here, we present a systematic study on the effect of a range of background ions on the crystallization of solid phases in the $CaCO_3$-H_2O system, which has been thoroughly studied due to its technical and mineralogical importance, and is known to undergo non-classical crystallization pathways. The induction time for the onset of calcium carbonate nucleation and effective critical supersaturation are systematically higher in the presence of background ions with decreasing ionic radii. We propose that the stabilization of water molecules in the pre-nucleation clusters by background ions can explain these results. The stabilization of solvation water hinders cluster dehydration, which is an essential step for precipitation. This hypothesis is corroborated by the observed correlation between parameters such as the macroscopic equilibrium constant for the formation of calcium/carbonate ion associates, the induction time, and the ionic radius of the background ions in the solution. Overall, these results provide new evidence supporting the hypothesis that pre-nucleation cluster dehydration is the rate-controlling step for calcium carbonate precipitation.

Keywords: background electrolytes; dehydration kinetics; calcium carbonate; calcite; clusters; nucleation; vaterite; ACC

---

1. Introduction

Calcium carbonate precipitation has been widely studied due to the extensive distribution of carbonates, predominantly calcium carbonate, in surface rocks of the earth and scale formation in industrial processes. It presents a relatively simple model system to work with, and its wide occurrence in many biominerals provides interdisciplinary significance [1,2]. It is known that many organic additives play a key role in nucleation and growth processes, with either enhancing or hindering effects [3–8]. The latter frequently leads to the stabilization of more soluble metastable phases, such as amorphous phases (i.e., amorphous calcium carbonate, or ACC), that are known to play a key role in biomineralization processes [9–12]. However, not only organic molecules have been recognized to either modify crystal morphology or stabilize more soluble precursor phases. Other ions present in solution, such as Mg^{2+}, may also influence mineral formation processes [13–17]. The role that background ions play in calcium carbonate precipitation, either from experimental reagents, or additionally dosed or naturally present in aqueous environments, has been frequently neglected or underestimated. Recent experimental and computational results [18,19] suggest that Ca^{2+} and CO_3^{2-} ions can associate into stable complexes prior to the onset of liquid or solid $CaCO_3$ formation and that, when conditions are favorable, this process of solute clustering is primarily controlled by the release of water molecules from ion hydration layers [20]. Therefore, any factor affecting cluster solvation should influence the stability of pre-nucleation ion associates, and thus also affect nucleation.

Background electrolytes can affect nucleation kinetics, crystal growth, dissolution, crystal size distribution, and the purity of precipitates by inducing changes in the aqueous solvation environment [21–25]. Ions in solution are able to modify water structure dynamics in their local environment as a result of effects associated with their hydration shells [26], which immobilize and electrostrict water [27]. The dehydration kinetics of ions (or clusters) in solution will be a competition between ion (or clusters)-water and water-water interactions [28,29], which can be significantly modified by the presence of background ions in solution [30]. At low ionic strength, the effect of background electrolytes on ion (or clusters)-water electrostatic interactions will be dominant [26].

We present a systematic study on the effect of a series of 1:1 background electrolytes on the precipitation of calcium carbonate. Experiments were performed under conditions of low ionic strength. Our main goal was to test the basic hypotheses that (i) the effect of electrolytes on the stability of pre-nucleation species and the onset of nucleation can be related to the influence of background electrolytes on the solvation of the ions building the crystals; and (ii) the systematic trends observed in the stability of pre-nucleation species and the onset of nucleation for the different background salts are due to the intrinsic properties of the background ions. To do so, we performed $CaCO_3$ precipitation (titration) experiments in the presence of different background electrolytes and at different ionic strengths, continuously monitoring pH, free-Ca^{2+} concentration, conductivity, and solution transmittance in batch reactors. At the same time, we studied the particle size distribution and the structural and textural features of precipitated phases.

2. Materials and Methods

2.1. Titration Experiments

Two types of experiments were performed in order to study the influence of different background electrolytes on calcium carbonate precipitation. In both types of experiments, a 10 mM aqueous calcium solution was continuously added, at a rate of 2 μL/s, into a reactor containing 100 mL of a 10 mM carbonate solution. The first type of experiment (Type I) was performed by changing the counter-ion of the salts used for calcium carbonate precipitation. For this purpose, Li_2CO_3, Na_2CO_3, K_2CO_3 and Cs_2CO_3 were used as carbonate sources, and $CaCl_2$, $CaBr_2$ and CaI_2 were used as calcium sources. This allowed us to study the influence of different background ions (both anions and cations) at a very low ionic strength (IS) of 0.026, defined according to Equation (1).

$$IS = \frac{1}{2} \sum_{i=1}^{n} c_i \cdot z_i^2 \tag{1}$$

where c_i is the molar concentration of "i" ion and z_i the charge of each ion.

The second type of experiment (Type II) was performed by selecting Na_2CO_3 and $CaCl_2$ as the carbonate and calcium sources, respectively, for calcium carbonate precipitation. Different background ions were introduced as foreign salts, in addition to the NaCl already present in the growth solution (i.e., LiCl, NaCl, NaBr, NaI, KCl and CsCl) at two different concentrations, 10 and 25 mM, with the aim to study both the effect of the background ions themselves, and the influence of two different ionic strengths (i.e., IS = 0.035 and 0.049, respectively).

Both types of experiments were performed using a 200 mL jacketed glass reactor coupled to a thermostatic bath, in order to maintain a constant T of 25 °C inside the reactor, which included a stirrer module (module 801, Metrohm, Gallen, Switzerland). The pH was measured using a glass electrode from Metrohm, conductivity with an 856 conductivity module (Metrohm, Gallen, Switzerland), transmittance with an Optrode sensor (Metrohm, Gallen, Switzerland) using a wavelength of 610 nm, and free calcium in solution (Ca^{2+}) with an ion-selective electrode (ISE, Mettler-Toledo, DX240-Ca, Columbus, OH, USA), using the pH electrode as a reference electrode. Sensors and dosing devices were coupled to a Titrando 905 module from Metrohm controlled with the software Tiamo v2.5 (Metrohm, Gallen, Switzerland). The above parameters (pH, transmittance, conductivity, and free-Ca) were recorded continuously during precipitation experiments.

Inorganic salts used in the two types of experiments (Li_2CO_3, Na_2CO_3, K_2CO_3, Cs_2CO_3, $CaCl_2$, $CaBr_2$, CaI_2, LiCl, NaCl, NaBr, NaI, KCl and CsCl) were purchased from Sigma-Aldrich (Merck Darmstadt, Germany) with purity of at least 99%. Solutions used in titration experiments were prepared with ultrapure water Type I + (resistivity $\geq$ 18.2 MΩ·cm). The initial pH values inside the reactor for the different type of experiments and salts were 11.05 $\pm$ 0.05 (IS = 0.026, Type I), 10.93 $\pm$ 0.04 (IS = 0.035, Type II), and 10.91 $\pm$ 0.05 (IS = 0.049, Type II). pH adjustment was avoided in order not to introduce more foreign ions into solutions, which would also modify the ionic strength (IS). In any case, measured pH changes (i.e., reduction) over the course of the titration experiments never exceeded 0.1 pH units. Higher reactant concentrations were avoided because at low IS, changes in activity coefficients for the different background ions tested could be considered as negligible [31,32].

2.2. Dynamic Light Scattering

During titration experiments, the controlled reference heterodyne method and in situ dynamic light scattering (DLS) were used to evaluate the time evolution of the precipitate particle size distribution as it referred to equivalent sphere diameter. Measurements were conducted at a scattering angle of 180° using a Microtrac NANO-flex particle size analyzer (MTB, Madrid, Spain) equipped with a diode laser (λ = 780 nm, 5 mW) and a 1 m-long flexible measuring probe (diameter = 8 mm) with sapphire window as the sample interface. Scattering was continuously monitored in situ during titration experiments, with an acquisition time of 30 s per run during 120 consecutive runs. DLS measurements started immediately before the slope of free-Ca^{2+} begun to flatten during titration experiments. Particle size distributions (PSDs) were computed with the Microtrac FLEX application software package (v.11.1.0.2, Microtrac, Montgomeryville, PA, USA). The presented PSD graphs show average values of particle size for the 120 consecutive measurements. Background scattering for pure water, as well as for solutions of each individual salt tested were collected. However, no particles within the analyzed size range were detected in these control runs.

2.3. Electron Microscopy

An Auriga field emission-scanning electron microscope (FE-SEM, ZEISS, Jena, Germany) was used for morphology examinations of calcium carbonate precipitates formed both at the early stages of precipitation and at the end of titration experiments. Prior to observations, samples were carbon coated. Secondary electron (SE) images were acquired using a SE-InLens detector. Observations were carried out at an accelerating voltage of 3 kV.

2.4. X-ray Diffraction Analysis

Solid phases formed during the different precipitation tests were analyzed by X-ray diffraction (XRD, PANalytica, Eindhoven, The Netherlands) using a Panalytical X'Pert PRO diffractometer. The following working conditions were used: radiation CuKα (λ = 1.5405 Å), voltage 45 kV, current 40 mA, scanning angle (2θ) 10°–70° and goniometer speed 0.016° 2θ s^{-1}.

3. Results

3.1. Type I Experiments

3.1.1. Effect of Different Counter-Ions in the Carbonate or Calcium Sources

Changing the cation (M^+) in the carbonate source led to appreciable changes in both the maximum free calcium concentration (Ca^{2+}_{max}) reached in titration experiments, and the corresponding induction time for the onset of precipitation (Figure 1). To correlate experimental values from titration experiments, the ionic radius was selected as the most simple and straightforward variable to fit our results. However, other parameters, such as the hydration enthalpy of an ion, could be used to obtain similar trends (see Figure S1). As depicted in Figure 1a,b, for the same calcium salt (i.e., $CaCl_2$), the lower the ionic radius of the cation (M^+) in the carbonate source (M_2CO_3), the higher the induction time and corresponding concentration of Ca^{2+}_{max} reached before the onset of spontaneous precipitation. In parallel, using Na_2CO_3 as the carbonate source, and a calcium source (CaX_2) with different anions (X^-), a slight decrease in Ca^{2+}_{max} concentration and induction time with increasing ionic radius of anions was observed (Figure 1c,d). Although differences between electrolytes are in some cases (e.g., K^+ and Cs^+) within errors, there is a clear observable trend in both the induction time and maximum free calcium concentration for the onset of nucleation, which in both cases (X^- and M^+) decreases with increasing ionic size. Nonetheless, it should be taken into account that the radius of a hydrated Cs^+ ion is slightly larger than K^+, whereas the hydrated radius of ions tends to decrease with increasing atomic number (i.e., $r + \Delta r$: $Li^+ > Na^+ > Cs^+ > K^+$). This suggests that the effective hydrated radius could be a relevant parameter to consider in the particular case of K^+ and Cs^+. On the other hand, a lower effect of the background electrolyte is observable in the case of anions. This appears to be related to the less structured solvation shell of anions (e.g., the hydration enthalpy of the anions is typically smaller than that of the cations, see Figure S1).

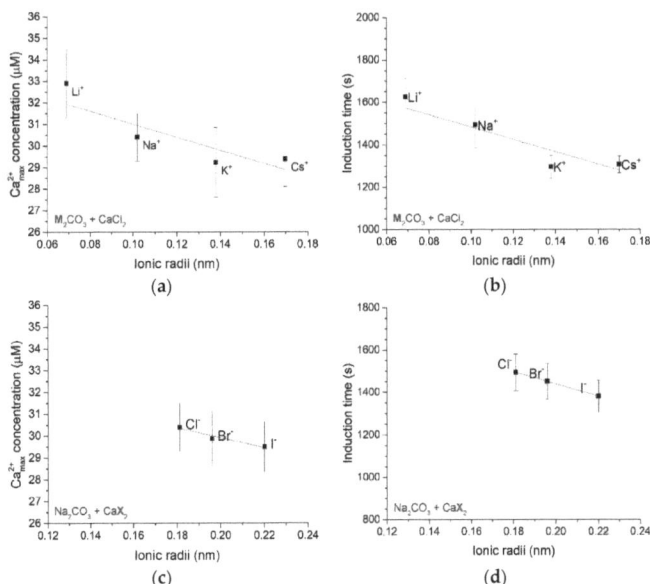

Figure 1. Maximum free Ca^{2+} concentration and corresponding induction time for type I titration experiments. (**a,b**) $CaCl_2$ was added to solutions of different carbonate sources (Li_2CO_3, Na_2CO_3, K_2CO_3, Cs_2CO_3) placed in the reactor. (**c,d**) Na_2CO_3 was always present in the reactor, and different calcium sources ($CaCl_2$, $CaBr_2$, CaI_2) were continuously added to the reactor. Errors bars show $2\sigma_N$.

3.1.2. Dynamics of Calcium Carbonate Binding

From titration experiments, the free Ca^{2+} concentration inside the reactor could be plotted vs. time (Figure 2a,b). Before precipitation, the amount of total calcium added to the reactor was always higher than the free calcium concentration measured by the ion-selective electrode. The latter is related to ion pairing and clustering phenomena in the pre-nucleation regime [20]. On the one hand, as depicted in Figure 2, larger background ions, both in the case of different background anions in the calcium source or cations in the carbonate source, seem to slightly increase the slope of the measured free calcium concentration. This indicates a lower ion pairing/clustering. On the other hand, in the case of Li_2CO_3 as the carbonate source, it was also observed that the Ca^{2+} concentration at the steady state (Figure 2a), corresponding to the solubility product of the precipitating phase (the more soluble one, if two or more phases form after the initial precipitated phase [20]), was higher compared to the other carbonate sources. The latter suggests a lower stability of the precipitating phase, as stated by Gebauer et al. [20]. In this case, it is likely that a more soluble ACC phase, possibly similar to the ACC II phase suggested by Gebauer et al. [20], was the first to precipitate in this system. Alternatively, the latter could be also explained by a different water content in the precipitating ACC, as suggested by Rodriguez-Navarro et al. [33] (i.e., a higher water content results in higher solubilities or size-related solubility effects; [34] see below). In contrast, it was systematically observed that when using CaI_2 as the calcium source, the free calcium concentration decreased linearly after precipitation (no actual steady state was observed), which suggested a faster transition, possibly by means of a dissolution–reprecipitation process, to produce a less soluble phase (Figure 2b).

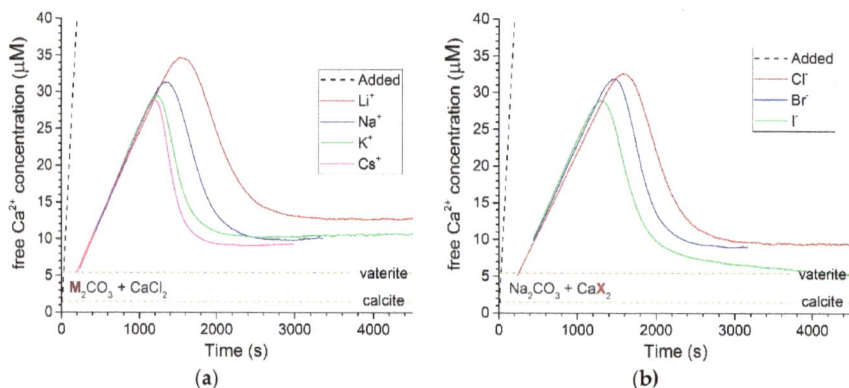

Figure 2. Free-Ca^{2+} concentration measured in type I titration experiments for (**a**) different background cations in the carbonate source and (**b**) different background anions in the calcium source. Horizontal orange dashed lines in (**a,b**) show the expected free-Ca^{2+} concentration for the different phases according to calculations using the geochemical computer code (PHREEQc) [35]. The nearly vertical black dashed line shows the time evolution of the total amount of calcium dosed during the titration experiment.

3.1.3. Free Energy of Ion Associates

The free energy of Ca^{2+} and CO_3^{2-} ions associated into stable complexes, present before the onset of $CaCO_3$ solid phase formation, could be calculated using the thermodynamic expression $\Delta G_{binding} = -RT \ln K'$, where K' can be estimated through the measurement of Ca^{2+} concentrations according to the method proposed by Gebauer et al. [20]. From these calculations, it can be concluded that the smaller the counterion, either from the carbonate or the calcium sources, the greater the stabilization of the pre-nucleation species (i.e., more negative $\Delta G_{binding}$), as depicted in Figure 3a,b. This stabilizing effect is more clearly observed in the case of different anions in the calcium source (i.e., CaX_2 where $X = Cl^-$, Br^- or I^-). However, in the cases of both anions and cations, changes in $\Delta G_{binding}$ were limited, and almost within error values, in comparison with reported values for other additives, such as organics, e.g., polyacrylate on $CaCO_3$ [36], citrate on calcium oxalate [37], or apatite [38].

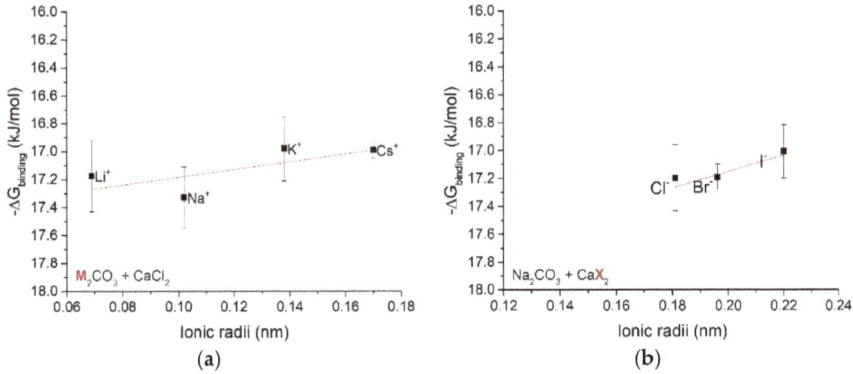

Figure 3. $\Delta G_{binding}$ calculated for (**a**) different cations in the carbonate source, and (**b**) different anions in the calcium source. Errors bars show $2\sigma_N$.

3.1.4. Particle Size and Phase Evolution

Dynamic light scattering measurements show particle sizes ranging from 20 nm to more than 1 µm (Figure 4). It is likely that particles are a mixture of all detected phases (ACC, vaterite and calcite; see phase analysis below), the smallest ones being ACC (more abundant at the early stages), and the larger ones vaterite and calcite (more abundant at later stages). It was also observed that the smaller the size of the background ion present in the solution, the broader the size distribution and the higher the amount of the smallest particles (i.e., left tail of PSD plot). This held true for both cations (especially in the case of Li^+, Figure 4a) and anions (especially in the case of Cl^-, Figure 4b). These results show that smaller precipitates were achieved using smaller background ions.

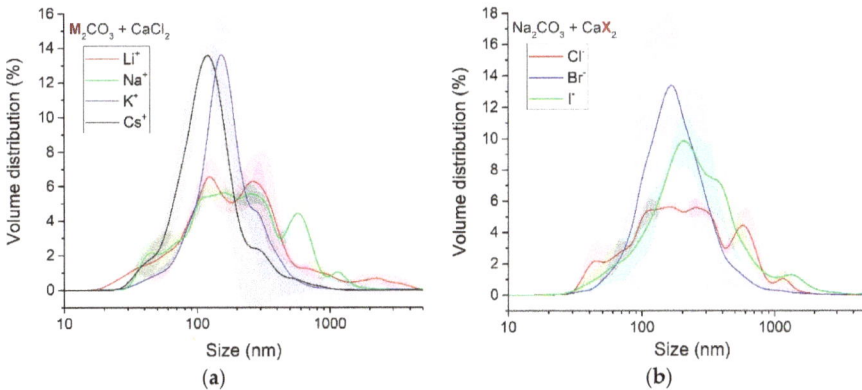

Figure 4. Particle size distribution (average of 120 consecutive measurements), expressed as volume percentage, of the precipitates during titration experiments (**a**) using different cations in the carbonate source and (**b**) using different anions in the calcium source. Shaded areas represent standard deviation from at least five titration experiments.

The morphology of the initial and final precipitates, and their mineralogy, were determined using FE-SEM images (Figure 5a–d) and X-ray diffraction analysis (Figure 5e,f). Irrespective of the type of background ions, FE-SEM observations showed spherical ACC nanoparticles from the early stages of nucleation that were consistent with the sizes detected from DLS measurements (Figure 5a). XRD analyses confirmed the amorphous nature of these early precipitates (Figure 5e). Samples collected at the very end of the titration experiments showed the presence of ~90% of vaterite and ~10% of calcite irrespective of the background electrolyte used in the titration experiment (see XRD results in Figure 5f). FE-SEM observations showed the presence of vaterite structures in close contact with calcite rhombohedra, which suggests that the latter formed after the (partial) dissolution of the former (Figure 5b). The existence of such a dissolution–precipitation process [39] was confirmed by observations of vaterite casts on (104) faces of calcite crystals (Figure S2). Interestingly, FE-SEM observations showed that vaterite structures displayed an almost perfect hexagonal plate-shaped morphology (Figure 5c). At a higher magnification, FE-SEM imaging disclosed a nanogranular structure made up of oriented vaterite nanoparticles (Figure 5d). The latter observation points to a structural development via (oriented) nanoparticle aggregation (Figure 5d), as reported by Jiang et al. [40].

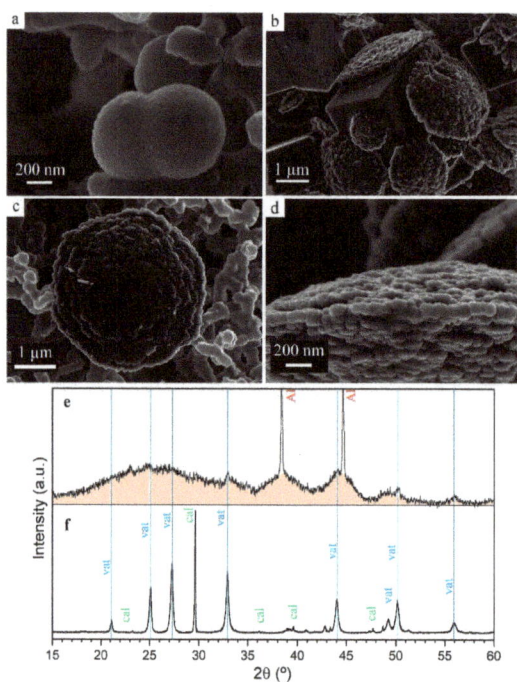

Figure 5. Representative field emission-scanning electron microscope FE-SEM images showing (**a**) spherical amorphous calcium carbonate (ACC) nanoparticles consistent in size with dynamic light scattering (DLS) measurements; (**b**) coupling of vaterite dissolution and rhombohedral calcite growth; (**c**) vaterite showing an almost perfect hexagonal plate shape; (**d**) detail of a vaterite crystal surface showing a nanogranular structure; (**e**) X-ray diffraction (XRD) analysis of precipitates obtained during titration experiments at the early stages of nucleation; and (**f**) representative diffraction pattern of samples collected at the very end of the titration experiments. Legend: cal: calcite, vat: vaterite, Al: aluminum sample holder (used as internal standard). The orange shaded area in (**e**) indicates the characteristic region of amorphous calcium carbonate (see Figure S2 for additional FE-SEM images showing the transformation of vaterite into calcite via a dissolution-precipitation process).

3.2. Type II Experiments

The main objective of this second type of experiment was to investigate the influence that background ions added as foreign electrolytes, as well as their concentration, exert on $CaCO_3$ precipitation. Titration experiments showed that in the case of the different cations (M = Li^+, Na^+, K^+ or Cs^+) in the background salt (MCl), and for the lowest ionic strength tested (IS = 0.035, [MCl] = 10 mM), a slight reduction in both the Ca^{2+}_{max} concentration (Figure 6a) and induction time (Figure 6b) were found with increasing size of background cations. However, for the highest background salt concentration tested (IS = 0.049, [MCl] = 25 mM), this trend seemed to flatten and/or slightly invert. However, note that measured values in this latter case were within error values.

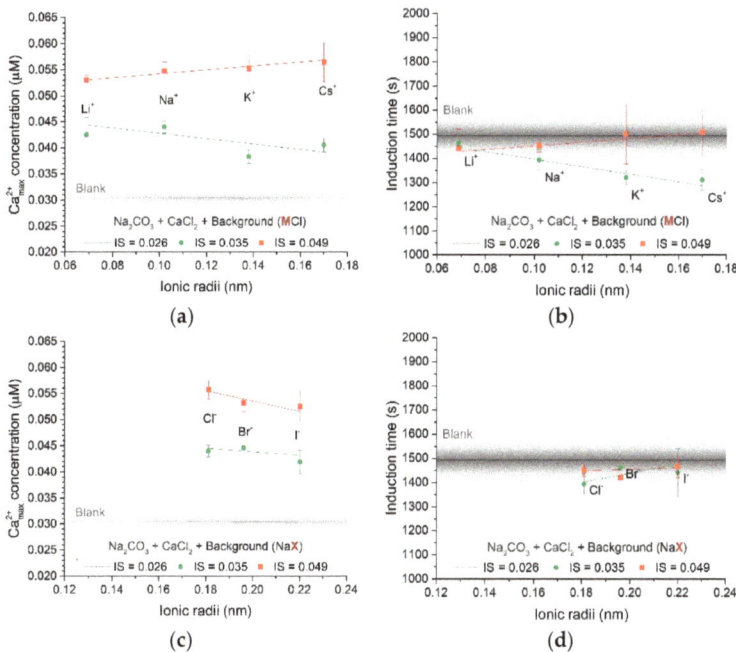

Figure 6. Maximum free Ca^{2+} concentration and elapsed time before the onset of $CaCO_3$ precipitation for type II experiments. $CaCl_2$ was used as titrant, and Na_2CO_3 as carbonate source in the reactor together with different external background salts (**a,b**) using different chloride salts, (**c,d**) using different sodium salts. Two different background electrolyte concentrations (10 and 25 mM) were used, increasing the ionic strength (IS) to 0.035 (green lines) and 0.049 (red lines), respectively. Errors bars show $2\sigma_N$. The blank black line and grey band show the average value and $2\sigma_N$, respectively, from titration experiments without any external background salt.

In the case of the different anions (X = Cl^-, Br^- or I^-) in the background salt (NaX), an apparent slight reduction in Ca^{2+}_{max} with increasing size of the background anions, and an increase in Ca^{2+}_{max} with increasing background salt concentrations were observed (Figure 6c). Induction times were within error values and similar in all cases (Figure 6d).

4. Discussion

Background electrolytes have been shown to modify both the growth and the dissolution of many minerals, e.g., gypsum, calcite, barite, whewellite, among others [21,22,24,25,41–50]. Ions at the mineral-solution interface are continuously attaching to and detaching from step edges and, consequently, mineral dissolution or growth kinetics depend on energy barriers involving these ions [51]. However, nucleation is a precondition for mineral growth, and is interpreted as an energetic event in which a system tends towards a reduction in its total free energy once activation energy barriers are overcome [52,53]. In this energetic scenario, several different interactions have to be considered (i.e., water-water, ion-water, and ion-ion interactions) [54], which will affect both the pre-nucleation and nucleation regimes.

4.1. Ion-Water Interactions

As a result of their charge, ions in solution promote the development of two distinct hydration regions around themselves: one closer to the ion, in which water is tightly bound and electrostricted as a hydration (inner) shell; and (outer shell) water that is under the influence of the electric field of the ions within the bulk water [27]. Ions interact with water molecules around them, forming a "cavity in the water" of radius r + Δr (i.e., an ion plus its hydration shell), which allows them to interact with the bulk water as if it were uncharged due to charge dispersion and dipole-induced forces between water molecules and the ion [55]. The mobility of water molecules in the vicinity of ions has been well-studied through numerous experimental and theoretical investigations [56]. This water mobility controls the diffusion of ions in aqueous solutions. Frequently, it has been presented as the ratio between the residence time of a water molecule in the solvation shell of the ion and in pure water (i.e., τ_i/τ_0) [57]. The latter depends on the competition between the tendency of an ion to orient water molecules in its solvation shell and the opposition of water to disrupt its hydrogen-bonded network [58]. Additionally, Samoïlov [59] pointed out that residence times are also due to differences in the activation energies (ΔE_i) between removing a water molecule from the ion solvation shell (E_i), and the activation energy required for transferring a water molecule from the first to the next coordination shell of another water molecule (E_0). Therefore, the aforementioned parameters allow us to define ions as positively or negatively hydrated, according to a retarded (i.e., $\tau_i > \tau_0$; $\Delta E_i > 0$) or an increased ($\tau_i < \tau_0$; $\Delta E_i < 0$) mobility of water from the solvation shell, respectively [30].

Small ions present a high charge density (i.e., high ionic potential), that results in strong hydration due to a closer approximation between the point charge of the ion and the point charge of the opposite charge in the water molecules [54]. This results in a higher activation energy needed to remove a water molecule from the ion solvation shell (E_i) and a longer residence time (τ_i) of a water molecule in the ion solvation shell. Therefore, hydration effects in aqueous solutions are highly dependent on the size and charge of the ions present. This inevitably will profoundly influence both the dissolution and growth of a mineral, as the characteristics of the aqueous solutions from which growth or dissolution occurs change depending on the ions in the solution.

4.2. Calcium Carbonate Pre-Nucleation Clusters

Gebauer et al. have reported on the existence of stable calcium carbonate pre-nucleation clusters (PNC) based on ion potential measurements in combination with analytical ultracentrifugation results [60]. This has also been confirmed by cryogenic transmission electron microscopy [60] and, additionally, through atomistic computational simulations [18]. The size of calcium carbonate PNC has been estimated to be below 2 nm; however, discrepancies on the exact size value exist due to the effect of surrounding hydration layers. Such hydration layers could be influenced by the electrostatic environment related to the nature of the background ionic species present in the solution.

Hydration of both Ca^{2+} and CO_3^{2-} ions, and/or PNC, which are considered solute species [20,52], in the presence of background ions will be controlled by the dynamics of water molecules in such solutions (E_{BCKG_0}), apart from the potential mineral structure-building ions and/or PNC interactions with solvent molecules (E_{BCKG_i}). In the presence of background ions, water–water interactions would be different with respect to pure water, due to the electrostatic environment related to background ions present in the solution. The stronger the background ion–water interactions (X^--H_2O and M^+-H_2O) resulting from higher charge density and lower ionic size, the higher the activation energy barrier of expelling water from the solvation shells of background ions [24,61,62]. Consequently, the more structured the water network, the lower the activation energy required to break water–water interactions in the presence of background ions (E_{BCKG_0}). Restricted water movement at the PNC-solution interface (i.e., Li$^+$ or Cl$^-$) would hinder phase separation, because desolvation would be the rate-limiting step for nucleation [10,61].

Also, the presence of background ions (M$^+$ and X$^-$) influences the required energy to strip off a water molecule from the solvation shell of a structure-forming ion (such as Ca^{2+} and CO$_3{}^{2-}$) and/or from PNC solute species. At low ionic strength, similar to our working conditions, the average distance between ions building the crystal and background counterions (Ca^{2+}//X$^-$ and/or CO$_3{}^{2-}$//M$^+$) is smaller than between ions building the crystal and background ions with the same charge sign [63]. Background ions with a different charge sign (counterions) with respect to Ca^{2+} or CO$_3{}^{2-}$ ions reduce the potential energy of water molecules in the Ca^{2+} or CO$_3{}^{2-}$ solvation shells, as a result of attractive interactions between the partial charge of the water dipole and the oppositely charged electric field of the counterions [24,64,65]. The latter results in stabilization and, therefore, increasing residence times of these water molecules (τ_{BCKG_i}) in the Ca^{2+} or CO$_3{}^{2-}$ solvation shells.

The stabilization of water molecules both surrounding and incorporated within PNC, by a strengthening of the electric field emanating from background ions, could decrease the dynamics of the clusters' equilibrium and stabilize pre-nucleation species. However, their dehydration and subsequent transformation into solid species would be also hampered. According to our results, the smaller the size of the background ion, and therefore the stronger its interaction with water, the larger the overall reduction in ΔG_{PNC}, resulting in a stabilization of PNC and a less favorable calcium carbonate precipitation. The latter is in agreement with the higher free Ca^{2+} concentration reached, and the resulting retardation of the onset of nucleation.

In summary, the solvation shell stability of either Ca^{2+}, CO$_3{}^{2-}$, ion pairs and/or PNC will depend, among other variables, on the characteristics of the background ions present in the solution. According to our experimental results, background ions induce stabilization of PNC, as deduced from the overall values of $\Delta G_{binding}$ (see Figure 3), according to the following trend Li$^+$ > Na$^+$ > K$^+$ ≥ Cs$^+$ for the cations, and Cl$^-$ > Br$^-$ > I$^-$ for the anions (i.e., higher stabilization is achieved with decreasing ionic radius). More negative $\Delta G_{binding}$ values (such as in the case of Li$^+$ as a carbonate source or Cl$^-$ as a calcium source), would limit dehydration kinetics of PNC by increasing the residence time of water molecules entering into their structure. Thus, longer induction times and higher supersaturation values would be required before the onset of nucleation occurs. The higher supersaturation values required for the onset of nucleation would lead to lower particle sizes [10], which is in agreement with our DLS experimental results (as seen in the left tails in the PSD plots shown in Figure 4).

4.3. Effects on ACC Solubility and Polymorph Selection

Two different ACC "polymorphs" have been reported in the literature, ACC I and ACC II [20], with different solubilities attributed to their proto-calcite and proto-vaterite structure, respectively. However, we have not found any clear relationship between observed lower or higher solubilities determined following titration experiments and the final polymorph selection. Actually, calcite-vaterite ratios were similar, within error, for all the different experimental precipitation runs. In our experiments, it is not clear whether or not a specific ACC proto-structure determines the phase selection. In all cases, the first phase formed after ACC (irrespective of ACC solubility) is vaterite, that during a dissolution–precipitation process transforms into calcite, although such a replacement is incomplete within the time-span of our titration experiments.

An alternative explanation for the existence of ACC nanoparticles with a higher solubility (i.e., runs including Li$^+$ or Cl$^-$, Figure 2) might be related to particle size effects. Zou et al. [34] have shown that there is a clear relationship between ACC particle size and solubility, which increases with decreasing ACC particle diameter (for particles with diameter <200 nm). This is fully consistent with our DLS measurements (Figure 4) showing that the left-tails (i.e., smallest particle size) of PSD plots expand to smaller sizes in the presence of Li^+, as well as for Cl^-. It should be also considered that another explanation for the measured solubility could be related to variations in the water content of ACC. Rodriguez-Navarro et al. have shown that there is a linear correlation between water content in ACC and solubility (i.e., lower solubility in anhydrous ACC and increased solubility as the water content in ACC increases) [33]. However, in all cases, the dissolution of ACC particles resulted in a solution with a sufficiently high supersaturation with respect to vaterite, as to enable its precipitation (following Ostwald's step rule) [10].

4.4. Influence of Background Electrolyte Concentration

Our results show that in type II experiments, differences in both induction time and $Ca^{2+}{}_{max}$ concentration for $CaCO_3$ nucleation diminished with increasing concentration of background electrolytes, and therefore the solution ionic strength, through the addition of different 1:1 salts. This suggests that at low ionic strength (i.e., low background electrolyte concentration), the electrostatic environment created by background ions controls ion–water interactions, while also depending substantially on the nature of the background electrolyte [26,54]. With the addition of background electrolytes with smaller ionic sizes, Ca^{2+}, $CO_3{}^{2-}$, ion pairs and/or PNCs will be stabilized, as stated above, and subsequently the time required for nucleation will be higher (see Figure 6).

However, at higher background electrolyte concentrations (and higher ionic strengths), the destabilizing effect, induced by the presence of like-charge ions on PNC hydration water, will become increasingly important, due to the reduced average distance between ionic species. This is because of like-charge ions in the solution, and the effect of background ions on the PNC hydration, which will stop controlling ion–water interactions (τ_{BCKG_i}) [63,65]. However, background electrolytes may still influence bulk solvent structure dynamics (τ_{BCKG_0}). The stronger the interaction between background ions and water, the weaker the hydration of structure-forming ions (or PNC). Therefore, clustering and dehydration would be favored, enhancing the nucleation of a solid phase. However, no reduction in the induction times for nucleation was observed when an enhancement of $CaCO_3$ precipitation with increasing IS should be expected. The latter could be attributed to our experimental design, due to the impossibility to fix both IS and pH. When increasing the background electrolyte concentration, the pH slightly decreased, which, according to Gebauer et al. [20], results in an increase in both the induction time and $Ca^{2+}{}_{max}$ for $CaCO_3$ precipitation that may counteract the theoretically expected decrease in nucleation time.

Finally, we stress that the observed background electrolyte effects should not be constrained to the nucleation and growth of calcium carbonate; they also should play a role in the dissolution of this phase. This is consistent with differences in both dissolution rates, measured from in situ flow-through calcite dissolution experiments using atomic force microscopy (AFM) [22,49,66], and atypical surface morphological features development, as in the presence of NH_4Cl background ions, which lead to spicule formation on {104} calcite surfaces [67].

5. Conclusions

First, it should be considered that water molecules in the solvation shell of a solute ion are stabilized by the presence of background counterions in the solution [65,68]. The increased affinity of water molecules for solute ions hampers precipitation in the case of strongly hydrated background ions (e.g., Li^+ and Cl^-). This is related to the structuring effect of water molecules when they are ordered around strongly hydrated ions and/or clusters. If the hydration shells of ions and/or clusters are stabilized due to the presence of background salts, the overall precipitation process would be hampered.

Our results suggest that the ACC solubility differences found in experiments with specific background ions (e.g., in the case of Li^+ in the carbonate source or Cl^- in the calcium source) might not be related to a specific ACC protostructure, but rather to the particle size of ACC formed at the very early stages of precipitation and/or the degree of hydration of ACC. Such solubility (and stability) differences are related to the ionic radius of background electrolyte ions. Due to the presence of strongly hydrated background electrolytes, the stabilization of pre-nucleation species results in higher supersaturation values and longer induction times before the onset of nucleation occurs. The latter is also in agreement with the observed lower particle sizes.

These results have important implications not only for a better understanding of non-classical crystallization of calcium carbonate in laboratory experiments, but also in natural environments. It should be pointed out that a comparison between calcium carbonate precipitation experiments by different researchers may not be straightforward because of the use of different reactants and IS. We also suggest that stabilization of PNC by specific background electrolytes can be selectively used to favor or hinder the precipitation or transport in solution of calcium carbonate in industrial and pharmaceutical processes.

Supplementary Materials: The following are available online at www.mdpi.com/2075-163X/7/7/126/s1, Figure S1: (a) Enthalpy of hydration vs. ionic radii and (b) Standard molar Gibbs energy of hydration of an ion vs. ionic radii. Note the quasilinear relationship between both variables with decreasing ionic radii; Figure S2: FESEM images of vaterite to calcite transition from the same titration run as seen in Figure 5b showing (a) vaterite attached to calcite. The development of growth steps is observed on the calcite surface; and (b) calcite crystals with vaterite casts.

Acknowledgments: This research was done within the grants MAT2012-37584, CGL2015-70642-R and P11-RNM-7550, funded by the Spanish Government, European Commission (ERDF funds) and the Junta de Andalucía. Additional funding was provided by the research group RNM-179 of the Junta de Andalucía and the Unidad Científica de Excelencia UCE-PP2016-05 of the University of Granada. Encarnacion Ruiz-Agudo acknowledges the receipt of a Ramón y Cajal grant from the Spanish Government (Ministerio de Economía y Competitividad) and CVP acknowledges funding from the EU Marie Curie initial training networks: Minsc, CO_2 React and Flowtrans as well as an Australian Research Council (ARC) grant awarded to Julian Gale at Curtin University, Perth Australia.

Author Contributions: Encarnacion Ruiz-Agudo conceived and designed the experiments; Alejandro Burgos-Cara performed the experiments; Alejandro Burgos-Cara and Christine V. Putnis analyzed the data; Carlos Rodriguez-Navarro contributed analysis tools; Alejandro Burgos-Cara and Carlos Rodriguez-Navarro wrote the paper, with contributions from all authors.

Conflicts of Interest: The authors declare no conflict of interest.

References

1. Gower, L.B. Biomimetic model systems for investigating the amorphous precursor pathway and its role in biomineralization. *Chem. Rev.* **2008**, *108*, 4551–4627. [CrossRef] [PubMed]

2. Cam, N.; Georgelin, T.; Jaber, M.; Lambert, J.F.; Benzerara, K. In vitro synthesis of amorphous Mg-, Ca-, Sr- and Ba-carbonates: What do we learn about intracellular calcification by cyanobacteria? *Geochim. Cosmochim. Acta* **2015**, *161*, 36–49. [CrossRef]

3. Zhong, C.; Chu, C.C. Acid polysaccharide-induced amorphous calcium carbonate (ACC) films: Colloidal nanoparticle self-organization process. *Langmuir* **2009**, *25*, 3045–3049. [CrossRef] [PubMed]

4. Tobler, D.J.; Blanco, J.D.R.; Dideriksen, K.; Sand, K.K.; Bovet, N.; Benning, L.G.; Stipp, S.L.S. The effect of aspartic acid and glycine on amorphous calcium carbonate (ACC) structure, stability and crystallization. *Proced. Earth Planet. Sci.* **2014**, *10*, 143–148. [CrossRef]

5. Bentov, S.; Weil, S.; Glazer, L.; Sagi, A.; Berman, A. Stabilization of amorphous calcium carbonate by phosphate rich organic matrix proteins and by single phosphoamino acids. *J. Struct. Biol.* **2010**, *171*, 207–215. [CrossRef] [PubMed]

6. De Yoreo, J.J.; Dove, P.M. Shaping crystals with biomolecules. *Science* **2004**, *306*, 1301–1302. [CrossRef] [PubMed]

7. Zeng, J.; Yin, Z.; Chen, Q. Effect of tetracarbon additives on gibbsite precipitation from seeded sodium aluminate liquor. *J. Cent. South Univ. Technol.* **2008**, *15*, 622–626. [CrossRef]

8. Orshesh, Z.; Hesaraki, S.; Khanlarkhani, A. Blooming gelatin: An individual additive for enhancing nanoapatite precipitation, physical properties, and osteoblastic responses of nanostructured macroporous calcium phosphate bone cements. *Int. J. Nanomed.* **2017**, *12*, 745–758. [CrossRef] [PubMed]

9. Wolf, S.E.; Böhm, C.F.; Harris, J.; Demmert, B.; Jacob, D.E.; Mondeshki, M.; Ruiz-Agudo, E.; Rodríguez-Navarro, C. Nonclassical crystallization in vivo et in vitro (I): Process-structure-property relationships of nanogranular biominerals. *J. Struct. Biol.* **2016**, *196*, 244–259. [CrossRef] [PubMed]

10. Rodríguez-Navarro, C.; Ruiz-Agudo, E.; Harris, J.; Wolf, S.E. Nonclassical crystallization in vivo et in vitro (II): Nanogranular features in biomimetic minerals disclose a general colloid-mediated crystal growth mechanism. *J. Struct. Biol.* **2016**, *196*, 260–287. [CrossRef] [PubMed]

11. Lowenstam, H.A.; Weiner, S. *On Biomineralization*; Oxford University Press: Oxford, UK, 1989.

12. Gago-Duport, L.; Briones, M.J.I.; Rodríguez, J.B.; Covelo, B. Amorphous calcium carbonate biomineralization in the earthworm's calciferous gland: Pathways to the formation of crystalline phases. *J. Struct. Biol.* **2008**, *162*, 422–435. [CrossRef] [PubMed]

13. Loste, E.; Wilson, R.M.; Seshadri, R.; Meldrum, F.C. The role of magnesium in stabilising amorphous calcium carbonate and controlling calcite morphologies. *J. Cryst. Growth* **2003**, *254*, 206–218. [CrossRef]

14. Gonzalez-Munoz, M.T.; Ben Chekroun, K.; Ben Aboud, A.; Arias, J.M.; Rodriguez-Gallego, M. Bacterially induced Mg-calcite formation: Role of Mg^{2+} in development of crystal morphology. *J. Sediment. Res.* **2000**, *70*, 559–564. [CrossRef]

15. Davis, K.J. The Role of Mg^{2+} as an Impurity in Calcite Growth. *Science* **2000**, *290*, 1134–1137. [CrossRef] [PubMed]

16. Zhang, Y.; Dawe, R.A. Influence of Mg^{2+} on the kinetics of calcite precipitation and calcite crystal morphology. *Chem. Geol.* **2000**, *163*, 129–138. [CrossRef]

17. Berg, J.K.; Jordan, T.; Binder, Y.; Börner, H.G.; Gebauer, D. Mg^{2+} Tunes the Wettability of Liquid Precursors of $CaCO_3$: Toward Controlling Mineralization Sites in Hybrid Materials. *J. Am. Chem. Soc.* **2013**, *135*, 12512–12515. [CrossRef] [PubMed]

18. Demichelis, R.; Raiteri, P.; Gale, J.D. Structure of hydrated calcium carbonates: A first-principles study. *J. Cryst. Growth* **2013**. [CrossRef]

19. Demichelis, R.; Raiteri, P.; Gale, J.D.; Quigley, D.; Gebauer, D. Stable prenucleation mineral clusters are liquid-like ionic polymers. *Nat. Commun.* **2011**, *2*, 590. [CrossRef] [PubMed]

20. Gebauer, D.; Volkel, A.; Colfen, H. Stable prenucleation calcium carbonate clusters. *Science* **2008**, *322*, 1819–1822. [CrossRef] [PubMed]

21. Burgos-Cara, A.; Putnis, C.V.; Rodriguez-Navarro, C.; Ruiz-Agudo, E. Hydration effects on gypsum dissolution revealed by in situ nanoscale atomic force microscopy observations. *Geochim. Cosmochim. Acta* **2016**, *179*, 110–122. [CrossRef]

22. Ruiz-Agudo, E.; Kowacz, M.; Putnis, C.V.; Putnis, A. The role of background electrolytes on the kinetics and mechanism of calcite dissolution. *Geochim. Cosmochim. Acta* **2010**, *74*, 1256–1267. [CrossRef]

23. Ruiz-Agudo, E.; Putnis, C.V.; Wang, L.; Putnis, A. Specific effects of background electrolytes on the kinetics of step propagation during calcite growth. *Geochim. Cosmochim. Acta* **2011**, *75*, 3803–3814. [CrossRef]

24. Kowacz, M.; Putnis, A. The effect of specific background electrolytes on water structure and solute hydration: Consequences for crystal dissolution and growth. *Geochim. Cosmochim. Acta* **2008**, *72*, 4476–4487. [CrossRef]

25. Ruiz-Agudo, E.; Urosevic, M.; Putnis, C.V.; Rodríguez-Navarro, C.; Cardell, C.; Putnis, A. Ion-specific effects on the kinetics of mineral dissolution. *Chem. Geol.* **2011**, *281*, 364–371. [CrossRef]

26. Collins, K.D.; Neilson, G.W.; Enderby, J.E. Ions in water: Characterizing the forces that control chemical processes and biological structure. *Biophys. Chem.* **2007**, *128*, 95–104. [CrossRef] [PubMed]

27. Marcus, Y. Thermodynamics of solvation of ions. Part 5.—Gibbs free energy of hydration at 298.15 K. *J. Chem. Soc. Faraday Trans.* **1991**, *87*, 2995–2999. [CrossRef]

28. Saharay, M.; Yazaydin, A.O.; Kirkpatrick, R.J. Dehydration-induced amorphous phases of calcium carbonate. *J. Phys. Chem. B* **2013**, *117*, 3328–3336. [CrossRef] [PubMed]

29. Saharay, M.; James Kirkpatrick, R. Onset of orientational order in amorphous calcium carbonate (ACC) upon dehydration. *Chem. Phys. Lett.* **2014**, *591*, 287–291. [CrossRef]

30. Kowacz, M.; Prieto, M.; Putnis, A. Kinetics of crystal nucleation in ionic solutions: Electrostatics and hydration forces. *Geochim. Cosmochim. Acta* **2010**, *74*, 469–481. [CrossRef]

31. Kielland, J. Individual activity coefficients of ions in aqueous solutions. *J. Am. Chem. Soc.* **1937**, *59*, 1675–1678. [CrossRef]

32. Zhuo, K.; Dong, W.; Wang, W.; Wang, J. Activity coefficients of individual ions in aqueous solutions of sodium halides at 298.15 K. *Fluid Phase Equilib.* **2008**, *274*, 80–84. [CrossRef]

33. Rodriguez-Navarro, C.; Kudlacz, K.; Cizer, O.; Ruiz-Agudo, E.; Kudłacz, K.; Cizer, Ö.; Ruiz-Agudo, E.; Kudlacz, K.; Cizer, O.; Ruiz-Agudo, E. Formation of amorphous calcium carbonate and its transformation into mesostructured calcite. *CrystEngComm* **2015**, *17*, 58–72. [CrossRef]

34. Zou, Z.; Bertinetti, L.; Politi, Y.; Jensen, A.C.S.; Weiner, S.; Addadi, L.; Fratzl, P.; Habraken, W.J.E.M. Opposite particle size effect on amorphous calcium carbonate crystallization in water and during heating in air. *Chem. Mater.* **2015**, *27*, 4237–4246. [CrossRef]

35. Parkhurst, D.L.; Appelo, C.A.J. *Description of Input and Examples for PHREEQC Version 3–A Computer Program for Speciation, Batch-Reaction, One-Dimensional Transport, and Inverse Geochemical Calculations*; U.S. Geological Survey: Reston, VA, USA, 2013.

36. Verch, A.; Gebauer, D.; Antonietti, M.; Cölfen, H. How to control the scaling of $CaCO_3$: A "fingerprinting technique" to classify additives. *Phys. Chem. Chem. Phys.* **2011**, *13*, 16811–16820. [CrossRef] [PubMed]

37. Qiu, S.R.; Wierzbicki, A.; Salter, E.A.; Zepeda, S.; Orme, C.A.; Hoyer, J.R.; Nancollas, G.H.; Cody, A.M.; De Yoreo, J.J. Modulation of calcium oxalate monohydrate crystallization by citrate through selective binding to atomic steps. *J. Am. Chem. Soc.* **2005**, *127*, 9036–9044. [CrossRef] [PubMed]

38. Hu, Y.-Y.; Rawal, A.; Schmidt-Rohr, K. Strongly bound citrate stabilizes the apatite nanocrystals in bone. *Proc. Natl. Acad. Sci. USA* **2010**, *107*, 22425–22429. [CrossRef] [PubMed]

39. Rodriguez-Blanco, J.D.; Shaw, S.; Benning, L.G. The kinetics and mechanisms of amorphous calcium carbonate (ACC) crystallization to calcite, via vaterite. *Nanoscale* **2011**, *3*, 265–271. [CrossRef] [PubMed]

40. Jiang, W.; Pacella, M.S.; Athanasiadou, D.; Nelea, V.; Vali, H.; Hazen, R.M.; Gray, J.J.; McKee, M.D. Chiral acidic amino acids induce chiral hierarchical structure in calcium carbonate. *Nat. Commun.* **2017**, *8*, 15066. [CrossRef] [PubMed]

41. Bosbach, D.; Junta-Rosso, J.L.; Becker, U.; Hochella, M.F. Gypsum growth in the presence of background electrolytes studied by scanning force microscopy. *Geochim. Cosmochim. Acta* **1996**, *60*, 3295–3304. [CrossRef]

42. Wolthers, M.; Nehrke, G.; Gustafsson, J.P.; Van Cappellen, P. Calcite growth kinetics: Modeling the effect of solution stoichiometry. *Geochim. Cosmochim. Acta* **2012**, *77*, 121–134. [CrossRef]

43. Weaver, M.L.; Qiu, S.R.; Hoyer, J.R.; Casey, W.H.; Nancollas, G.H.; De Yoreo, J.J. Inhibition of calcium oxalate monohydrate growth by citrate and the effect of the background electrolyte. *J. Cryst. Growth* **2007**, *306*, 135–145. [CrossRef]

44. Putnis, C.V.; Ruiz-Agudo, E. The mineral-water interface: Where minerals react with the environment. *Elements* **2013**, *9*, 177–182. [CrossRef]

45. Ruiz-Agudo, E.; Putnis, C.V.; Rodriguez-Navarro, C. Reactions between minerals and aqueous solutions. In *Mineral Reaction Kinetics: Microstructures, Textures, Chemical and Isotopic Signatures*; Mineralogical Society of Great Britain & Ireland: Middlesex, UK, 2017; Volume 16, pp. 419–467.

46. Dove, P.M.; Czank, C.A. Crystal chemical controls on the dissolution kinetics of the isostructural sulfates: Celestite, anglesite, and barite. *Geochim. Cosmochim. Acta* **1995**, *59*, 1907–1915. [CrossRef]

47. Pokrovsky, O.S.; Schott, J. Iron colloids/organic matter associated transport of major and trace elements in small boreal rivers and their estuaries (NW Russia). *Chem. Geol.* **2002**, *190*, 141–179. [CrossRef]

48. Kowacz, M.; Groves, P.; Esperança, J.M.S.S.; Rebelo, L.P.N. On the use of ionic liquids to tune crystallization. *Cryst. Growth Des.* **2011**, *11*, 684–691. [CrossRef]

49. Ruiz-Agudo, E.; Putnis, C.V. Direct observations of mineral fluid reactions using atomic force microscopy: The specific example of calcite. *Mineral. Mag.* **2012**, *76*, 227–253. [CrossRef]

50. Aoba, T. The effect of fluoride on apatite structure and growth. *Crit. Rev. Oral Biol. Med.* **1997**, *8*, 136–153. [CrossRef] [PubMed]

51. DeYoreo, J.J.; Vekilov, P.G. Principles of crystal nucleation and growth. *Rev. Mineral. Geochem.* **2003**, *54*, 57–93. [CrossRef]

52. Gebauer, D.; Kellermeier, M.; Gale, J.D.; Bergström, L.; Cölfen, H. Pre-nucleation clusters as solute precursors in crystallisation. *Chem. Soc. Rev.* **2014**, *43*, 2348–2371. [CrossRef] [PubMed]

53. Sear, R.P. Nucleation: Theory and applications to protein solutions and colloidal suspensions. *J. Phys. Condens. Matter* **2007**, *19*, 33101. [CrossRef]

54. Collins, K.D. Charge density-dependent strength of hydration and biological structure. *Biophys. J.* **1997**, *72*, 65–76. [CrossRef]

55. Collins, K.D. Ion hydration: Implications for cellular function, polyelectrolytes, and protein crystallization. *Biophys. Chem.* **2006**, *119*, 271–281. [CrossRef] [PubMed]

56. Marcus, Y. Effect of ions on the structure of water: Structure making and breaking. *Chem. Rev.* **2009**, *109*, 1346–1370. [CrossRef] [PubMed]

57. Marcus, Y. *Ions in Water and Biophysical Implications*; Springer: Dordrecht, The Netherlands, 2012; Volume 53.

58. Hribar, B.; Southall, N.T.; Vlachy, V.; Dill, K.A. How ions affect the structure of water. *J. Am. Chem. Soc.* **2002**, *124*, 12302–12311. [CrossRef] [PubMed]

59. Samoïlov, O.I. *Structure of Aqueous Electrolyte Solutions and the Hydration of Ions*; Consultants Bureau: London, UK, 1965.

60. Pouget, E.M.; Bomans, P.H.H.; Goos, J.A.C.M.; Frederik, P.M.; de With, G.; Sommerdijk, N.A.J.M. The initial stages of template-controlled $CaCO_3$ formation revealed by cryo-TEM. *Science* **2009**, *323*, 1455–1458. [CrossRef] [PubMed]

61. Dorvee, J.R.; Veis, A. Water in the formation of biogenic minerals: Peeling away the hydration layers. *J. Struct. Biol.* **2013**, *183*, 278–303. [CrossRef] [PubMed]

62. Malenkov, G.G.; Samoïlov, O.I. Electrostatic interaction and coordination of molecules in water. *J. Struct. Chem.* **1965**, *6*, 6–10. [CrossRef]

63. Kinoshita, M.; Harano, Y. Potential of mean force between solute atoms in salt solution: Effects due to salt species and relevance to conformational transition of biomolecules. *Bull. Chem. Soc. Jpn.* **2005**, *78*, 1431–1441. [CrossRef]

64. Samoïlov, O.I. The theory of the salting out in aqueous solutions-II. Dependence of dehydration and hydration on the initial degree of hydration of a cation undergoing salting out. *J. Struct. Chem.* **1967**, *7*, 177–180. [CrossRef]

65. Samoïlov, O.I. Theory of salting out from aqueous solutions-III. Dependence of salting out on characteristics of ions of salting-out agent. *J. Struct. Chem.* **1971**, *11*, 929–931. [CrossRef]

66. Buhmann, D.; Dreybrodt, W. Calcite dissolution kinetics in the system H_2O-CO_2-$CaCO_3$ with participation of foreign ions. *Chem. Geol.* **1987**, *64*, 89–102. [CrossRef]

67. Long, X.; Meng, R.; Wu, W.; Ma, Y.; Yang, D.; Qi, L. Calcite microneedle arrays produced by inorganic ion-assisted anisotropic dissolution of bulk calcite crystal. *Chem.-A Eur. J.* **2014**, *20*, 4264–4272. [CrossRef] [PubMed]

68. Samoïlov, O.I. The theory of salting out from aqueous solutions-I. General problems. *J. Struct. Chem.* **1967**, *7*, 12–19. [CrossRef]

minerals

MDPI

Article

Liquid Cell Transmission Electron Microscopy and the Impact of Confinement on the Precipitation from Supersaturated Solutions

Roland Kröger * and Andreas Verch

Department of Physics, University of York, York, YO10 5DD, UK; andreas.verch@leibniz-inm.de
* Correspondence: roland.kroger@york.ac.uk

Received: 4 September 2017; Accepted: 12 January 2018; Published: 15 January 2018

Abstract: The study of nucleation and growth from supersaturated ion solutions is a key area of interest in biomineralization research and beyond with high-resolution in situ imaging techniques such as liquid cell transmission electron microscopy (LCTEM) attracting substantial attention. However, there is increasing experimental evidence that experiments performed with this technique differ from those performed in bulk solutions due to the spatial restriction, which is a prerequisite for LCTEM to provide electron transparent samples. We have performed 2D Finite Elements (FE) simulations to study the impact of confinement on the steady state concentration profiles around a nanoparticle in a supersaturated solution of the constituent ions. We find that confinement below a critical value significantly reduces the concentration of available ions in solutions and hence the stability of the precipitates. These findings could explain the necessity to substantially increase ion activities of Ca^{2+} and CO_3^{2-} to induce precipitation in LCTEM.

Keywords: liquid cell TEM; precipitation; confinement; crystallization

---

1. Introduction

Precipitation in confinement is a phenomenon of significant interest in a number of areas such as geology (e.g., the impact of salt crystallization in rocks) [1], environmental research (e.g., for the remediation of contaminants) [2] as well as biomineralization and biomorphism [3,4]. Hence, it has been studied with a variety of techniques such as electron microscopy and X-ray spectroscopy to reveal the details of the transformation process from ion solutions to solid, often crystalline, precipitates. Moreover, the recent advent of liquid cell transmission electron microscopy techniques raises the prospect of investigating crystal growth in situ at the nanometer level [5,6] but also the question of how the confinement, which is a prerequisite for the application of this technique, affects the dynamics of crystallization.

An open aspect is here the impact of confinement on the concentration of the ions in solutions upon precipitation, which is the focus of this work. We are concentrating on the transport of ions to the growing particle and how it is affected in the presence of an already formed precipitate particle by the reduction of dimension.

The application of LCTEM for the investigation of particle precipitation and dissolution dynamics in polar solvents such as water, where the precipitates are either of amorphous or crystalline nature and a key problem is the mechanism of ion transport and its impact on the stability of precipitates. This is particularly important when taking into account the spatial boundary conditions defining the problem, namely the confinement of the liquid by thin membranes to create an electron transparent sample. Inspired by previous precipitation experiments using LCTEM we developed a MATLAB code to solve the involved ion transport equation numerically using Finite Element (FE) simulations for different confinement scenarios.

2. Experimental and Simulation Details

The precipitation experiments used a three-port Poseidon 210 (Protochips Inc., Morrisville, NC, USA) liquid cell holder in a JEOL 2200 FS (JEOL (UK) Ltd, Herts, UK) in scanning transmission electron microscopy (STEM) mode. Hence will use the acronym LCSTEM from here. Two ports were used to inject the cation and anion solutions separately to be mixed in the observation area of the holder as described in the text and the used solution was extracted through an exit port. Standard pump rate was 300 μL/min and the experiment was performed at room temperature.

Two-dimensional (2D) simulations of ion concentration profiles were performed by finite-element calculations of the respective transport equations using a self-written MATLAB code (version R2017b, MathWorks Inc., Natick, MA, USA) approximating the respective differential equation by finite elements on 2D domains as discussed in the text [7].

3. CaCO$_3$ Formation in LCSTEM

The inspiration for this work originates from observations made for supersaturated aqueous solutions of calcium and carbonate using liquid cell transmission electron microscopy (LCSTEM). As has been reported the precipitation of crystalline calcium carbonate phases is only possible under standard conditions using a membrane spacing of 100–5000 nm for concentrations well exceeding 10 mM, which is more than 10 times above the supersaturation limit. To illustrate this we have studied the dynamics of calcium carbonate precipitation from supersaturated solutions with LCSTEM using a 500 nm spacer and 10 mM calcium chloride (CaCl$_2$) and sodium carbonate concentration (Na$_2$CO$_3$). This corresponds to a significant supersaturation of 170 in bulk solution with respect to calcite at pH = 8 and a temperature of 300 K.

The growth or dissolution of a particle depends on the saturation level of the constituent ions. Since the standard liquid cell holders operate with membranes as depicted in Figure 1 the confined area essentially depends on the lateral size of the smaller of the two encasing membrane-bearing chips generally in the order of several mm, and the choice of spacers [8]. To create the electron transparent volume, spacers of varying thickness are used and are typically in the range between 0.1 and 5 μm. This results in an aspect ratio of the distance between the membranes and the cross-section of the flow channels surrounding the membrane area of approx. 10^{-3} to 10^{-2}. These values result effectively in a convective flow barrier. Hence, in a two or three port liquid cell holder the supplied liquid can only arrive at the membrane area by diffusion [9].

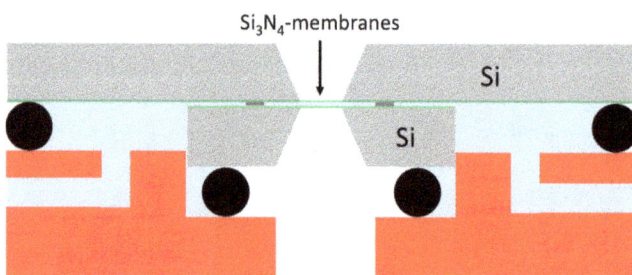

Figure 1. Schematic of LCTEM holder design (not to scale) in cross section. Black circles indicate O-rings sealing the water bearing tubes from the microscope vacuum. The electron transparent volume is given by the lateral extensions of the silicon nitride membranes and the spacers separating the two chips.

Imaging conditions were: 1.5 pA beam current, 200 kV acceleration voltage and 1 μs dwell time resulting in 0.5 s for each frame. For the lowest magnification and hence lowest electron dose (note, that we use "dose" in this context not in the strict radiochemistry sense, which considers the electron stopping power) no precipitation was observed throughout the field of view.

At magnifications of 20 k corresponding to approx. 3 e/Å^2 even for long time periods of >30 min no precipitation could be observed in the observable membrane area of approx. 400 × 400 μm.

Upon the increase of magnification a spontaneous precipitation was triggered by the electron beam and, upon dose reduction, i.e., by lowering the magnification, the formed particles dissolved (see image sequence in Figure 2). An inspection of the whole electron transparent sample area after the recording of the video sequence proved that no precipitation had occurred throughout in the absence of electron beam radiation. From the morphology of the nanoparticles we speculate that the formed particles were calcite rather than aragonite, vaterite or amorphous calcium carbonate.

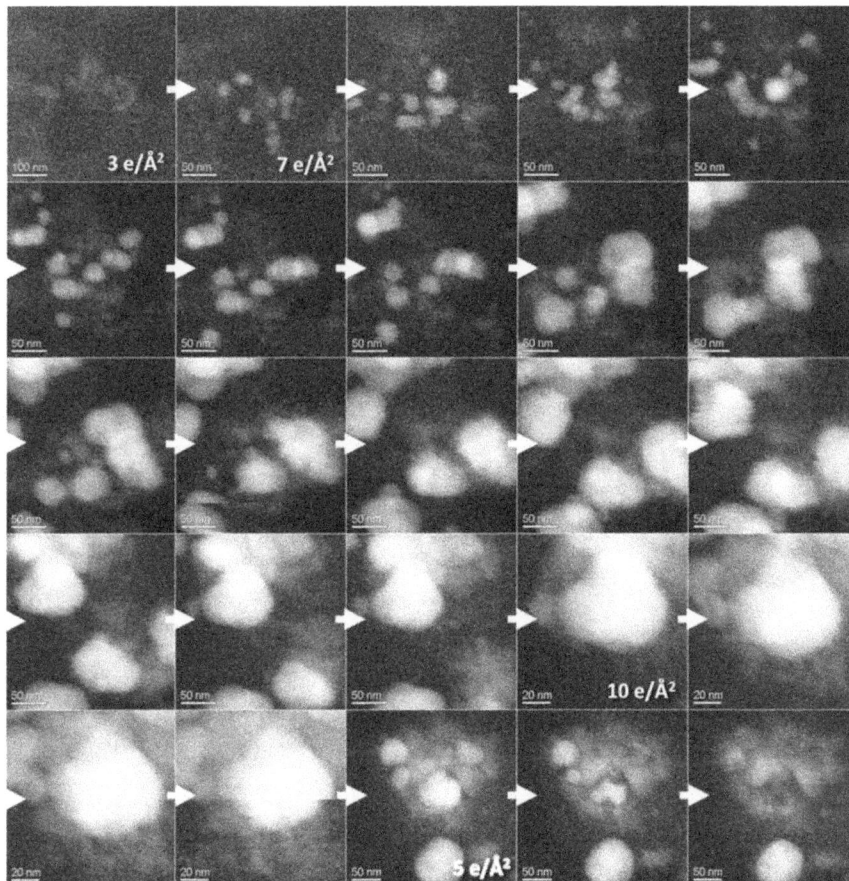

Figure 2. Sequence of LCSTEM stills from a video recording electron beam induced calcium carbonate formation from 10 mM supersaturated solution of Ca^{2+} and CO_3^{2-} ions and subsequent dissolution of most of the nanoparticles upon reduction of electron dose. Electron dose values were between 3e/Å and 10 e/Å and are indicated in the stills when changed. The whole frame frequency was recorded within a time interval of approx. 120 s, hence each frame was recorded at increment time steps of approx. 5 s with a recording time of 0.5 s.

A similar observation could be made when studying calcium carbonate formation in the presence of proteins where the nacre derived protein AP7 was found to lead to calcium binding and enhanced protein induced local ion concentration [10]. These observations do not exclude a possible role of radiolysis in lowering the pH as has been reported elsewhere [11]. However, we did not observe any bubble formation as to be expected from the CO_2 production accompanying the calcite particle dissolution resulting from a potential electron beam induced lowering of the pH value. Furthermore, the electron dose was maximal (10 e/$Å^2$) when the particle formation was observed in contrast to the reported pH reduction, which would lead to inhibition of nucleation and growth rather than to an enhancement. We further regard the previously reported finding that calcium and carbonate ion solution concentrations of a minimum of 20 mM were required to obtain precipitation with LCTEM as supportive for our observations [12]. These observations raise the question of what role confinement plays in the control of crystallization. To quantitatively understand the impact of confinement on the concentration profiles established upon crystallization we therefore undertook Finite Element (FE) simulations.

4. 2D Finite Element Calculations of Supersaturation in Confinement

To study how confinement affects the dynamics of crystal evolution we assume that the precipitate consists of a bi-ionic compound constituted by divalent ions following our experimental example using Ca^{2+} and CO_3^{2-}. If a critical nucleus of the compound AB has been formed in the center between the membranes its further growth will depend on the supply by the constituent ions A^{2+} and B^{2-}. As long as the ion concentration in the vicinity of the seed precipitate is above supersaturation relative to the AB composite the crystal will continue to grow, if the concentration falls below the supersaturation level the crystal will dissolve.

Two transport determining processes need to be considered here: (i) the diffusion of ions from the boundaries of the confined region to the precipitate and (ii) the reaction rate at the precipitate surface. The first process will be driven by Fick's second law of diffusion resulting in a second order linear partial differential equation (PDE).

$$\frac{\partial c_i}{\partial t} = D_i \Delta c_i \tag{1}$$

where i refers to either of the ions constituting the compound.

We employed finite element (FE) simulations to calculate the impact of constrains on the ion transport in the presence of a reactive surface emulating a precipitate in a supersaturated solution of ions. This approach was chosen since it allows for the numerical solution of the respective second order diffusion equation identifying appropriate boundary conditions at the membrane surface and entrance sites as discussed below.

For simplicity we are concentrating on the 2D solution for this PDE and steady state $D_i \Delta c_i = 0$ with equal concentrations of A^{2+} and B^{2-}. Our work uses a simplified system and does not aim to fully reproduce the geometry, the surface chemistry occurring in LCSTEM tips or the quantitative dependence of the precipitation on the ion activities. We neglect here the reported phenomenon of membrane bowing [13] resulting from the pressure difference between the liquid and the surrounding vacuum. As a consequence of our simplifications we point out that the results of the simulations do not provide a complete quantitative analysis, which would require a full implementation of the (not fully available) nucleation and growth kinetics as well as the geometry of the liquid cell. Simplifications include the assumption of parallel membranes rather than the reported bowed membranes [13], a first order surface reaction for the precipitate and a constant ion concentration at the membrane entrances.

The growth or dissolution of the precipitate is incorporated by the choice of boundary condition on the precipitate surface and can be expressed as:

$$D\frac{\partial c_i}{\partial x}\bigg|_{surf} = k_g \Omega \tag{2}$$

where k_g is the surface reaction rate and the relative supersaturation Ω is given by

$$\Omega = \frac{c_i}{c_{eq}} - 1 \tag{3}$$

with c_{eq} denoting the supersaturation concentration of the respective ion. We assume equal concentrations of the anions and cations so that c_{eq} can be determined from the square root of the solubility product K_s (for calcite at 25 °C: $K_s = 3.36 \times 10^{-9}$ mol^2/dm^3) [14].

The characteristic parameters for this problem are shown in Figure 3 where h is the spacing between the membranes and L the lateral extension of the smaller of the two chips as shown in Figure 1. We have used a square shaped precipitate geometry with a lateral size of 1 nm.

The boundary conditions applicable to this problem are no-flux conditions at the confining walls (the SiN membranes at $z = \pm h/2$) and constant concentrations $c_i = 10$ mM at $x = \pm L/2$. We assume $L = 5$ mm and that the precipitate is much smaller than h Hence we do not take the change in precipitate size into account effectively assuming that steady-state is reached faster than the time scale associated with particle growth or dissolution.

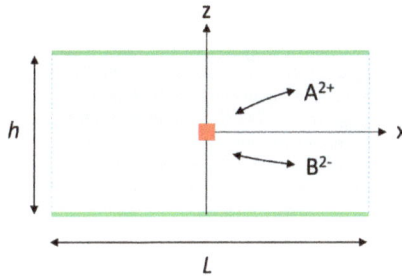

Figure 3. Schematic of the setup used for the finite element simulation.

To calculate the resulting concentration profiles a mesh was created with a refined node density around the precipitate as shown in Figure 4. The number of nodes number is adapted to the values of h and L.

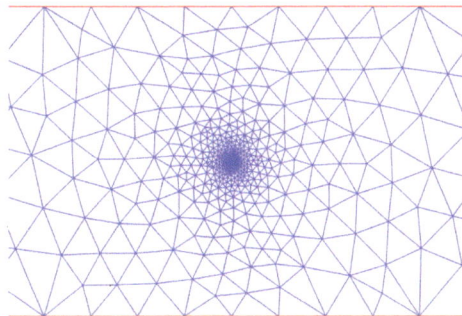

Figure 4. Example of a mesh created for the 2D Finite Element calculations presented in this work with a nanometer-sized precipitate in the center.

We approximated the dependence of growth and dissolution rate on supersaturation using a rate constant of $k_g = 1.7 \times 10^{15}$ m^2/s in accordance with values reported in the literature for the precipitation of CaCO$_3$ from supersaturated solutions of calcium and carbonate ions [15]. 2D solutions for the partial differential equation were obtained choosing following parameters: $D = 9.0 \times 10^{-10}$ m^2/s [8] and $C_{eq} = \sqrt{K_S}$ assuming equal concentrations for cations and anions. Calculations were performed for membrane separations between 100 nm and 10 mm with the latter case effectively representing the absence of confinement.

The results of the 2D numerical calculations of Equation (1) using the mentioned boundary conditions show how the concentration profile around a precipitate is affected by increasing confinement under steady state conditions. Figure 5 shows three concentration maps with a steep recovery of the bulk ion concentration in the vicinity of the precipitate if essentially no confinement is applied (Figure 5a), an increasing depletion of ions around the precipitate when a confinement of 10 μm is applied (Figure 5b) and a significant ion depletion for a confinement of 1 μm (Figure 5c).

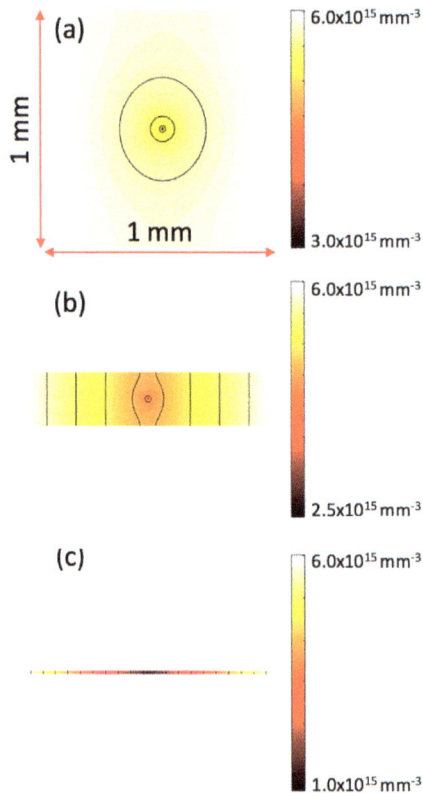

Figure 5. 2D Finite Element simulation of steady state concentration profile as a function of increasing confinement for (**a**) 1 mm, (**b**) 10 μm and (**c**) 1 μm spacing. Yellow indicates high, red/purple low concentrations. Lines indicate iso-concentration contours.

Concentration line profiles for different values for h from $x = \pm L/2$ across the precipitate at $(0, 0)$ are plotted in Figure 6. Due to the diffusion transport and the steric limitations the concentration profile shows a decrease of ion concentration at the precipitate position depending on the extent of confinement. The change in supersaturation is small for spacer values below 1 μm and above 1 mm (which is not experimentally realized). These calculations do not take into account temporal variations since we are focusing here on the steady state situation. However, from this we can already derive important information regarding the impact of confinement and growth rates on the concentration profiles and hence ion transport properties.

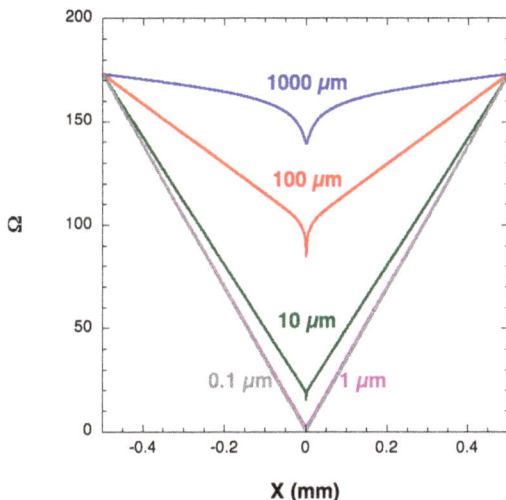

Figure 6. Concentration profiles across the precipitate at $x = 0$ in the FE simulation for a membrane spacing between 100 nm and 1 mm.

The incorporation of ions from solutions into the growing precipitate leads to a decrease of ion concentration and hence supersaturation in the immediate vicinity of the precipitate. The resulting concentration gradient drives the continued flux of ions from the solution to the precipitate. As mentioned, our simulations only consider a steady state between ion incorporation and supply and do not take into account the impact of supersaturation on nucleation. However, we can assume that simultaneous nucleation of neighboring precipitates will lead to an enhanced ion-depletion in solution and hence our problem provides an upper limit for the impact of confinement on supersaturation.

The change of supersaturation at the precipitate position as a function of h is shown in Figure 7 and shows a sharp drop of supersaturation to values close to zero for h values below 1 μm and a steep increase and saturation near values of 150 above 1 mm spacing between the membranes.

These calculations provide insights into the impact of spatial confinement on the ion concentration and hence supersaturation levels in the presence of stable nuclei. Since the nucleation rate scales with $\exp(-A\gamma^3/\ln(\Omega)^2)$ [16], where A is a constant and γ denotes the interfacial energy (i.e., between precipitate and the membrane), confinement also forms a barrier for nucleation [17]. Hence, both nucleation and growth are significantly diminished in this case.

The simulated supersaturation at the precipitate position as a function of h was fitted in the range $0 < h < 10^4$ μm by the relation.

$$\Omega_P = \Omega(0,0) \approx \frac{1}{6.7 \times 10^{-3} + \frac{0.47\,\mu m}{h}} \tag{4}$$

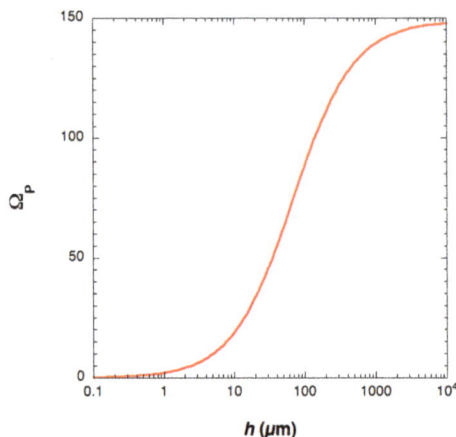

Figure 7. Finite element calculation of the supersaturation Ω_P at the precipitate surface as a function of membrane spacing h.

Equation (4) evinces a significant reduction of the supersaturation for a confinement of 70 µm and lower in our case. For typical spacer values of 5 µm and below Ω decreases to <5 and hence to less than 3% of the bulk value leading to reduced growth rates or, if Ω becomes negative, depending on pH and concentration, can result in a dissolution of the precipitate. We conclude that the confinement can considerably affect the stability of precipitates and hence leads to a reduction of both nucleation and growth if ion activities fall below critical levels.

We have ignored in our considerations the impact of complexation and hence ion removal from solution at the given supersaturation levels, which would lead to a further reduction of ion concentrations around the precipitate. However, this would primarily impact the density of available ions overall without affecting the qualitative validity of our observation.

As previously reported the ion transport to the crystal growth front in 10 mM calcium and carbonate ion solutions occurs through an ion depletion zone resulting from the ion incorporation into the growing crystal [18]. The zone width has been found to be of the order of 4–5 µm [10] under standard conditions. These experiments had been performed using atmospheric SEM observing precipitation in a petri dish through a Si_xN_y membrane in the bottom of the dish with no restriction of dimensions. In that case, the ion supply occurs from the bulk of the solution. We regard the width of the depletion zone as the limiting factor for the growth in confinement coinciding with our observed reduction in supersaturation if the confinement is reduced to similar values. It has also been shown that confinement in one dimension to below 10 µm has a strong impact e.g., on the morphology and polymorphism of calcium carbonate [19]. Our observations indicate that a reduction below 1 µm has further repercussions for the overall stability of a solid phase.

5. Conclusions

We show that confinement has a strong impact on the ion transport affecting the mineralization process, which is highly important also in biological systems where mineral growth generally occurs in confinement affecting the polymorph selection, shape evolution and composition of many compounds in nature. Ion incorporation rates and ion flux from solution due to diffusion are the controlling factors for phase stabilities in these confined volumes and their understanding and quantification is key for a detailed understanding of crystal growth under these conditions.

Acknowledgments: We acknowledge funding by the European Research Council in the framework of the project SMILEY (FP7-NMP-2012-SMALL-6-310637) and the Engineering and Physical Sciences Research Council (EP/I001514/1).

Author Contributions: R.K. conceived and designed the experiment and performed the FE calculations; A.V. performed the LCSTEM experiment jointly with R.K.; R.K. performed data analysis; R.K. analyzed the data and wrote the paper.

Conflicts of Interest: The authors declare no conflict of interest.

References

1. Desarnaud, J.; Bonn, D.; Shahidzadeh, N. The Pressure Induced by Salt Crystallization in Confinement. *Sci. Rep.* **2017**, *6*, 30856. [CrossRef] [PubMed]
2. Stack, A.G.; Fernandez-Martinez, A.; Allard, L.F.; Bañuelos, J.L.; Rother, G.; Anovitz, L.M.; Cole, D.R.; Waychunas, G.A. Pore-Size-Dependent Calcium Carbonate Precipitation Controlled by Surface Chemistry. *Environ. Sci. Technol.* **2014**, *48*, 6177–6183. [CrossRef] [PubMed]
3. Nakouzi, E.; Steinbock, O. Self-organization in precipitation reactions far from the equilibrium. *Sci. Adv.* **2016**, *2*, e1601144. [CrossRef] [PubMed]
4. Loste, E.; Park, R.J.; Warren, J.; Meldrum, F.C. Precipitation of Calcium Carbonate in Confinement. *Adv. Funct. Mater.* **2004**, *14*, 1211–1220. [CrossRef]
5. Smeets, P.J.M.; Cho, K.R.; Kempen, R.G.E.; Sommerdijk, N.A.J.M.; de Yoreo, J.J. Calcium carbonate nucleation driven by ion binding in a biomimetic matrix revealed by in situ electron microscopy. *Nat. Mater.* **2015**, *14*, 394–399. [CrossRef] [PubMed]
6. Kröger, R. Ion-binding and Biomineralization. *Nat. Mater.* **2015**, *14*, 369–370. [CrossRef] [PubMed]
7. Strang, G.; Fix, G. *An Analysis of the Finite Element Method*, 2nd ed.; Wellesley-Cambridge Press: Wellesley, MA, USA, 2008.
8. Ross, F. (Ed.) *Liquid Cell Electron Microscopy (Advances in Microscopy and Microanalysis)*; Cambridge University Press: Cambridge, UK, 2017.
9. Van Driessche, A.E.S.; Kellermeier, M.; Benning, L.G.; Gebauer, D. Liquid Phase TEM Investigations of Crystal Nucleation, Growth, and Transformation. In *New Perspectives on Mineral Nucleation and Growth: From Solution Precursors to Solid Materials*; Springer International Publishing: Cham, Switzerland, 2017; pp. 353–374.
10. Perovic, I.; Verch, A.; Chang, E.P.; Rao, A.; Cölfen, H.; Kröger, R.; Evans, J.S. An Oligomeric C-RING Nacre Protein Influences Prenucleation Events and Organizes Mineral Nanoparticles. *Biochemistry* **2014**, *53*, 7259–7268. [CrossRef] [PubMed]
11. Schneider, N.M.; Norton, M.M.; Mendel, B.J.; Grogan, J.M.; Ross, F.M.; Bau, H.H. Electron-Water Interactions and Implications for Liquid Cell Electron Microscopy. *J. Phys. Chem. C* **2014**, *118*, 22373–22382. [CrossRef]
12. Nielsen, M.H.; Aloni, S.; de Yoreo, J.J. In situ TEM imaging of CaCO$_3$ nucleation reveals coexistence of direct and indirect pathways. *Science* **2014**, *345*, 1158–1162. [CrossRef] [PubMed]
13. Holtz, M.; Yu, Y.; Gao, J.; Abruña, H.; Muller, D. In Situ Electron Energy-Loss Spectroscopy in Liquids. *Microsc. Microanal.* **2013**, *19*, 1027–1035. [CrossRef] [PubMed]
14. Benjamin, M.M. *Water Chemistry*; McGraw-Hill: New York, NY, USA, 2002; ISBN 0-07-238390-9.
15. Morse, J.W.; Arvidson, R.S.; Lüttge, A. Calcium Carbonate Formation and Dissolution. *Chem. Rev.* **2007**, *107*, 342–381. [CrossRef] [PubMed]
16. De Yoreo, J.J.; Vekilov, P.G. Principles of Crystal Nucleation and Growth. *Rev. Mineral. Geochem.* **2003**, *54*, 57–93. [CrossRef]
17. Lioliou, M.G.; Paraskeva, C.A.; Koutsoukos, P.G.; Payatakes, A.C. Heterogeneous nucleation and growth of calcium carbonate on calcite and quartz. *J. Colloid Interface Sci.* **2007**, *308*, 421–428. [CrossRef] [PubMed]
18. Verch, A.; Morrison, I.E.G.; van de Locht, R.; Kröger, R. Electron microscopy studies of calcium carbonate precipitation from aqueous solution with and without organic additives. *J. Struct. Biol.* **2013**, *183*, 270–277. [CrossRef] [PubMed]
19. Stephens, C.J.; Christopher; Ladden, S.F.; Meldrum, F.C.; Christenson, H.K. Amorphous Calcium Carbonate is Stabilized in Confinement. *Adv. Funct. Mater.* **2010**, *20*, 1616–3028. [CrossRef]

Review

In Situ Atomic Force Microscopy Studies on Nucleation and Self-Assembly of Biogenic and Bio-Inspired Materials

Cheng Zeng [1], Caitlin Vitale-Sullivan [2] and Xiang Ma [2,3,*]

[1] Department of Chemistry, Indiana University, Bloomington, IN 47403, USA; chzeng@umail.iu.edu
[2] Department of Chemistry, Idaho State University, Pocatello, ID 83209, USA; vitacait@isu.edu
[3] Department of Chemistry, Grand View University, Des Moines, IA 50316, USA
* Correspondence: max@umail.iu.edu; Tel.: +1-515-263-2951

Received: 30 May 2017; Accepted: 16 August 2017; Published: 31 August 2017

Abstract: Through billions of years of evolution, nature has been able to create highly sophisticated and ordered structures in living systems, including cells, cellular components and viruses. The formation of these structures involves nucleation and self-assembly, which are fundamental physical processes associated with the formation of any ordered structure. It is important to understand how biogenic materials self-assemble into functional and highly ordered structures in order to determine the mechanisms of biological systems, as well as design and produce new classes of materials which are inspired by nature but equipped with better physiochemical properties for our purposes. An ideal tool for the study of nucleation and self-assembly is *in situ* atomic force microscopy (AFM), which has been widely used in this field and further developed for different applications in recent years. The main aim of this work is to review the latest contributions that have been reported on studies of nucleation and self-assembly of biogenic and bio-inspired materials using *in situ* AFM. We will address this topic by introducing the background of AFM, and discussing recent *in situ* AFM studies on nucleation and self-assembly of soft biogenic, soft bioinspired and hard materials.

Keywords: *in situ* atomic force microscopy; nucleation; self-assembly; biomaterials; biomimetic; bioinspired

1. Introduction

Nature has always been a fascinating source of inspiration for scientists and engineers. During billions of years of evolution, highly sophisticated and ordered structures in living systems with remarkable properties have been created [1,2]. These properties, such as physical mechanics, are often far better than those of the equivalent synthetic materials with similar chemical compositions. For example, spider silk [3] has a tensile strength comparable to steel, but with a higher flexibility and much less density [4–7], which makes it an ideal material for a wide variety of military, industrial, and consumer applications [8–11]. Moreover, biological systems are made with significantly weak components, such as soft proteins, brittle minerals and water. Finally, these highly ordered structures can form at mild temperature and pressure conditions spontaneously or with relatively low energy consumption, mainly through self-assembly [12,13]. Incorrect or undesired self-assembly processes can lead to malfunction or even diseases, such as Alzheimer's and Parkinson's [14]. Therefore, it is fundamentally important and interesting to determine nucleation and self-assembly pathways of biogenic materials in order to understand the mechanisms of living systems and provide a rationale for the design and production of new classes of materials, such as biochips, biosensors, and novel drugs [15].

Bio-inspired materials are a class of materials whose structure, function and/or property mimic those of biogenic materials or living matter [16–22]. The field of bio-inspired materials is highly interdisciplinary. It focuses on the understanding of biological synthesis, self-assembly and hierarchical organization of biogenic materials, and uses this understanding to design and produce new "bio-inspired" synthetic materials for diverse applications [23–30]. A few examples of bio-inspired materials include peptoid [31–33] inspired by peptide, light-harvesting photonic materials [34–44] that mimic photosynthesis, water collecting surfaces [45–52] inspired by the head-stander beetle, and organic-inorganic hybrid structures that mimic the hierarchical architecture of nacre or bone [53–60]. One critical limitation to the mimicking of biological structures in materials, however, lies in our insufficient understanding of how biogenic and bio-inspired materials self-assemble into sophisticated structures.

Atomic force microscopy (AFM) is an imaging tool which utilizes a nanometer-sized tip as a localized force sensor. Since its invention [61], it has been widely used to study biology specimens including nucleic acid [62–64], proteins [65], viruses [66,67], cells [68], etc. AFM generates topographic images by raster scanning over a sample surface. The most commonly applied imaging modes [69] are contact mode [70–72], tapping (semi-contact) mode [73–77], non-contact mode [78–81], and the newly developed peak-force tapping mode [82–86]. Contact mode generates high lateral forces while the tip is scanning over the surface, and therefore, is normally being used to image hard materials such as minerals, metals, glass and ceramics, whose Young's moduli (E) are >10 GPa. Tapping mode greatly reduces the lateral forces by oscillating the tip to realize only intermittent contacts with the sample [74,76], and therefore, is used for soft materials such as cell membranes, biomolecules and some polymers (E < 10 GPa). However, the imaging force is difficult to accurately estimate. In peak force tapping mode, the tip-sample forces are directly controlled at ultralow levels while at the same time the lateral forces are minimized, providing a suitable imaging environment for soft biological samples [85]. By continuously imaging the same area of interest, AFM allows researchers to take snapshots of various processes, thus directly visualizing the dynamics. The observation of nucleation and self-assembly by AFM can reveal *in situ* single-particle dynamics with the potential of mechanical manipulation.

For studies of nucleation and self-assembly, *in situ* AFM (Scheme 1) is an ideal tool and has unique advantages over several other structural techniques (e.g., transmission electron microscope, scanning electron microscope, optical microscope, light scattering, and small angle X-ray scattering) [69,87]. AFM tip apex radius is normally on the order of a few to tens of nanometers. Thanks to the very small size of tip, AFM can achieve extremely high spatial resolution in both vertical (sub-Å) and lateral (sub-nanometer) directions, which enables people to directly monitor the nucleation and growth of single assemblies, rather than bulk behavior. Besides, AFM requires simple sample preparation. It is a label free method. Unlike electron microscopy (EM) which requires freezing or drying and staining, samples can be imaged by AFM at physiological or near-reacting conditions, which creates a suitable environment for studies of living systems and synthetic materials. Furthermore, recent development of AFM enables simultaneous collections of topographical, mechanical and electronic/electrochemical properties of samples, generating multi-parametrical data to examine the system of interest. Lastly, high-speed AFM (HS-AFM) [65,88–92] has been applied to visualize processes at an imaging rate of ~20 frames per second, which provides new opportunities for researchers to further capture and comprehend nucleation and self-assembly processes with much higher temporal resolution [93–99]. One also has the option to greatly improve temporal resolution by sacrificing spatial information. As AFM uses a raster scanning approach to generate images, there are two perpendicular scan directions: fast scan direction and slow scan direction. In a normal operation, the AFM tip will scan in the fast scan direction to complete imaging all pixels in one line before it moves by one pixel along the slow scan direction to the next line. The slow scan can be disabled, which allows the tip to repeatedly scan over the same line. The recorded image will show the evolution of a line profile over time and is very good for studying dynamics of processes such as nucleation.

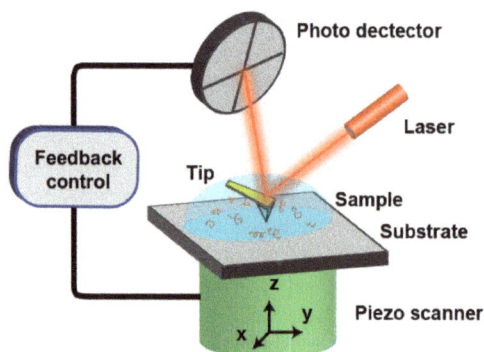

Scheme 1. Schematic illustration of *in situ* atomic force microscopy (**AFM**). The AFM tip is scanning on the surface to collect continuous images, while ions/molecules self-assemble into ordered structures in solution.

Besides imaging, AFM can also provide other physical information of the sample by measuring the force. For example, by (nano) indenting or stretching materials, their mechanical properties, such as adhesion and elasticity, can be precisely measured and mapped on the surface [100–104]. AFM-based nanoindentation and force spectroscopy has been applied to study many biological and bio-inspired systems, including bacterial adhesive material [105], viruses [106–118], proteins [119–125], nucleic acids [126–130], cells [131–133], and plant cell walls [134]. The mechanical properties of the materials can be connected to the processes of material nucleation and self-assembly [100–103].

In this review, a series of examples will be presented to illustrate the applications of *in situ* AFM and HS-AFM in the field of nucleation and self-assembly. These recent reports will be categorized into three groups based on their subjects: soft ($E < 10$ GPa) biogenic materials, soft bio-inspired materials, and hard ($E > 10$ GPa) materials. Since many previous review articles [135–144] have summarized AFM applications in nucleation of hard materials, this review will focus on soft materials.

2. Soft Biogenic Materials

2.1. Peptide and Protein

Alzheimer's disease is partially caused by the self-assembly and accumulation of fibrillar amyloid β (Aβ) peptides in the brain [145]. Biophysical studies related to the self-assembly of Aβ peptides have significantly improved our understanding of the mechanism of amyloid formation and its role in pathogenesis of Alzheimer's disease. The formation of amyloid fibril is believed to be a surface-mediated event [146]. As a surface sensitive technique, AFM is thus a very suitable candidate to study this process as the surface chemistry of substrate can be very well controlled. Lashuel and Dietler et al. investigated the mechanism of Aβ fibril formation using *in situ* AFM [147]. They demonstrated that the structure and polymorphism of Aβ fibrils are critically influenced by the oligomeric state of the starting materials, the ratio of monomeric-to-aggregated forms of Aβ (oligomers and protofibrils), and the occurrence of secondary nucleation. They also demonstrated that monomeric Aβ plays an important role in mediating structural transitions in the amyloid pathways. For the first time, they provided evidence for the existence of secondary-nucleation sites on the Aβ fibrils. Their research findings have implications for understanding the molecular and structural basis of amyloid formation and development of therapeutic strategies for the treatment of Alzheimer's disease.

The requirement of using a nanometer-sized mechanical probe in AFM imaging brings in a unique way of tuning assembly: mechanical manipulation. Although AFM imaging force is normally kept small to prevent sample damage, people can also apply a higher force to initiate new events. Liu's group used an AFM probe to mechanically cut the amyloid fibrils, which are formed by mucin 1 peptide fused with Q11 (MUC1-Q11), and observed growth of fibrils at the newly exposed termini using *in situ* AFM [148]. They showed that orientation and length of branched fibrils can be controlled by the nuclei orientation and reaction time. Therefore, local mechanical force can be used to affect the fibril formation and assembly of MUC1-Q11. Their approach offers a pure physical and label-free means to control the growth of MUC1 epitopes. In a similar case, Hu et al. demonstrated that a new peptide nanofilament can be introduced by mechanically generating an active end at designated positions on an existing filament [149]. The self-assembly of new filaments is specifically guided by the direction of new active ends.

Many soft bio-inspired materials are inherently connected to soft/hard interface or interphases (transitions of material properties, e.g., mechanical gradients) [150,151]. Disassembly of amelogenin, an enamel protein, onto hydroxyapatite (HAP) surfaces has been studied using high-resolution *in situ* AFM by Tao's group [152]. Their study showed that the amelogenin nanospheres disassemble onto the HAP surface, and then break down into oligomeric (25-mer) subunits of the larger nanosphere. The binding energy of the protein to a specific face of HAP (100) was extracted from quantification of the adsorbate amounts by size analysis. The kinetics of disassembly indicated a time-dependent increase in oligomer-oligomer binding interactions within the nanosphere. Their study elucidates the mechanism of supramolecular protein interactions and break down at surfaces.

Two dimensional (2D) assembly of S-layer proteins has been studied using *in situ* AFM on supported lipid bilayers by De Yoreo et al. Their results suggested that proteins undergo conformational transformations which direct the pathway of assembly. Monomers with an extended conformation first form a mobile adsorbed phase, and then condense into amorphous clusters, which undergo a phase transition through S-layer folding into crystalline clusters composed of compact tetramers [153]. They also showed that the system of S-layer proteins on mica possesses a kinetic trap associated with conformational differences between a long-lived transient state, which exhibits the characteristics of a kinetic trap in a folding funnel, and the final stable state [154].

HS-AFM has been utilized to visualize the fast dynamics of growth and assembly of lithostathine, a protein produced by pancreas acinar cells, at a high imaging rate (one frame/s). Formation of fibrils via protofibril lateral association and stacking follows a zipper-like mechanism of association. They also demonstrated that two lithostathine protofibrils can associate to form helical fibrils (Figure 1) [99]. Their study provided new insights into lithostathine protofibril elongation and assembly.

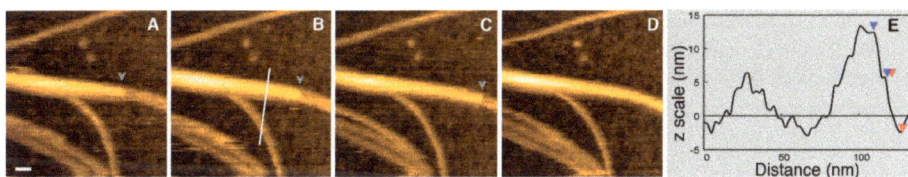

Figure 1. (**A–D**) Time lapse of the protofibril elongation stacked on top of a fibril. The grey arrowheads indicate the edge of the elongating protofibril on top of a fibril. (**E**) The profile (corresponding to the white line in (**B**)) indicates that only two fibers are stacked on top of each other. Images were collected successively separated by 1 s. Scale bar: 30 nm. Adapted from Ref. [99].

2.2. Nucleic Acid

Direct visualization of nucleic acids in aqueous solution has always been a challenging topic due to relatively weak interactions between nucleic acids and substrates. Appropriate modification of substrate surfaces and selection of force detection method can significantly enhance imaging quality and resolution [64]. For example, Yamada's group used an ultra-low-noise AFM, frequency modulation AFM (FM-AFM), to successfully observe local structures of native plasmid DNA in water. Even individual functional groups within DNA were clearly discerned [155]. In another case, Ye et al.

utilized a carboxyl-terminated self-assembled monolayer modified gold surface where single-stranded DNA probes were covalently anchored for visualization of single DNA hybridization events using *in situ* AFM. By introducing Ni^{2+} ions (or some other divalent cations), the surface-DNA interaction can be switched between a strong state and a weak state, which accommodates for both high-resolution imaging and DNA hybridization [156,157].

2.3. Lipid

Lipid-protein/peptide interactions have been studied using *in situ* AFM to understand self-assembly of lipids and health-related mechanisms. Legleiter et al. have used *in situ* AFM to directly monitor the interaction between huntingtin (htt) exon 1 protein and brain lipid extract [158]. They observed that the exon1 fragments accumulated on the lipid membranes comprised of total brain lipid extract, which caused disruption of the membrane. This disruption is dependent on length of polyQ in exon1. They also observed that adding an N-terminal myc-tag to the htt exon1 fragments can prevent the interaction of htt with the bilayer. The interaction between melittin, a 26-residue amphipathic peptide, and lipid membranes has been studied using *in situ* AFM by Pan et al. [159]. Their results showed that melittin induced defects in lipid bilayers, which can be delayed by introducing cholesterol. They also proposed a kinetic defect growth model based on their AFM observations.

Kobayashi et al. used HS-AFM to investigate interactions between self-assembled sphingomyelin (SM)-cholesterol (Chol) lipid membranes and a SM-binding pore-forming toxin, lysenin. Lysenin recognized SM-rich domain and self-assembled into hexagonal close-packed (hcp) structure on the SM-rich domain, which did not perturb the phase boundary between SM and Chol. After the SM-rich domain is fully covered by lysenin, the hcp assembly started to occur in Chol liquid-disordered phase and eventually covered the entire membrane [160].

3. Soft Bio-Inspired Materials

3.1. Peptoid

Peptoid is a bio-inspired material, which has a structure similar to peptide but lacking hydrogen bonds between backbones. Because of this difference, peptoid has several advantages over peptide, such as high stability against temperature and protease degradation, flexibility and simplicity. Not until very recently, researchers started to appreciate the simplicity of peptoid systems and employ the system for studies on nucleation and self-assembly.

Ma et al. specifically designed a peptoid system to fundamentally understand the relationship between molecular sequence and structure [31]. They determined the dynamics and kinetics of nucleation and self-assembly of two peptoid molecules (Pep_c vs. Pep_b) with similar sequence using *in situ* AFM. Ma's team, for the first time, provided direct evidence showing that nucleation pathways are indeed sequence dependent: a slight difference in sequence can switch nucleation pathway from classical to a two-step process (Figure 2, Scheme 2). They also proposed a mathematical model to explicitly explain and predict sequence-dependent selection of self-assembly pathways. Their research findings shed new light on non-classical nucleation pathways, which has implications in biology, chemistry, geology and biomedicine.

The sequence of peptoid can be tailored to self-assemble into 2D membranes [33]. These peptoid membranes self-assembled on mica through a process of heterogeneous nucleation and lateral growth. Details about kinetics and dynamics of self-assembly of 2D membranes have been studied. *In situ* AFM images showed that peptoid membranes exhibit self-repair properties and are highly stable. These membranes can provide a platform to incorporate and pattern a diverse range of functional objects either before assembly or by co-crystallization approaches.

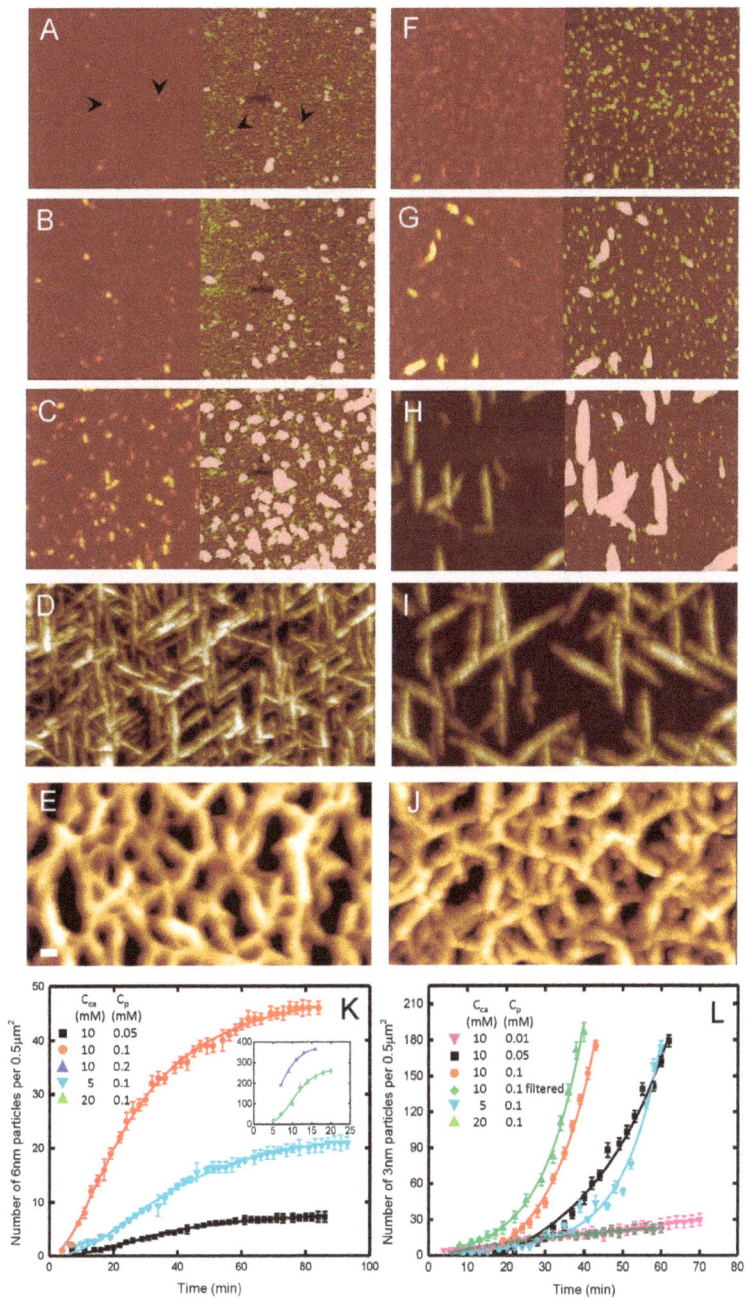

Figure 2. *In situ* view and kinetics of peptoid assembly. Nucleation and self-assembly of Pepc (**A–E**), and Pepb (**F–J**). (**A–D,F–I**): *In situ* AFM images showing assembly pathway of Pepc at *t* = 24.2 (**A**); 32.3 (**B**); 50.1 (**C**) and 275.1 (**D**) min; and Pepb at *t* = 4.3 (**F**); 17.2 (**G**); 69.1 (**H**); and 111.2 (**I**) min. (**E,J**): *Ex situ* AFM images showing self-assembled porous networks of Pepc (**E**) and Pepb (**J**). (**K,L**): Nuclei number density versus time for Pepc (**K**) and Pepb (**L**). Adapted from Ref. [31].

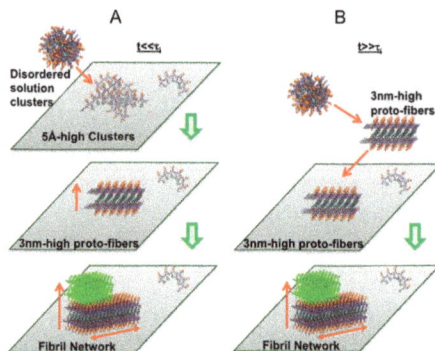

Scheme 2. Proposed model for the peptoid assembly process at early (**A**) and late (**B**) stages showing the effect of hydrophobic conjugate on the propensity for peptoid aggregation. Adapted from Ref. [31].

3.2. Other Polymers

The nucleation and growth of several bio-inspired polymers has been studied using *in situ* AFM. Interestingly, it was found that a carefully chosen surface can be used to facilitate self-assembly in such systems. For instance, Yip et al. revealed a distinct behavior of elastin-like peptides on difference surfaces: they form aggregates on mica, yet self-assemble into ordered fibrils on HOPG (highly ordered pyrolytic graphite) surface [14]. Such a difference shows that the hydrophobic peptide-substrate interactions are crucial for the formation of ordered structures. Their results provided direct evidence of frustrated fibril nuclei and oriented growth of independent fibril domains [161]. Similarly, Cohen Stuart et al. reported the self-assembly of a biosynthetic amino acid into fibrils on a charge silica surface, which would not happen in solution under the same buffer conditions due to the electrostatic repulsion between monomers carrying the same charges [162]. The presence of the substrate, which is negatively charged, stabilizes the positive charged monomers on surface and thus allows for heterogeneous nucleation of fibers. Zhou et al. observed "upright" fibrilar conformation for an amyloid-like peptide on mica surface, which is different from its "flat" conformation on the HOPG surface [163]. Zhou et al. also examined the effect of salt on self-assembly of amyloid-like peptides on mica [164].

As discussed above, self-assembly can be induced by a localized mechanical force applied by the AFM tip [15,165,166]. For instance, He et al. have shown an interesting *in situ* AFM based approach to pattern silk fibroin (SF) proteins on surface [15]. Specifically, they used an AFM tip to scan over a mica surface in a solution of SF monomers. After repeated scans on the same area, SF proteins are deposited onto the surface. Interestingly, such features can only be observed in areas where scanning was performed. On the contrary, the rest of the surface which was not extensively scanned by the AFM tip stayed clean. Such results suggest that the surface patterning was achieved upon contact of the tip and surface. This process was observed to occur in both contact mode and tapping mode. He et al. proposed that a sol-gel transition of the SF molecule is key to their deposition. In contact mode, this transition occurs at the tip-substrate interface as a result of shear force; whereas in tapping mode, the sol-gel transition of SF molecules is proposed to first occur on the tip due to tip oscillation, and the accumulated SF molecules are subsequently transferred to the surface upon tip-surface contact. The patterning process is governed by mechanical contact; thus, it can be manipulated by altering the scanning speed, which determines the frequency of tip-surface contact, and the force applied in each contact. In addition, the SF concentration is also very important. This approach is advantageous because, unlike conventional dip-pen lithography methods performed in air, patterning is achieved in liquid which possibly helps to retain the native structure of proteins.

Seog et al. reported a study of silk-elastin-like peptide (SELP), which is another example of mechanically induced self-assembly on surface with possibly a different mechanism [165]. Formation of SELP amyloid fibers was observed after repeated scanning the AFM tip over a single line, which is achieved by disabling the scan in the slow scan direction. In contrast to the previous example, fibers grow exclusively in the direction perpendicular to the scan direction of the tip. This directional bias suggests some specificity in nucleation. The proposed mechanism is that the AFM tip can stretch the SELP monomer to expose a domain which can interact with other peptides and lead to subsequent assembly.

Although the density of spontaneously formed fibers over the entire surface can be adjusted by concentration and buffer conditions, additional control on fiber growth over a local region can be achieved by nanomechanical manipulation. In a study of peptide EAK 16-II, which assemble into fibers on mica surface, Chen et al. reported that such fibers can be broken up by the imaging force applied in tapping mode AFM. Such fragments can then initiate new nucleation and growth into new fibers [166].

4. Hard Materials

4.1. Calcium Phosphate

It is important to determine the kinetics and dynamics during nucleation and growth of calcium phosphate (Ca-P) in order to understand several health-related issues, such as bone and tooth biomineralization [167], and pathological calcification in stones and in cardiovascular disease [168,169]. Very recently, Wang et al. examined nucleation of hydroxyapatite (HAP, $Ca_{10}(PO_4)_6(OH)_2$), a major component in bone and teeth, at different supersaturations using *in situ* AFM with long imaging times (~30 h) to capture complete processes of HAP surface growth. They showed the morphology evolution during HAP surface growth, where a coexistence of a classical spiral growth by adding monomeric chemical species (molecules or ions) and nonclassical particle attachment by aggregating nanoparticles to form triangular solids which then transformed into hexagonal aggregates was observed [170] (Scheme 3). In another recent report, Wang's group studied the kinetics and mechanisms of Ca-P surface crystallization modulated by amelogenin [171]. They observed that, during *in situ* growth via a nonclassical particle attachment pathway, the assembly of amelogenin's C-termini induced an elongated aggregation of Ca-P nanoparticles, which directed Ca-P mineralization. They further determined the binding free energy of the C-terminal fragment absorbed to the (100) face of Ca-P using force spectroscopy to reveal mechanisms of shape evolution from spherical particles to elongated nanorods.

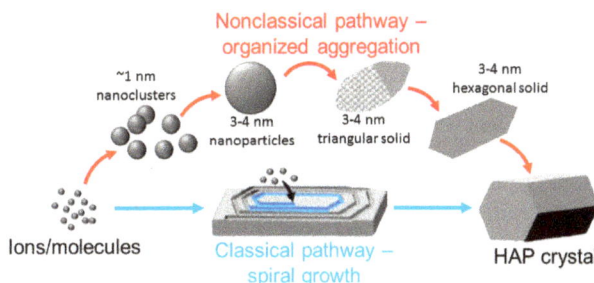

Scheme 3. Classical and nonclassical pathways for hydroxyapatite (HAP) crystallization. In the classical pathway, monomeric species (ions or molecules) are added at step edges during spiral growth, while in the nonclassical pathway, HAP crystals grow by the attachment of different size particles (~1 nm or 3–4 nm in height) to form organized, but may be amorphous, assemblies/aggregates with triangular and hexagonal morphologies.

Nucleation of HAP and amorphous calcium phosphate (ACP) on collagen has been studied using *in situ* AFM [172]. By measuring the rate of Ca-P nucleation as a function of supersaturation, they showed that ACP formation cannot be directly reconciled with classical nucleation theory. However, when taking into account the existence of pre-nucleation complexes and the particle size dependence of the interfacial free energy, the thermodynamic barrier for nucleation is dramatically lowered, which unites classical and non-classical nucleation theories—the observed non-classical route to ACP formation can be explained using classical theory.

4.2. Calcite

Biomineralization of Calcite ($CaCO_3$), the most abundant carbonate mineral on Earth and a common biomineral, has attracted researchers' interest for decades. The involvement of nonclassical nucleation pathways has been implied by previous studies, but only recently, direct evidence regarding the actual mechanism of calcite growth via an amorphous phase was reported. Rodriguez-Navarro et al. used *in situ* AFM to show that calcite can grow via a nonclassical particle-mediated colloidal crystal growth mechanism involving a layer-by-layer attachment of amorphous calcium carbonate (ACC) precursors, followed by restructuring and fusion with the calcite substrate in perfect crystallographic registry [173]. The transformation of ACC to calcite occurs through interface-coupled dissolution-reprecipitation and affects the nanogranular texture of the colloidal growth layer regulated by organic molecules. Another recent report by Colfen et al. indicates that polymer-induced liquid precursor phases play a role in nonclassical calcite crystallization and growth processes. They observed gel-like precursors, which spread out on the surface suggesting a liquid character, for the crystal growth on a calcite surface using *in situ* AFM [174].

The effects of inorganic and organic molecules/ions on calcite nucleation and growth have also been studied by *in situ* AFM. For example, Voitchovsky and Schmidt et al. investigated calcite (104)-water interface in the presence of $NaNO_3$ [175]. *In situ* AFM images confirmed the alteration of crystallographic characteristics, and AFM force spectroscopy indicated the ability of dissolved $NaNO_3$ to modify the structure of interfacial water. Wang and Putnis et al. showed that the presence of glucose-6-phosphate (G6P) changes the morphology of etch pits from the typical rhombohedral to a fan-shaped form, which can be explained by a site-selective mechanism of G6P-calcite surface interactions stabilizing the energetically unfavorable (0001) or (0112) faces through step-specific adsorption of G6P [176]. Wahl et al. observed specific binding of an acidic cement protein to step edge atoms on (1014) calcite surfaces. The protein then assembles to form one-dimensional nanofibrils, which affects the morphology of calcite [177].

4.3. Calcium Oxalate Monohydrate

Due to the clinical importance as the main inorganic component in human kidney stones, calcium oxalate monohydrate (COM) and regulation of its nucleation and growth has been of great interest for researchers. *In situ* AFM has been employed to investigate the effect of biomolecules, such as peptide, protein and nucleic acid, on COM growth. Huang and Qiu et al. performed a series of *in situ* AFM studies to understand the inhibitory effect of linear enantiomers of L- and D-Asp_6 on the growth of COM crystal [178]. They showed that D-Asp_6 has a larger inhibitory effect on the growth of the [100] step on the (010) face than L-Asp_6. They further showed that these enantiomers create the impurity pinning along the steps, and the major impact of Asp_6 is to block active kink sites. The inhibition of COM nucleation and aggregation by osteopontin (OPN) proteins has been investigated by Wang et al. Their data indicated that phosphorylated OPN peptides affect the step retreat rates via step-specific interactions, which in turn regulates the kinetics of COM nucleation and aggregation at the expense of brushite crystals by means of the interfacial mineral replacement reactions. Their research findings offer general insights concerning the control of kidney stone formation and the mechanisms through which aberrant crystallization kinetics is inhibited [179].

5. Conclusions

In this review, we summarized recent applications of *in situ* AFM to studies of nucleation and self-assembly of biogenic and bio-inspired materials, including peptide, protein, nucleic acid, lipid, peptoid, Ca-P, calcite and COM. Development of AFM techniques, such as *in situ* and HS-AFM, has enabled direct characterization of many biological and physicochemical processes which have never been revealed before.

Despite these key advancements, current AFM still has a number of limitations and technological issues, such as limited scan range (up to ~1 μm), uncontrollable forces applied on the sample, and the lack of (bio)chemical specificity. The scan range has recently been increased to over ~40 $\times$ 40 μm^2, due to an implementation of an inversion-based feedforward control technique [180] and an enhanced iterative inverse control technique [181,182], although z-scanner still suffers from bandwidth limitation [65]. The peak-force tapping mode has been developed to directly control the force applied on the sample during scanning, which is beneficial for biological and soft material studies. The use of spatially resolved single-molecule force spectroscopy with AFM tips that are coated with specific chemical groups has enabled the detection and localization of single functional molecules on the surfaces of cells and other materials [183].

In the future, the applications of AFM will be further broadened into the fields of energy storage and conversion, catalysis, geology and biomedicine, especially due to the development of multimodal *in situ* imaging capabilities. For example, nanomechanical imaging can provide mechanical properties simultaneously; tip-enhanced Raman imaging can resolve chemical details spatially; current/conductive AFM imaging can give electronic and electrochemical properties of the samples; and AFM-based nanomanipulation can build three-dimensional nanostructures and nanodevices. Integrating AFM with other imaging, characterizing and manipulation techniques opens new avenues to understanding biological/chemical materials and processes comprehensively.

Acknowledgments: We thank the support from Idaho State University.

Author Contributions: C.V.-S. collected literature; C.Z. and X.M. wrote the paper.

Conflicts of Interest: The authors declare no conflict of interest.

References

1. Wainwright, S.A. *Mechanical Design in Organisms*; Princeton University Press: Princeton, NJ, USA, 1982.
2. Vincent, J.F. *Structural Biomaterials*; Princeton University Press: Princeton, NJ, USA, 2012.
3. Vollrath, F. Biology of spider silk. *Int. J. Biol. Macromol.* **1999**, *24*, 81–88. [CrossRef]
4. Heim, M.; Keerl, D.; Scheibel, T. Spider silk: From soluble protein to extraordinary fiber. *Angew. Chem. Int. Ed.* **2009**, *48*, 3584–3596. [CrossRef] [PubMed]
5. Gosline, J.M.; DeMont, M.E.; Denny, M.W. The structure and properties of spider silk. *Endeavour* **1986**, *10*, 37–43. [CrossRef]
6. Rammensee, S.; Slotta, U.; Scheibel, T.; Bausch, A. Assembly mechanism of recombinant spider silk proteins. *Proc. Natl. Acad. Sci. USA* **2008**, *105*, 6590–6595. [CrossRef] [PubMed]
7. Shao, Z.; Vollrath, F. Materials: Surprising strength of silkworm silk. *Nature* **2002**, *418*, 741. [CrossRef] [PubMed]
8. Hinman, M.B.; Jones, J.A.; Lewis, R.V. Synthetic spider silk: A modular fiber. *Trends Biotechnol.* **2000**, *18*, 374–379. [CrossRef]
9. Lee, S.-M.; Pippel, E.; Gösele, U.; Dresbach, C.; Qin, Y.; Chandran, C.V.; Bräuniger, T.; Hause, G.; Knez, M. Greatly increased toughness of infiltrated spider silk. *Science* **2009**, *324*, 488–492. [CrossRef] [PubMed]
10. O'Brien, J.P.; Fahnestock, S.R.; Termonia, Y.; Gardner, K.H. Nylons from nature: Synthetic analogs to spider silk. *Adv. Mater.* **1998**, *10*, 1185–1195. [CrossRef]
11. Lewis, R.V. Spider silk: Ancient ideas for new biomaterials. *Chem. Rev.* **2006**, *106*, 3762–3774. [CrossRef] [PubMed]
12. Whitesides, G.M.; Boncheva, M. Beyond molecules: Self-assembly of mesoscopic and macroscopic components. *Proc. Natl. Acad. Sci. USA* **2002**, *99*, 4769–4774. [CrossRef] [PubMed]

13. Zhang, S. Fabrication of novel biomaterials through molecular self-assembly. *Nat. Biotechnol.* **2003**, *21*, 1171–1178. [CrossRef] [PubMed]

14. Yang, G.; Woodhouse, K.A.; Yip, C.M. Substrate-facilitated assembly of elastin-like peptides: Studies by variable-temperature in situ atomic force microscopy. *J. Am. Chem. Soc.* **2002**, *124*, 10648–10649. [CrossRef] [PubMed]

15. Zhong, J.; Ma, M.; Zhou, J.; Wei, D.; Yan, Z.; He, D. Tip-induced micropatterning of silk fibroin protein using in situ solution atomic force microscopy. *ACS Appl. Mater. Interfaces* **2013**, *5*, 737–746. [CrossRef] [PubMed]

16. Zhao, N.; Wang, Z.; Cai, C.; Shen, H.; Liang, F.; Wang, D.; Wang, C.; Zhu, T.; Guo, J.; Wang, Y. Bioinspired materials: From low to high dimensional structure. *Adv. Mater.* **2014**, *26*, 6994–7017. [CrossRef] [PubMed]

17. Chworos, A.; Smitthipong, W. Bio-inspired materials. In *Bio-Based Composites for High-Performance Materials: From Strategy to Industrial Application*; CRC Press: Boca Raton, FL, USA, 2014; Volume 43.

18. Barron, A.E.; Zuckerman, R.N. Bioinspired polymeric materials: In-between proteins and plastics. *Curr. Opin. Chem. Biol.* **1999**, *3*, 681–687. [CrossRef]

19. Sanchez, C.; Arribart, H.; Guille, M.M.G. Biomimetism and bioinspiration as tools for the design of innovative materials and systems. *Nat. Mater.* **2005**, *4*, 277–288. [CrossRef] [PubMed]

20. Wegst, U.G.; Bai, H.; Saiz, E.; Tomsia, A.P.; Ritchie, R.O. Bioinspired structural materials. *Nat. Mater.* **2015**, *14*, 23–36. [CrossRef] [PubMed]

21. Dujardin, E.; Mann, S. Bio-inspired materials chemistry. *Adv. Mater.* **2002**, *14*, 775. [CrossRef]

22. Aizenberg, J.; Fratzl, P. Biological and biomimetic materials. *Adv. Mater.* **2009**, *21*, 387–388. [CrossRef]

23. Sarikaya, M.; Tamerler, C.; Jen, A.K.-Y.; Schulten, K.; Baneyx, F. Molecular biomimetics: Nanotechnology through biology. *Nat. Mater.* **2003**, *2*, 577–585. [CrossRef] [PubMed]

24. Lindsey, J.S. Self-assembly in synthetic routes to molecular devices. Biological principles and chemical perspectives: A review. *New J. Chem.* **1991**, *15*, 153–179. [CrossRef]

25. Mayer, M.; Tebbe, M.; Kuttner, C.; Schnepf, M.J.; König, T.A.; Fery, A. Template-assisted colloidal self-assembly of macroscopic magnetic metasurfaces. *Faraday Discuss.* **2016**, *191*, 159–176. [CrossRef] [PubMed]

26. Pan, H.M.; Seuss, M.; Neubauer, M.P.; Trau, D.W.; Fery, A. Tuning the mechanical properties of hydrogel core–shell particles by inwards interweaving self-assembly. *ACS Appl. Mater. Interfaces* **2016**, *8*, 1493–1500. [CrossRef] [PubMed]

27. Kuttner, C.; Hanisch, A.; Schmalz, H.; Eder, M.; Schlaad, H.; Burgert, I.; Fery, A. Influence of the polymeric interphase design on the interfacial properties of (fiber-reinforced) composites. *ACS Appl. Mater. Interfaces* **2013**, *5*, 2469–2478. [CrossRef] [PubMed]

28. Kuttner, C.; Tebbe, M.; Schlaad, H.; Burgert, I.; Fery, A. Photochemical synthesis of polymeric fiber coatings and their embedding in matrix material: Morphology and nanomechanical properties at the fiber-matrix interface. *ACS Appl. Mater. Interfaces* **2012**, *4*, 3484–3492. [CrossRef] [PubMed]

29. Fratzl, P.; Burgert, I.; Gupta, H.S. On the role of interface polymers for the mechanics of natural polymeric composites. *Phys. Chem. Chem. Phys.* **2004**, *6*, 5575–5579. [CrossRef]

30. Kuttner, C. *Macromolecular Interphases and Interfaces in Composite Materials*; Verlag Dr. Hut: München, Germany, 2014.

31. Ma, X.; Zhang, S.; Jiao, F.; Newcomb, C.J.; Zhang, Y.; Prakash, A.; Liao, Z.; Baer, M.D.; Mundy, C.J.; Pfaendtner, J. Tuning crystallization pathways through sequence engineering of biomimetic polymers. *Nat. Mater.* **2017**. [CrossRef] [PubMed]

32. Robertson, E.J.; Battigelli, A.; Proulx, C.; Mannige, R.V.; Haxton, T.K.; Yun, L.; Whitelam, S.; Zuckermann, R.N. Design, synthesis, assembly, and engineering of peptoid nanosheets. *Acc. Chem. Res.* **2016**, *49*, 379–389. [CrossRef] [PubMed]

33. Jin, H.; Jiao, F.; Daily, M.D.; Chen, Y.; Yan, F.; Ding, Y.-H.; Zhang, X.; Robertson, E.J.; Baer, M.D.; Chen, C.-L. Highly stable and self-repairing membrane-mimetic 2d nanomaterials assembled from lipid-like peptoids. *Nat. Commun.* **2016**, *7*, 12252. [CrossRef] [PubMed]

34. Wagner, R.W.; Lindsey, J.S. A molecular photonic wire. *J. Am. Chem. Soc.* **1994**, *116*, 9759–9760. [CrossRef]

35. Nam, Y.S.; Shin, T.; Park, H.; Magyar, A.P.; Choi, K.; Fantner, G.; Nelson, K.A.; Belcher, A.M. Virus-templated assembly of porphyrins into light-harvesting nanoantennae. *J. Am. Chem. Soc.* **2010**, *132*, 1462–1463. [CrossRef] [PubMed]

36. Calzaferri, G.; Bossart, O.; Brühwiler, D.; Huber, S.; Leiggener, C.; Van Veen, M.K.; Ruiz, A.Z. Light-harvesting host–guest antenna materials for quantum solar energy conversion devices. *C. R. Chim.* **2006**, *9*, 214–225. [CrossRef]

37. Choi, M.S.; Yamazaki, T.; Yamazaki, I.; Aida, T. Bioinspired molecular design of light—Harvesting multiporphyrin arrays. *Angew. Chem. Int. Ed.* **2004**, *43*, 150–158. [CrossRef] [PubMed]

38. Xu, J.; Guo, Z. Biomimetic photonic materials with tunable structural colors. *J. Coll. Interface Sci.* **2013**, *406*, 1–17. [CrossRef] [PubMed]

39. Andrews, D.L. *Energy Harvesting Materials*; World Scientific: Singapore, 2005.

40. Jacobs, M.; Lopez-Garcia, M.; Phrathep, O.-P.; Lawson, T.; Oulton, R.; Whitney, H.M. Photonic multilayer structure of begonia chloroplasts enhances photosynthetic efficiency. *Nat. Plants* **2016**, *2*, 16162. [CrossRef] [PubMed]

41. Calver, C.F.; Schanze, K.S.; Cosa, G. Biomimetic light-harvesting antenna based on the self-assembly of conjugated polyelectrolytes embedded within lipid membranes. *ACS Nano* **2016**, *10*, 10598–10605. [CrossRef] [PubMed]

42. Henry, S.L.; Withers, J.M.; Singh, I.; Cooper, J.M.; Clark, A.W.; Burley, G.A.; Cogdell, R.J. DNA-directed spatial assembly of photosynthetic light-harvesting proteins. *Organ. Biomol. Chem.* **2016**, *14*, 1359–1362. [CrossRef] [PubMed]

43. Hemmig, E.A.; Creatore, C.; Wünsch, B.; Hecker, L.; Mair, P.; Parker, M.A.; Emmott, S.; Tinnefeld, P.; Keyser, U.F.; Chin, A.W. Programming light-harvesting efficiency using DNA origami. *Nano Lett.* **2016**, *16*, 2369–2374. [CrossRef] [PubMed]

44. Kundu, S.; Patra, A. Nanoscale strategies for light harvesting. *Chem. Rev.* **2017**, *117*, 712–757. [CrossRef] [PubMed]

45. Parker, A.R.; Lawrence, C.R. Water capture by a desert beetle. *Nature* **2001**, *414*, 33–34. [CrossRef] [PubMed]

46. White, B.; Sarkar, A.; Kietzig, A.-M. Fog-harvesting inspired by the stenocara beetle—An analysis of drop collection and removal from biomimetic samples with wetting contrast. *Appl. Surface Sci.* **2013**, *284*, 826–836. [CrossRef]

47. Garrod, R.; Harris, L.; Schofield, W.; McGettrick, J.; Ward, L.; Teare, D.; Badyal, J. Mimicking a stenocara beetle's back for microcondensation using plasmachemical patterned superhydrophobic-superhydrophilic surfaces. *Langmuir* **2007**, *23*, 689–693. [CrossRef] [PubMed]

48. Pang, C.; Kim, S.M.; Rahmawan, Y.; Suh, K.-Y. Beetle-inspired bidirectional, asymmetric interlocking using geometry-tunable nanohairs. *ACS Appl. Mater. Interfaces* **2012**, *4*, 4225–4230. [CrossRef] [PubMed]

49. Zhu, H.; Guo, Z.; Liu, W. Biomimetic water-collecting materials inspired by nature. *Chem. Commun.* **2016**, *52*, 3863–3879. [CrossRef] [PubMed]

50. Zhu, H.; Guo, Z. Hybrid engineered materials with high water-collecting efficiency inspired by namib desert beetles. *Chem. Commun.* **2016**, *52*, 6809–6812. [CrossRef] [PubMed]

51. Zeng, X.; Qian, L.; Yuan, X.; Zhou, C.; Li, Z.; Cheng, J.; Xu, S.; Wang, S.; Pi, P.; Wen, X. Inspired by stenocara beetles: From water collection to high-efficiency water-in-oil emulsion separation. *ACS Nano* **2016**, *11*, 760–769. [CrossRef] [PubMed]

52. Zhang, S.; Huang, J.; Chen, Z.; Lai, Y. Bioinspired special wettability surfaces: From fundamental research to water harvesting applications. *Small* **2016**. [CrossRef] [PubMed]

53. Espinosa, H.D.; Rim, J.E.; Barthelat, F.; Buehler, M.J. Merger of structure and material in nacre and bone—Perspectives on de novo biomimetic materials. *Prog. Mater. Sci.* **2009**, *54*, 1059–1100. [CrossRef]

54. Finnemore, A.; Cunha, P.; Shean, T.; Vignolini, S.; Guldin, S.; Oyen, M.; Steiner, U. Biomimetic layer-by-layer assembly of artificial nacre. *Nat. Commun.* **2012**, *3*, 966. [CrossRef] [PubMed]

55. Tang, Z.; Kotov, N.A.; Magonov, S.; Ozturk, B. Nanostructured artificial nacre. *Nat. Mater.* **2003**, *2*, 413–418. [CrossRef] [PubMed]

56. Katti, K.S.; Mohanty, B.; Katti, D.R. *Biomimetic Lessons Learnt from Nacre*; INTECH Open Access: Rijeka, Croatia, 2010.

57. Li, C.; Born, A.K.; Schweizer, T.; Zenobi-Wong, M.; Cerruti, M.; Mezzenga, R. Amyloid-hydroxyapatite bone biomimetic composites. *Adv. Mater.* **2014**, *26*, 3207–3212. [CrossRef] [PubMed]

58. Thein-Han, W.; Misra, R. Biomimetic chitosan-nanohydroxyapatite composite scaffolds for bone tissue engineering. *Acta Biomater.* **2009**, *5*, 1182–1197. [CrossRef] [PubMed]

59. Ng, J.; Spiller, K.; Bernhard, J.; Vunjak-Novakovic, G. Biomimetic approaches for bone tissue engineering. *Tissue Eng. Part B Rev.* **2017**. [CrossRef] [PubMed]

60. Lopez-Heredia, M.A.; Łapa, A.; Mendes, A.C.; Balcaen, L.; Samal, S.K.; Chai, F.; Van der Voort, P.; Stevens, C.V.; Parakhonskiy, B.V.; Chronakis, I.S. Bioinspired, biomimetic, double-enzymatic mineralization of hydrogels for bone regeneration with calcium carbonate. *Mater. Lett.* **2017**, *190*, 13–16. [CrossRef]

61. Binnig, G.; Quate, C.F.; Gerber, C. Atomic force microscope. *Phys. Rev. Lett.* **1986**, *56*, 930–933. [CrossRef] [PubMed]

62. Bustamante, C.; Vesenka, J.; Tang, C.L.; Rees, W.; Guthold, M.; Keller, R. Circular DNA molecules imaged in air by scanning force microscopy. *Biochemistry* **1992**, *31*, 22–26. [CrossRef] [PubMed]

63. Hansma, H.G.; Sinsheimer, R.L.; Li, M.-Q.; Hansma, P.K. Atomic force microscopy of single-and double-stranded DNA. *Nucleic Acids Res.* **1992**, *20*, 3585–3590. [CrossRef] [PubMed]

64. Hansma, H.G.; Laney, D.E. DNA binding to mica correlates with cationic radius: Assay by atomic force microscopy. *Biophys. J.* **1996**, *70*, 1933–1939. [CrossRef]

65. Ando, T.; Uchihashi, T.; Kodera, N. High-speed afm and applications to biomolecular systems. *Annu. Rev. Biophys.* **2013**, *42*, 393–414. [CrossRef] [PubMed]

66. Kuznetsov, Y.G.; Malkin, A.J.; Lucas, R.W.; Plomp, M.; McPherson, A. Imaging of viruses by atomic force microscopy. *J. Gen. Virol.* **2001**, *82*, 2025–2034. [CrossRef] [PubMed]

67. De Pablo, P.J.; Carrión-Vázquez, M. Imaging biological samples with atomic force microscopy. *Cold Spring Harb. Protoc.* **2014**, *2014*, 167–177. [CrossRef] [PubMed]

68. Butt, H.J.; Wolff, E.K.; Gould, S.A.C.; Dixon Northern, B.; Peterson, C.M.; Hansma, P.K. Imaging cells with the atomic force microscope. *J. Struct. Biol.* **1990**, *105*, 54–61. [CrossRef]

69. Dufrêne, Y.F.; Ando, T.; Garcia, R.; Alsteens, D.; Martinez-Martin, D.; Engel, A.; Gerber, C.; Müller, D.J. Imaging modes of atomic force microscopy for application in molecular and cell biology. *Nat. Nanotechnol.* **2017**, *12*, 295–307. [CrossRef] [PubMed]

70. Rugar, D.; Hansma, P. Atomic force microscopy. *Phys. Today* **1990**, *43*, 23–30. [CrossRef]

71. Junno, T.; Anand, S.; Deppert, K.; Montelius, L.; Samuelson, L. Contact mode atomic force microscopy imaging of nanometer—Sized particles. *Appl. Phys. Lett.* **1995**, *66*, 3295–3297. [CrossRef]

72. Le Grimellec, C.; Lesniewska, E.; Giocondi, M.-C.; Finot, E.; Vié, V.; Goudonnet, J.-P. Imaging of the surface of living cells by low-force contact-mode atomic force microscopy. *Biophys. J.* **1998**, *75*, 695–703. [CrossRef]

73. Cleveland, J.; Anczykowski, B.; Schmid, A.; Elings, V. Energy dissipation in tapping-mode atomic force microscopy. *Appl. Phys. Lett.* **1998**, *72*, 2613–2615. [CrossRef]

74. Hansma, P.; Cleveland, J.; Radmacher, M.; Walters, D.; Hillner, P.; Bezanilla, M.; Fritz, M.; Vie, D.; Hansma, H.; Prater, C. Tapping mode atomic force microscopy in liquids. *Appl. Phys. Lett.* **1994**, *64*, 1738–1740. [CrossRef]

75. San Paulo, A.; Garcia, R. Unifying theory of tapping-mode atomic-force microscopy. *Phys. Rev. B* **2002**, *66*, 041406. [CrossRef]

76. Tamayo, J.; Garcia, R. Deformation, contact time, and phase contrast in tapping mode scanning force microscopy. *Langmuir* **1996**, *12*, 4430–4435. [CrossRef]

77. Kowalewski, T.; Holtzman, D.M. In situ atomic force microscopy study of alzheimer's β-amyloid peptide on different substrates: New insights into mechanism of β-sheet formation. *Proc. Natl. Acad. Sci. USA* **1999**, *96*, 3688–3693. [CrossRef] [PubMed]

78. Kitamura, S.-I.; Iwatsuki, M. Observation of 7×7 reconstructed structure on the silicon (111) surface using ultrahigh vacuum noncontact atomic force microscopy. *Jpn. J. Appl. Phys.* **1995**, *34*, L145. [CrossRef]

79. Morita, S.; Giessibl, F.J.; Meyer, E.; Wiesendanger, R. *Noncontact Atomic Force Microscopy*; Springer: Berlin, Germany, 2015; Volume 3.

80. Ramachandran, T.; Baur, C.; Bugacov, A.; Madhukar, A.; Koel, B.; Requicha, A.; Gazen, C. Direct and controlled manipulation of nanometer-sized particles using the non-contact atomic force microscope. *Nanotechnology* **1998**, *9*, 237. [CrossRef]

81. Sugimoto, Y.; Pou, P.; Abe, M.; Jelinek, P.; Pérez, R.; Morita, S.; Custance, O. Chemical identification of individual surface atoms by atomic force microscopy. *Nature* **2007**, *446*, 64–67. [CrossRef] [PubMed]

82. Trtik, P.; Kaufmann, J.; Volz, U. On the use of peak-force tapping atomic force microscopy for quantification of the local elastic modulus in hardened cement paste. *Cem. Concr. Res.* **2012**, *42*, 215–221. [CrossRef]

83. Alsteens, D.; Dupres, V.; Yunus, S.; Latgé, J.-P.; Heinisch, J.J.; Dufrêne, Y.F. High-resolution imaging of chemical and biological sites on living cells using peak force tapping atomic force microscopy. *Langmuir* **2012**, *28*, 16738–16744. [CrossRef] [PubMed]

84. Zhao, B.; Song, Y.; Wang, S.; Dai, B.; Zhang, L.; Dong, Y.; Lü, J.; Hu, J. Mechanical mapping of nanobubbles by peakforce atomic force microscopy. *Soft Matter* **2013**, *9*, 8837–8843. [CrossRef]

85. Foster, B. New atomic force microscopy(afm) approaches life sciences gently, quantitatively, and correctively. *Am. Lab.* **2012**, *44*, 24–28.

86. Glatz, B.A.; Tebbe, M.; Kaoui, B.; Aichele, R.; Kuttner, C.; Schedl, A.E.; Schmidt, H.-W.; Zimmermann, W.; Fery, A. Hierarchical line-defect patterns in wrinkled surfaces. *Soft Matter* **2015**, *11*, 3332–3339. [CrossRef] [PubMed]

87. De Yoreo, J.J.; Chung, S.; Friddle, R.W. In situ atomic force microscopy as a tool for investigating interactions and assembly dynamics in biomolecular and biomineral systems. *Adv. Funct. Mater.* **2013**, *23*, 2525–2538. [CrossRef]

88. Schitter, G.; Astrom, K.J.; DeMartini, B.E.; Thurner, P.J.; Turner, K.L.; Hansma, P.K. Design and modeling of a high-speed afm-scanner. *IEEE Trans. Control Syst. Technol.* **2007**, *15*, 906–915. [CrossRef]

89. Fantner, G.E.; Schitter, G.; Kindt, J.H.; Ivanov, T.; Ivanova, K.; Patel, R.; Holten-Andersen, N.; Adams, J.; Thurner, P.J.; Rangelow, I.W. Components for high speed atomic force microscopy. *Ultramicroscopy* **2006**, *106*, 881–887. [CrossRef] [PubMed]

90. Sulchek, T.; Hsieh, R.; Adams, J.; Minne, S.; Quate, C.; Adderton, D. High-speed atomic force microscopy in liquid. *Rev. Sci. Instrum.* **2000**, *71*, 2097–2099. [CrossRef]

91. Hartman, B.; Andersson, S.; Nagel, W.; Leang, K. Non-raster high-speed afm imaging of biopolymers. *Biophys. J.* **2017**, *112*, 587a. [CrossRef]

92. Takahashi, H.; Miyagi, A.; Redondo-Morata, L.; Scheuring, S. Development of temperature-controlled high-speed afm. *Biophys. J.* **2017**, *112*, 587a. [CrossRef]

93. Katan, A.J.; Dekker, C. High-speed afm reveals the dynamics of single biomolecules at the nanometer scale. *Cell* **2011**, *147*, 979–982. [CrossRef] [PubMed]

94. Ando, T.; Kodera, N.; Takai, E.; Maruyama, D.; Saito, K.; Toda, A. A high-speed atomic force microscope for studying biological macromolecules. *Proc. Natl. Acad. Sci. USA* **2001**, *98*, 12468–12472. [CrossRef] [PubMed]

95. Uchihashi, T.; Iino, R.; Ando, T.; Noji, H. High-speed atomic force microscopy reveals rotary catalysis of rotorless f1-atpase. *Science* **2011**, *333*, 755–758. [CrossRef] [PubMed]

96. Ruan, Y.; Miyagi, A.; Wang, X.; Chami, M.; Boudker, O.; Scheuring, S. Direct visualization of glutamate transporter elevator mechanism by high-speed afm. *Proc. Natl. Acad. Sci. USA* **2017**, *114*, 1584–1588. [CrossRef] [PubMed]

97. Kotani, N.; Kumaresan, R.; Kawamoto-Ozaki, Y.; Morii, T.; Okada, T. High-speed afm reveals advanced details on dynamic behavior of antibody. *Biophys. J.* **2017**, *112*, 587a. [CrossRef]

98. Watanabe-Nakayama, T.; Ono, K.; Itami, M.; Takahashi, R.; Teplow, D.B.; Yamada, M. High-speed atomic force microscopy reveals structural dynamics of amyloid β1–42 aggregates. *Proc. Natl. Acad. Sci. USA* **2016**, *113*, 5835–5840. [CrossRef] [PubMed]

99. Milhiet, P.-E.; Yamamoto, D.; Berthoumieu, O.; Dosset, P.; Le Grimellec, C.; Verdier, J.-M.; Marchal, S.; Ando, T. Deciphering the structure, growth and assembly of amyloid-like fibrils using high-speed atomic force microscopy. *PLoS ONE* **2010**, *5*, e13240. [CrossRef] [PubMed]

100. Zlotnikov, I.; Zolotoyabko, E.; Fratzl, P. Nano-scale modulus mapping of biological composite materials: Theory and practice. *Prog. Mater. Sci.* **2017**, *87*, 292–320. [CrossRef]

101. Gal, A.; Wirth, R.; Kopka, J.; Fratzl, P.; Faivre, D.; Scheffel, A. Macromolecular recognition directs calcium ions to coccolith mineralization sites. *Science* **2016**, *353*, 590–593. [CrossRef] [PubMed]

102. Casdorff, K.; Keplinger, T.; Bellanger, H.; Michen, B.; Schön, S.; Burgert, I. High-resolution adhesion mapping of the odd-even effect on a layer-by-layer coated biomaterial by atomic-force-microscopy. *ACS Appl. Mater. Interfaces* **2017**, *9*, 13793–13800. [CrossRef] [PubMed]

103. Li, Q.; Zhang, T.; Pan, Y.; Ciacchi, L.C.; Xu, B.; Wei, G. Afm-based force spectroscopy for bioimaging and biosensing. *RSC Adv.* **2016**, *6*, 12893–12912. [CrossRef]

104. Alsteens, D.; Newton, R.; Schubert, R.; Martinez-Martin, D.; Delguste, M.; Roska, B.; Müller, D.J. Nanomechanical mapping of first binding steps of a virus to animal cells. *Nat. Nanotechnol.* **2017**, *12*, 177–183. [CrossRef] [PubMed]

105. Berne, C.C.; Ma, X.; Licata, N.A.; Neves, B.R.; Setayeshgar, S.; Brun, Y.V.; Dragnea, B. Physiochemical properties of caulobacter crescentus holdfast: A localized bacterial adhesive. *J. Phys. Chem. B* **2013**, *117*, 10492–10503. [CrossRef] [PubMed]

106. Vaughan, R.; Tragesser, B.; Ni, P.; Ma, X.; Dragnea, B.; Kao, C.C. The tripartite virions of the brome mosaic virus have distinct physical properties that affect the timing of the infection process. *J. Virol.* **2014**, *88*, 6483–6491. [CrossRef] [PubMed]

107. Ni, P.; Wang, Z.; Ma, X.; Das, N.C.; Sokol, P.; Chiu, W.; Dragnea, B.; Hagan, M.; Kao, C.C. An examination of the electrostatic interactions between the n-terminal tail of the brome mosaic virus coat protein and encapsidated rnas. *J. Mol. Biol.* **2012**, *419*, 284–300. [CrossRef] [PubMed]

108. Kol, N.; Shi, Y.; Tsvitov, M.; Barlam, D.; Shneck, R.Z.; Kay, M.S.; Rousso, I. A stiffness switch in human immunodeficiency virus. *Biophys. J.* **2007**, *92*, 1777–1783. [CrossRef] [PubMed]

109. Castellanos, M.; Pérez, R.; Carrasco, C.; Hernando-Pérez, M.; Gómez-Herrero, J.; de Pablo, P.J.; Mateu, M.G. Mechanical elasticity as a physical signature of conformational dynamics in a virus particle. *Proc. Natl. Acad. Sci. USA* **2012**, *109*, 12028–12033. [CrossRef] [PubMed]

110. Marchetti, M.; Wuite, G.; Roos, W. Atomic force microscopy observation and characterization of single virions and virus-like particles by nano-indentation. *Curr. Opin. Virol.* **2016**, *18*, 82–88. [CrossRef] [PubMed]

111. Denning, D.; Wuite, G.J.; Roos, W.H. A combined imaging and force spectroscopy approach reveals the material properties of viral nanoparticles. *Biophys. J.* **2016**, *110*, 500a. [CrossRef]

112. Ramalho, R.; Rankovic, S.; Zhou, J.; Aiken, C.; Rousso, I. Analysis of the mechanical properties of wild type and hyperstable mutants of the hiv-1 capsid. *Retrovirology* **2016**, *13*, 17. [CrossRef] [PubMed]

113. Korneev, D.; Popova, A.; Generalov, V.; Zaitsev, B. Atomic force microscopy-based single virus particle spectroscopy. *Biophysics* **2016**, *61*, 413–419. [CrossRef]

114. Snijder, J.; Kononova, O.; Barbu, I.M.; Uetrecht, C.; Rurup, W.F.; Burnley, R.J.; Koay, M.S.; Cornelissen, J.J.; Roos, W.H.; Barsegov, V. Assembly and mechanical properties of the cargo-free and cargo-loaded bacterial nanocompartment encapsulin. *Biomacromolecules* **2016**, *17*, 2522–2529. [CrossRef] [PubMed]

115. Quintana-Cataño, C.A.; Vives-Flórez, M.J.; Forero-Shelton, M. Force spectroscopy of t4 bacteriophage adhesion during infection. *Biophys. J.* **2017**, *112*, 588a. [CrossRef]

116. Hernando-Perez, M.; Zeng, C.; Delalande, L.; Tsvetkova, I.; Bousquet, A.; Tayachi-Pigeonnat, M.; Temam, R.; Dragnea, B. Nanoindentation of isometric viruses on deterministically corrugated substrates. *J. Phys. Chem. B* **2016**, *120*, 340–347. [CrossRef] [PubMed]

117. Zeng, C.; Moller-Tank, S.; Asokan, A.; Dragnea, B. Probing the link among genomic cargo, contact mechanics, and nanoindentation in recombinant adeno-associated virus 2. *J. Phys. Chem. B* **2017**, *121*, 1843–1853. [CrossRef] [PubMed]

118. Delalande, L.; Tsvetkova, I.B.; Zeng, C.; Bond, K.; Jarrold, M.; Dragnea, B. Catching a virus in a molecular net. *Nanoscale* **2016**, *8*, 16221–16228. [CrossRef] [PubMed]

119. Eeftens, J.; Katan, A.; Kschonsak, M.; Hassler, M.; Dief, E.; de Wilde, L.; Haering, C.; Dekker, C. Single-molecule experiments to resolve structural and mechanical properties of condensin. *Biophys. J.* **2016**, *110*, 528a. [CrossRef]

120. Heinze, K.; Sasaki, E.; King, N.; Baker, D.; Hilvert, D.; Wuite, G.; Roos, W. Protein nanocontainers from nonviral origin: Testing the mechanics of artificial and natural protein cages by afm. *J. Phys. Chem. B* **2016**, *120*, 5945–5952. [CrossRef] [PubMed]

121. Moreno-Madrid, F.; Martín-González, N.; Llauró, A.; Ortega-Esteban, A.; Hernando-Pérez, M.; Douglas, T.; Schaap, I.A.; de Pablo, P.J. Atomic force microscopy of virus shells. *Biochem. Soc. Trans.* **2017**, *45*, 499–511. [CrossRef] [PubMed]

122. Zhang, X.F.; Xu, Y.; McKinnon, T.A.; Zhang, W. Biophysical mechanisms of von willebrand factor-collagen interactions. *Biophys. J.* **2017**, *112*, 455a. [CrossRef]

123. Van Pattten, W.J.; Walder, R.; Adhikari, A.; Ravichandran, R.; Tinberg, C.E.; Baker, D.; Perkins, T.T. A computationally designed protein-ligand interaction is mechanically robust. *Biophys. J.* **2017**, *112*, 455a. [CrossRef]

124. Yadav, A.; Paul, S.; Venkatramani, R.; Rama, S.; Ainavarapu, K. Examining the mechanical properties of copper binding azurin using single molecule force spectroscopy and steered molecular dynamics. *Biophys. J.* **2016**, *110*, 496a. [CrossRef]

125. Hughes, M.L.; Dougan, L. The physics of pulling polyproteins: A review of single molecule force spectroscopy using the afm to study protein unfolding. *Rep. Prog. Phys.* **2016**, *79*, 076601. [CrossRef] [PubMed]

126. Walder, R.; Van Patten, W.J.; Miller, T.W.; Perkins, T.T. Mechanical characterization of the hiv-1 rna hairpin using an atomic force microscope. *Biophys. J.* **2017**, *112*, 166a. [CrossRef]

127. Shlyakhtenko, L.S.; Dutta, S.; Li, M.; Harris, R.S.; Lyubchenko, Y.L. Single-molecule force spectroscopy studies of apobec3a-single-stranded DNA complexes. *Biochemistry* **2016**, *55*, 3102–3106. [CrossRef] [PubMed]

128. Fitzgibbon, C.J.; Josephs, E.A.; Marszalek, P.E. Resolving individual damage sites in DNA with afm using reengineered repair proteins. *Biophys. J.* **2016**, *110*, 496a. [CrossRef]

129. Dutta, S.; Armitage, B.A.; Lyubchenko, Y.L. Probing of minipegγ-pna–dna hybrid duplex stability with afm force spectroscopy. *Biochemistry* **2016**, *55*, 1523–1528. [CrossRef] [PubMed]

130. Camunas-Soler, J.; Ribezzi-Crivellari, M.; Ritort, F. Elastic properties of nucleic acids by single-molecule force spectroscopy. *Annu. Rev. Biophys.* **2016**, *45*, 65–84. [CrossRef] [PubMed]

131. Fisher, G.J.; Shao, Y.; He, T.; Qin, Z.; Perry, D.; Voorhees, J.J.; Quan, T. Reduction of fibroblast size/mechanical force down-regulates tgf-β type ii receptor: Implications for human skin aging. *Aging Cell* **2016**, *15*, 67–76. [CrossRef] [PubMed]

132. Rianna, C.; Radmacher, M. Cell Mechanics As a Marker for Diseases: Biomedical Applications of Afm. In *AIP Conference Proceedings*; AIP Publishing: Melville, NY, USA, 2016; p. 020057.

133. Ting, A.Y.P. Spatially resolved mapping of endogenous proteins and rna in living cells. *Biophys. J.* **2017**, *112*, 7a. [CrossRef]

134. Burgert, I.; Keplinger, T. Plant micro-and nanomechanics: Experimental techniques for plant cell-wall analysis. *J. Exp. Bot.* **2013**, *64*, 4635–4649. [CrossRef] [PubMed]

135. Wang, L.; Nancollas, G.H. Pathways to biomineralization and biodemineralization of calcium phosphates: The thermodynamic and kinetic controls. *Dalton Trans.* **2009**, 2665–2672. [CrossRef] [PubMed]

136. Putnis, A.; Pina, C.M.; Astilleros, J.M.; Fernández-Díaz, L.; Prieto, M. Nucleation of solid solutions crystallizing from aqueous solutions. *Phil. Trans. R. Soc. Lond. A* **2003**, *361*, 615–632. [CrossRef] [PubMed]

137. Gao, Y.; Wang, J.; Zhong, J.; Wang, Y.; Yin, Q.; Hou, B.; Hao, H. Application of atomic force microscopy in understanding crystallization process. *Sci. Adv. Mater.* **2017**, *9*, 89–101. [CrossRef]

138. Jun, Y.-S.; Kim, D.; Neil, C.W. Heterogeneous nucleation and growth of nanoparticles at environmental interfaces. *Acc. Chem. Res.* **2016**, *49*, 1681–1690. [CrossRef] [PubMed]

139. Wang, L.; Lu, J.; Xu, F.; Zhang, F. Dynamics of crystallization and dissolution of calcium orthophosphates at the near-molecular level. *Chin. Sci. Bull.* **2011**, *56*, 713–721. [CrossRef]

140. Motta, N.; Szkutnik, P.D.; Tomellini, M.; Sgarlata, A.; Fanfoni, M.; Patella, F.; Balzarotti, A. Role of patterning in islands nucleation on semiconductor surfaces. *C. R. Phys.* **2006**, *7*, 1046–1072. [CrossRef]

141. Wesson, J.A.; Ward, M.D. Role of crystal surface adhesion in kidney stone disease. *Curr. Opin. Nephrol. Hypertens.* **2006**, *15*, 386–393. [CrossRef] [PubMed]

142. Alamani, B.G.; Rimer, J.D. Molecular modifiers of kidney stones. *Curr. Opin. Nephrol. Hypertens.* **2017**, *26*, 256–265. [CrossRef] [PubMed]

143. Rimer, J.D.; Kolbach-Mandel, A.M.; Ward, M.D.; Wesson, J.A. The role of macromolecules in the formation of kidney stones. *Urolithiasis* **2017**, *45*, 57–74. [CrossRef] [PubMed]

144. Han, L.; Ohsuna, T.; Liu, Z.; Alfredsson, V.; Kjellman, T.; Asahina, S.; Suga, M.; Ma, Y.; Oleynikov, P.; Miyasaka, K. Structures of silica-based nanoporous materials revealed by microscopy. *Z. Anorg. Allg. Chem.* **2014**, *640*, 521–536. [CrossRef]

145. Glenner, G.G.; Wong, C.W. Alzheimer's disease: Initial report of the purification and characterization of a novel cerebrovascular amyloid protein. *Biochem. Biophys. Res. Commun.* **1984**, *120*, 885–890. [CrossRef]

146. Zhai, J.; Lee, T.-H.; Small, D.H.; Aguilar, M.-I. Characterization of early stage intermediates in the nucleation phase of aβ aggregation. *Biochemistry* **2012**, *51*, 1070–1078. [CrossRef] [PubMed]

147. Jeong, J.S.; Ansaloni, A.; Mezzenga, R.; Lashuel, H.A.; Dietler, G. Novel mechanistic insight into the molecular basis of amyloid polymorphism and secondary nucleation during amyloid formation. *J. Mol. Biol.* **2013**, *425*, 1765–1781. [CrossRef] [PubMed]

148. Karsai, A.; Slack, T.J.; Malekan, H.; Khoury, F.; Lin, W.F.; Tran, V.; Cox, D.; Toney, M.; Chen, X.; Liu, G.Y. Local mechanical perturbation provides an effective means to regulate the growth and assembly of functional peptide fibrils. *Small* **2016**, *12*, 6407–6415. [CrossRef] [PubMed]

149. Zhang, F.-C.; Zhang, F.; Su, H.-N.; Li, H.; Zhang, Y.; Hu, J. Mechanical manipulation assisted self-assembly to achieve defect repair and guided epitaxial growth of individual peptide nanofilaments. *ACS Nano* **2010**, *4*, 5791–5796. [CrossRef] [PubMed]

150. Dunlop, J.W.; Fratzl, P. Biological composites. *Annu. Rev. Mater. Res.* **2010**, *40*, 1–24. [CrossRef]

151. Dunlop, J.W.; Weinkamer, R.; Fratzl, P. Artful interfaces within biological materials. *Mater. Today* **2011**, *14*, 70–78. [CrossRef]

152. Tao, J.; Buchko, G.W.; Shaw, W.J.; De Yoreo, J.J.; Tarasevich, B.J. Sequence-defined energetic shifts control the disassembly kinetics and microstructure of amelogenin adsorbed onto hydroxyapatite (100). *Langmuir* **2015**, *31*, 10451. [CrossRef] [PubMed]

153. Chung, S.; Shin, S.-H.; Bertozzi, C.R.; De Yoreo, J.J. Self-catalyzed growth of s layers via an amorphous-to-crystalline transition limited by folding kinetics. *Proc. Natl. Acad. Sci. USA* **2010**, *107*, 16536–16541. [CrossRef] [PubMed]

154. Shin, S.-H.; Chung, S.; Sanii, B.; Comolli, L.R.; Bertozzi, C.R.; De Yoreo, J.J. Direct observation of kinetic traps associated with structural transformations leading to multiple pathways of s-layer assembly. *Proc. Natl. Acad. Sci. USA* **2012**, *109*, 12968–12973. [CrossRef] [PubMed]

155. Ido, S.; Kimura, K.; Oyabu, N.; Kobayashi, K.; Tsukada, M.; Matsushige, K.; Yamada, H. Beyond the helix pitch: Direct visualization of native DNA in aqueous solution. *ACS Nano* **2013**, *7*, 1817–1822. [CrossRef] [PubMed]

156. Abel, G.R., Jr.; Josephs, E.A.; Luong, N.; Ye, T. A switchable surface enables visualization of single DNA hybridization events with atomic force microscopy. *J. Am. Chem. Soc.* **2013**, *135*, 6399–6402. [CrossRef] [PubMed]

157. Thomson, N.H.; Kasas, S.; Smith, B.; Hansma, H.G.; Hansma, P.K. Reversible binding of DNA to mica for afm imaging. *Langmuir* **1996**, *12*, 5905–5908. [CrossRef]

158. Burke, K.A.; Hensal, K.M.; Umbaugh, C.S.; Chaibva, M.; Legleiter, J. Huntingtin disrupts lipid bilayers in a polyq-length dependent manner. *Biochim. Biophys. Acta (BBA)-Biomembr.* **2013**, *1828*, 1953–1961. [CrossRef] [PubMed]

159. Pan, J.; Khadka, N.K. Kinetic defects induced by melittin in model lipid membranes: A solution atomic force microscopy study. *J. Phys. Chem. B* **2016**, *120*, 4625–4634. [CrossRef] [PubMed]

160. Yilmaz, N.; Kobayashi, T. Visualization of lipid membrane reorganization induced by a pore-forming toxin using high-speed atomic force microscopy. *ACS Nano* **2015**, *9*, 7960–7967. [CrossRef] [PubMed]

161. Yang, G.; Wong, M.K.; Lin, L.E.; Yip, C.M. Nucleation and growth of elastin-like peptide fibril multilayers: An in situ atomic force microscopy study. *Nanotechnology* **2011**, *22*, 494018. [CrossRef] [PubMed]

162. Charbonneau, C.; Kleijn, J.M.; Cohen Stuart, M.A. Subtle charge balance controls surface-nucleated self-assembly of designed biopolymers. *ACS Nano* **2014**, *8*, 2328–2335. [CrossRef] [PubMed]

163. Kang, S.-G.; Li, H.; Huynh, T.; Zhang, F.; Xia, Z.; Zhang, Y.; Zhou, R. Molecular mechanism of surface-assisted epitaxial self-assembly of amyloid-like peptides. *ACS Nano* **2012**, *6*, 9276–9282. [CrossRef] [PubMed]

164. Dai, B.; Kang, S.-G.; Huynh, T.; Lei, H.; Castelli, M.; Hu, J.; Zhang, Y.; Zhou, R. Salts drive controllable multilayered upright assembly of amyloid-like peptides at mica/water interface. *Proc. Natl. Acad. Sci. USA* **2013**, *110*, 8543–8548. [CrossRef] [PubMed]

165. Varongchayakul, N.; Johnson, S.; Quabili, T.; Cappello, J.; Ghandehari, H.; Solares, S.D.J.; Hwang, W.; Seog, J. Direct observation of amyloid nucleation under nanomechanical stretching. *ACS Nano* **2013**, *7*, 7734–7743. [CrossRef] [PubMed]

166. Yang, H.; Fung, S.-Y.; Pritzker, M.; Chen, P. Mechanical-force-induced nucleation and growth of peptide nanofibers at liquid/solid interfaces. *Angew. Chem. Int. Ed.* **2008**, *47*, 4397–4400. [CrossRef] [PubMed]

167. LeGeros, R.Z. Calcium phosphate-based osteoinductive materials. *Chem. Rev.* **2008**, *108*, 4742–4753. [CrossRef] [PubMed]

168. Tiselius, H.-G. A hypothesis of calcium stone formation: An interpretation of stone research during the past decades. *Urol. Res.* **2011**, *39*, 231–243. [CrossRef] [PubMed]

169. Bazin, D.; Daudon, M.; Combes, C.; Rey, C. Characterization and some physicochemical aspects of pathological microcalcifications. *Chem. Rev.* **2012**, *112*, 5092–5120. [CrossRef] [PubMed]

170. Li, M.; Wang, L.; Zhang, W.; Putnis, C.V.; Putnis, A. Direct observation of spiral growth, particle attachment, and morphology evolution of hydroxyapatite. *Cryst. Growth Des.* **2016**, *16*, 4509–4518. [CrossRef]

171. Wu, S.; Yu, M.; Li, M.; Wang, L.; Putnis, C.V.; Putnis, A. In situ afm imaging of octacalcium phosphate crystallization and its modulation by amelogenin's c-terminus. *Cryst. Growth Des.* **2017**, *17*, 2194–2202. [CrossRef]

172. Habraken, W.J.; Tao, J.; Brylka, L.J.; Friedrich, H.; Bertinetti, L.; Schenk, A.S.; Verch, A.; Dmitrovic, V.; Bomans, P.H.; Frederik, P.M. Ion-association complexes unite classical and non-classical theories for the biomimetic nucleation of calcium phosphate. *Nat. Commun.* **2013**, *4*, 1507. [CrossRef] [PubMed]

173. Rodriguez-Navarro, C.; Burgos Cara, A.; Elert, K.; Putnis, C.V.; Ruiz-Agudo, E. Direct nanoscale imaging reveals the growth of calcite crystals via amorphous nanoparticles. *Cryst. Growth Des.* **2016**, *16*, 1850–1860. [CrossRef]

174. Wolf, S.L.P.; Caballero, L.; Melo, F.; Cölfen, H. Gel-like calcium carbonate precursors observed by in-situ afm. *Langmuir* **2016**, *33*, 158–163. [CrossRef] [PubMed]

175. Hofmann, S.; Voïtchovsky, K.; Spijker, P.; Schmidt, M.; Stumpf, T. Visualising the molecular alteration of the calcite (104)–water interface by sodium nitrate. *Sci. Rep.* **2016**, *6*, 21576. [CrossRef] [PubMed]

176. Wang, L.; Qin, L.; Putnis, C.V.; Ruiz-Agudo, E.; King, H.E.; Putnis, A. Visualizing organophosphate precipitation at the calcite–water interface by in situ atomic-force microscopy. *Environ. Sci. Technol.* **2015**, *50*, 259–268. [CrossRef] [PubMed]

177. So, C.R.; Liu, J.; Fears, K.P.; Leary, D.H.; Golden, J.P.; Wahl, K.J. Self-assembly of protein nanofibrils orchestrates calcite step movement through selective nonchiral interactions. *ACS Nano* **2015**, *9*, 5782–5791. [CrossRef] [PubMed]

178. Cho, K.R.; Salter, E.A.; De Yoreo, J.J.; Wierzbicki, A.; Elhadj, S.; Huang, Y.; Qiu, S.R. Growth inhibition of calcium oxalate monohydrate crystal by linear aspartic acid enantiomers investigated by in situ atomic force microscopy. *CrystEngComm* **2013**, *15*, 54–64. [CrossRef]

179. Li, S.; Zhang, W.; Wang, L. Direct nanoscale imaging of calcium oxalate crystallization on brushite reveals the mechanisms underlying stone formation. *Cryst. Growth Des.* **2015**, *15*, 3038–3045. [CrossRef]

180. Ando, T.; Uchihashi, T.; Fukuma, T. High-speed atomic force microscopy for nano-visualization of dynamic biomolecular processes. *Prog. Surf. Sci.* **2008**, *83*, 337–437. [CrossRef]

181. Li, Y.; Bechhoefer, J. Model-free iterative control of repetitive dynamics for high-speed scanning in atomic force microscopy. *Rev. Sci. Instrum.* **2009**, *80*, 013702. [CrossRef] [PubMed]

182. Yan, Y.; Wu, Y.; Zou, Q.; Su, C. An integrated approach to piezoactuator positioning in high-speed atomic force microscope imaging. *Rev. Sci. Instrum.* **2008**, *79*, 073704. [CrossRef] [PubMed]

183. Heinisch, J.J.; Lipke, P.N.; Beaussart, A.; Chatel, S.E.K.; Dupres, V.; Alsteens, D.; Dufrêne, Y.F. Atomic force microscopy—Looking at mechanosensors on the cell surface. *J. Cell Sci.* **2012**, *125*, 4189–4195. [CrossRef] [PubMed]

minerals

MDPI

Article

In Situ AFM Study of Crystal Growth on a Barite (001) Surface in BaSO$_4$ Solutions at 30 °C

Yoshihiro Kuwahara [1,*], **Wen Liu** [2], **Masato Makio** [2] **and Keisuke Otsuka** [2]

[1] Department of Environmental Changes, Faculty of Social and Cultural Studies, Kyushu University, Motooka, Fukuoka 819-0395, Japan

[2] Department of Comprehensive Earth Sciences, Graduate School of Integrated Sciences for Global Society, Kyushu University, Motooka, Fukuoka 819-0395, Japan; 2GS15057G@s.kyushu-u.ac.jp (W.L.); 3GS15017G@s.kyushu-u.ac.jp (M.M.); u77b22ced7pq96a@yahoo.co.jp (K.O.)

* Correspondence: ykuwa@scs.kyushu-u.ac.jp; Tel.: +81-92-802-5654

Academic Editor: Denis Gebauer
Received: 21 September 2016; Accepted: 24 October 2016; Published: 2 November 2016

Abstract: The growth behavior and kinetics of the barite (001) surface in supersaturated BaSO$_4$ solutions (supersaturation index (SI) = 1.1–4.1) at 30 °C were investigated using in situ atomic force microscopy (AFM). At the lowest supersaturation, the growth behavior was mainly the advancement of the initial step edges and filling in of the etch pits formed in the water before the BaSO$_4$ solution was injected. For solutions with higher supersaturation, the growth behavior was characterized by the advance of the <$uv0$> and [010] half-layer steps with two different advance rates and the formation of growth spirals with a rhombic to bow-shaped form and sector-shaped two-dimensional (2D) nuclei. The advance rates of the initial steps and the two steps of 2D nuclei were proportional to the SI. In contrast, the advance rates of the parallel steps with extremely short step spacing on growth spirals were proportional to SI^2, indicating that the lateral growth rates of growth spirals were directly proportional to the step separations. This dependence of the advance rate of every step on the growth spirals on the step separations predicts that the growth rates along the [001] direction of the growth spirals were proportional to SI^2 for lower supersaturations and to SI for higher supersaturations. The nucleation and growth rates of the 2D nuclei increased sharply for higher supersaturations using exponential functions. Using these kinetic equations, we predicted a critical supersaturation ($SI \approx 4.3$) at which the main growth mechanism of the (001) face would change from a spiral growth to a 2D nucleation growth mechanism: therefore, the morphology of bulk crystals would change.

Keywords: barite; atomic force microscopy (AFM); crystal growth; spiral growth; 2D nucleation growth

---

1. Introduction

Barite (BaSO$_4$) is the most abundant mineral of barium and occurs in a wide variety of geological environments that span geologic times from the Early Archean era to the present [1,2]. Barite is also one of the few marine authigenic minerals that are reported to form in the water column, as well as in marine sediments and locations around hydrothermal vents and cold seeps [3]. Due to its diverse modes of formation, barite can be utilized for paleoenvironmental, hydrogeological, and hydrothermal studies [3]. The precipitation and dissolution of barite control the concentration and mobility of Ba in the ground and surface water, due to its low solubility ($K_{sp} = 10^{-9.99}$ at 25 °C) in water [4,5]. The common scale mineral barite is also almost inevitable in industrial water, oil, and gas production systems due to its low solubility [4–9]. The uptake of radioactive Ra ions during barite formation [10] may occur from the systems from U mine wastes [2] or at a later stage of high-level waste-repository evolution [11] because Ra^{2+} and Ba^{2+} have a similar ionic radius and electronegativity.

Therefore, barite precipitation and dissolution reactions at the mineral–water interface are still substantial. Mineral growth in solutions preferentially occurs at kink sites, edges, steps, and defect outcrops, depending on the surface microtopography. Such microscopic growth features are also reflected in the bulk crystal appearance [12,13]. In situ Atomic Force Microscopy (AFM) allows direct observation of the growth and dissolution processes at the mineral–water interface at the site or step level (e.g., [9,14–16]). Several in situ AFM studies on barite growth, particularly on the (001) surface, which is a singular plane of barite, have been conducted to elucidate the processes involved and solve the problems cited above. The available AFM results mainly show two cases: one is growth on a singular interface with nucleus formation and spread (two-dimensional (2D) nucleation growth), and the other is growth on an imperfect singular interface, which controls the number of defect sites and the rate of lateral growth from these sites (spiral growth) (e.g., [4,8,13,17–23]). However, the available data and hypotheses regarding the growth kinetics of the two processes are still insufficient and unclear, although there are some reports regarding the lateral growth kinetics of steps or 2D islands {e.g., the lateral spreading rate of 2D islands is proportional to the supersaturation Ω (which is equivalent to IAP/K_{sp}, where IAP is the ion activity product of the ionic species in solution, $a(Ba^{2+})$ $a(SO_4^{2-})$, and K_{sp} is the solubility product of the growing mineral) [4,21]; the step velocity increases linearly with $S^{1/2}(S-1)$ where $S = \Omega^{1/2}$ [19]; and the 2D nucleation kinetics are shown by a linear correlation between $\ln N$ vs. $(\ln S)^{-1}$, where N is the density of the nuclei [21]}. In addition, the relationship between the anisotropic advance and retreat behaviors of the steps during barite growth and dissolution is not well understood.

In this study, we examined the growth behavior of the barite (001) surface in supersaturated $BaSO_4$ solutions at 30 °C using in situ AFM with an air/fluid heater/cooler system. The aims of this study were to reveal the microscopic growth behavior on the barite (001) surface at the step or site level and to estimate the growth kinetics of 2D nucleation and spiral growth that occur at the barite surface–aqueous solution interface, as well as the lateral advance rates of the initial steps, steps of the 2D nuclei, and steps on growth spirals.

2. Materials and Methods

2.1. Materials and Solutions

The barite (($Ba_{0.98}Ca_{0.01}Sr_{0.01})SO_4$) used for this study was obtained from the Stonehem Barite Deposit in Colorado, USA in the form of a single optically clear crystal. The barite crystal was cleaved parallel to the (001) cleavage plane with a sharp knife blade immediately before the AFM observations and fixed on an AFM mount with an adhesive tab (Ted Pella, Inc., Redding, CA, USA). The crystallographic direction of the cleaved samples was determined from the morphology of the etch pit and/or the 2D nucleus [4,8,15]. Various steps were mechanically formed on the (001) surfaces by cleaving with the knife blade (e.g., Figures 1a and 2a), as shown in our previous studies on barite dissolution [15,16,24]. The initial steps were generally parallel to <*uv*0> (mainly <120>, <130>, <140>, and so on) and [010], but they did not necessarily appear to be oriented parallel to an energetically favorable crystallographic direction, and slightly curved steps were also observed. The step heights corresponded to a half-*c* unit cell layer (3.6 Å), a single unit cell layer (7.2 Å), and multiple layers.

The supersaturated $BaSO_4$ aqueous solutions were prepared by mixing Na_2SO_4 and $Ba(NO_3)_2$ solutions consisting of analytical grade chemicals and deionized water immediately before the AFM observations. The concentrations of the $BaSO_4$ solutions were 20–100 μM ($[Ba^{2+}]/[SO_4^{2-}] = 1$). The degree of supersaturation was calculated using the program PHREEQC (the database used was minteq) [25]. The supersaturation index (*SI*) can be expressed as

$$SI = \Delta\mu/kT = \ln(a/a_e) = \ln(IAP/K_{sp}), \tag{1}$$

where $\Delta\mu$ is the difference in the chemical potential between the growth unit in the aqueous and solid phases, k is the Boltzmann constant, a is the activity of the ion species, and a_e is the equilibrium activity at the temperature T [26–28]. The *SI* values of the $BaSO_4$ solutions were 1.1 for 20 μM, 2.4 for 40 μM, 3.2 for 60 μM, 3.7 for 80 μM, and 4.1 for 100 μM at 30 °C.

2.2. Barite Growth Experiments and AFM Imaging

In situ observations of barite growth were performed using a Nanoscope III with a Multimode SPM unit (Digital Instruments/Bruker AXS, Yokohama, Japan) and a fluid cell with an air/fluid heater/cooler system (Veeco/Bruker AXS, Yokohama, Japan), which was operated in contact-mode AFM (CMAFM) on a vibration isolation platform in a temperature- and humidity-controlled room [22]. The cleaved barite crystals were first reacted with pure water at 30 ± 0.3 °C to ensure stable AFM scanning conditions and obtain reliable AFM images. We then replaced the water with a $BaSO_4$ solution in the fluid cell and began observing the growth process on the barite (001) surface at 30 ± 0.3 °C. The temperature was controlled by the heater/cooler-stage and a Thermal Applications Controller (TAC) (Veeco/Bruker AXS); it was also monitored by a thermocouple thermometer (Cole-Parmer International, Vernon Hills, IL, USA) [22,29]. The pure water and $BaSO_4$ solutions flowed through the fluid cell at a constant rate of 0.6 mL/h, which was controlled by a syringe pump. The solution residence time in the fluid cell at the flow rate of 0.6 mL/h was approximately 8 min [22]. (Incidentally, we also attempted the growth experiments in the lowest and higher supersaturated (20, 80, and 100 μM) $BaSO_4$ solutions at a flow rate of 1.2 mL/h to examine the effect of the solution flow rate on the barite crystal growth. The growth rates of the 2D nuclei and growth spirals did not differ for the two flow rate conditions.)

The AFM images were captured at scan rates of 1 to 4 Hz with 512×512 scan lines (and also with 256×256 scan lines in the 100 μM $BaSO_4$ solutions only) using a specialized heater/cooler, J-head piezoelectric scanner (125 μm X-Y and 5 μm Z scan size) and oxide-sharpened Si_3N_4 tip units. The setpoint and direct scanning duration on the observed surface were kept as small as possible to reduce the scan force and minimize the effect of the tip on the sample surface. All experimental runs lasted over 2 h, and the longest experiment was over 9 h. We collected parallel CMAFM height and deflection images. The AFM images were analyzed using the methods described in our earlier studies [15,16] to determine the step advance rate and growth rates of islands and spirals. We did not use the AFM images obtained immediately after the water was replaced with the $BaSO_4$ solution in the fluid cell to estimate the nucleation rate of the 2D nuclei because the images did not show steady formation of the 2D nuclei.

3. Results

The growth behavior on the barite (001) surface changed as the supersaturation increased. At the lowest supersaturation ($SI = 1.1$), the growth behavior appeared to begin and persist with the advancement of the initial step edges and the filling in of the etch pits formed in water before the $BaSO_4$ solution was injected (Figure 1a,b). Although the formation of growth spirals from a screw dislocation was observed, their growth was extremely slow. Sector-shaped 2D nuclei, which were formed at higher supersaturations, were not observed at the lowest supersaturation.

The growth behavior of the moderate supersaturations ($SI = 2.4$–3.7) was characterized by the advance of two half-layer steps with distinctly different advance rates, the relatively rapid formation of growth spirals and hillocks, and the relatively slow formation of sector-shaped 2D nuclei (Figures 2 and 3). Of course, the advance rates of the various steps increased with increasing supersaturation (Table 1). Figure 2a–c reveals that the initial [010] one-layer steps were immediately split into two half-layer [010] steps with different advance rates. Similar results were also observed for the two half-layer [120] steps (Figure 3). The large difference in the advance rates of the "f" and the "s" steps led to the formation of a new one-layer step (Figure 3d).

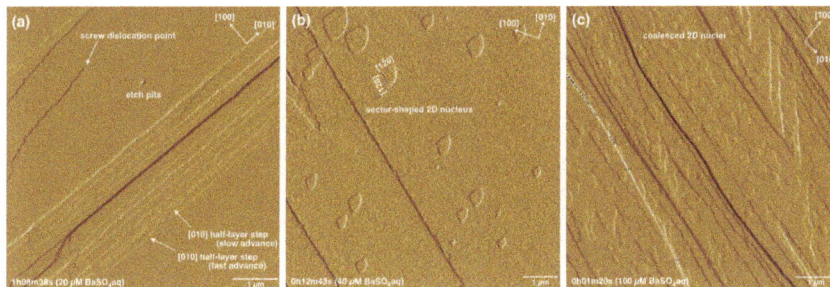

Figure 1. (a–c) CMAFM deflection images of a (001) surface in 20μM, 40 μM, and 100 μM BaSO$_4$ solutions at 30 °C, respectively. In (**b**), sector-shaped 2D nuclei formed sporadically, while in (**c**) these nuclei formed rapidly and coalesced.

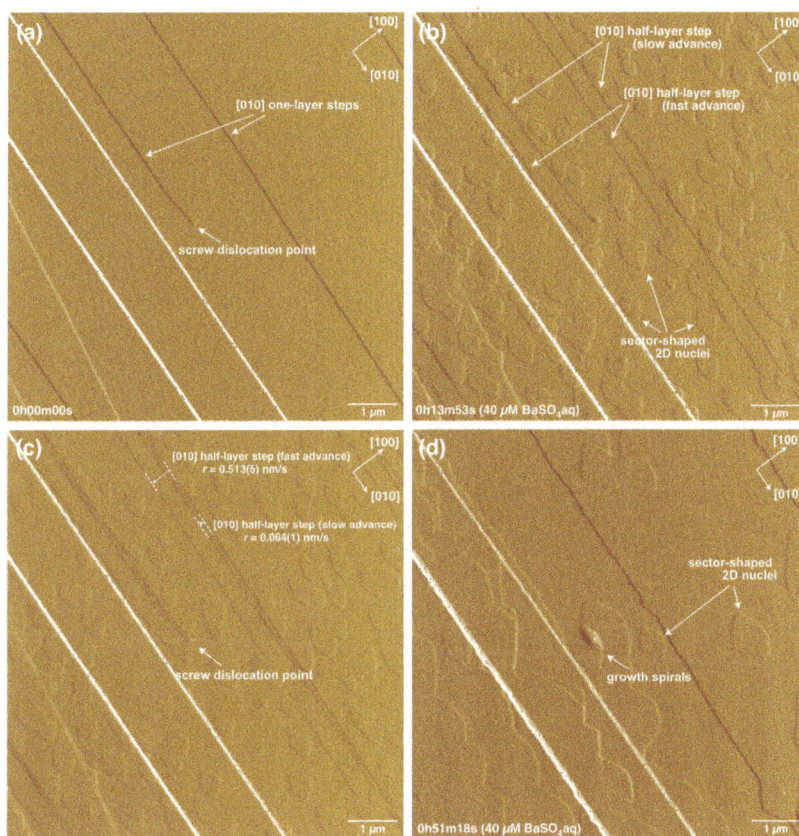

Figure 2. (**a**), (**b**), and (**d**) Sequential CMAFM deflection images of a (001) surface in a 40 μM BaSO$_4$ solution at 30 °C after 0 min, 14 min, and 51 min, respectively; each one-layer [010] step (step height = 7.2 Å) begins to split into two half-layer steps (step height = 3.6 Å) with different advance rates. Sector-shaped 2D nuclei were randomly formed, independent of the microtopography. In (**d**), the birth of a growth spirals that formed from a screw dislocation point is shown. (**c**) A comparison of (**a**) and (**b**), which are overlapped. The splitting process and advance rates of the steps are clearly demonstrated.

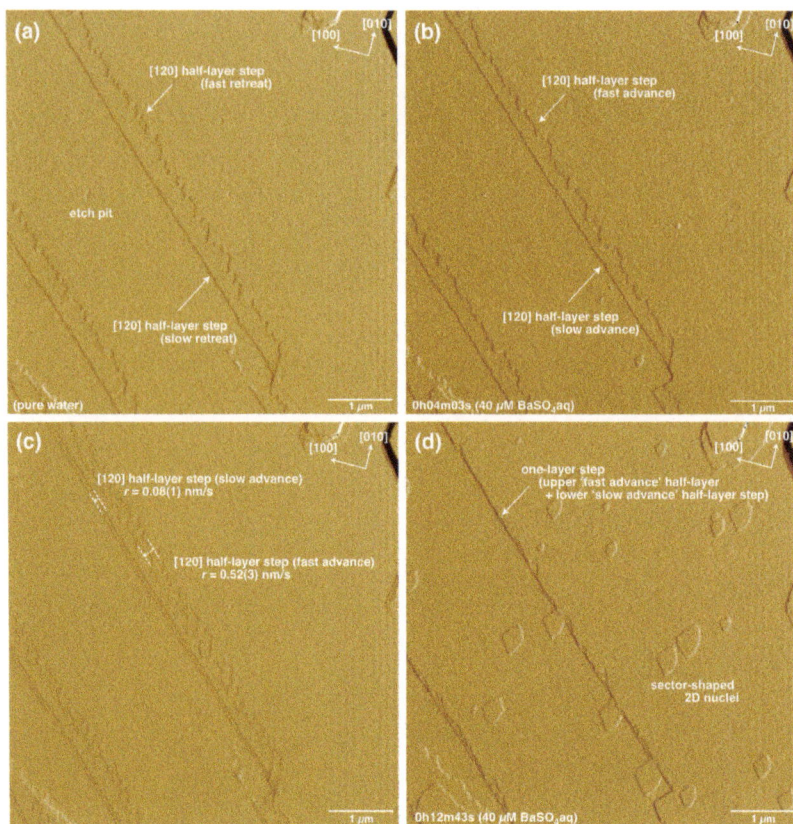

Figure 3. (a), (b), and (d) Sequential CMAFM deflection images of a (001) surface in pure water and in a 40 μM BaSO$_4$ solution at 30 °C after 4 min and 12 min, respectively. The [120] "fast retreat" and "slow retreat" half-layer steps during dissolution showed "fast advance" and "slow advance" behaviors, respectively, during growth. In (d), a new one-layer step was formed because the front of the half-layer step with the fast advance rate caught up with that of the immediately underlying half-layer step with the slow advance rate. (c) A comparison of (a) and (b), which are overlapped. The difference in the advance rate between the two [120] half-layer steps is clearly revealed.

Figure 2 also reveals the birth of growth spirals from a screw dislocation point at a moderate supersaturation (SI = 2.4). The morphology of the growth spirals was a rhombic form elongated along the [010] direction. The growth spirals tended to have a more curved outline with increasing supersaturation and showed mainly one-layer step sequences with approximately regular, narrow step spacings at each supersaturation condition (Figure 4). The growth rates of growth spirals also increased with increasing supersaturation (Table 2). Our AFM observations further captured the formation of growth hillocks where sector-shaped, half-layer 2D islands formed from a specific point were piled up and pointing in opposite orientations due to the presence of the 2_1 screw axis normal to the (001) plane, although these structures could only be observed at a moderate supersaturation (SI = 2.4) (Figure 4a). The growth hillocks may have formed from an edge dislocation point and grew slowly, while the growth spirals formed from a screw dislocation point grew rapidly.

The sector-shaped 2D nuclei were defined by half-layer steps parallel to [120], [1$\bar{2}$0], and a curved step edge tangent to [010] (Figures 2 and 3). Such sector-shaped 2D nuclei have been formed in various supersaturated $BaSO_4$ solutions at room temperature {e.g., *SI* = 3 [8]; *SI* = 3.3 and 4.3 [18]; and *SI* = 4.0 and 4.5 [4]}. The [120] and curved steps of the 2D nuclei advanced at constant rates in each supersaturated solution for the duration of the experiment (i.e., the 2D nuclei grew at a constant rate) (Figures 5 and 6, Table 3).

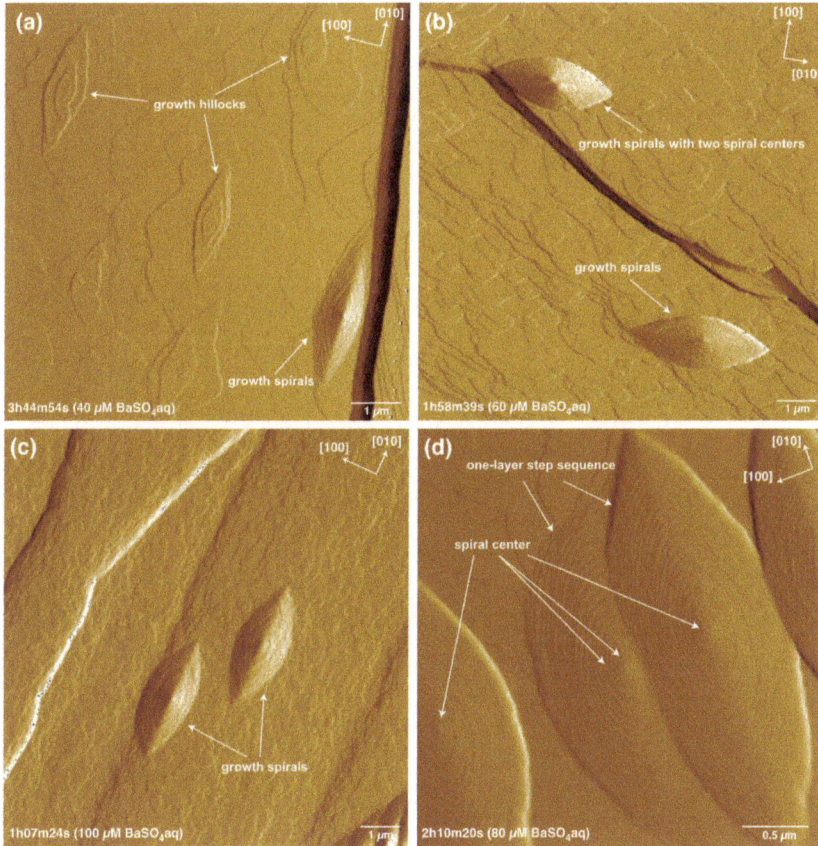

Figure 4. (**a–c**) CMAFM deflection images of a (001) surface in 40 µM, 60 µM, and 100 µM $BaSO_4$ solutions at 30 °C, respectively, showing the growth spirals that formed from screw dislocations. The growth hillocks that likely formed from edge dislocations are also shown in (**a**). (**d**) CMAFM deflection image of a (001) surface in an 80 µM $BaSO_4$ solution at 30 °C showing growth spirals with a one-layer step sequence and regular step spacing.

The growth behavior at the highest supersaturation (*SI* = 4.1) was characterized by the rapid formation and growth of sector-shaped 2D nuclei and growth spirals immediately after the solutions were injected into the AFM fluid cell (Figures 1d and 4c). We could not estimate the advance rates of initial steps and the [120] and curved steps of the 2D nuclei at higher supersaturations because the 2D nuclei formed and grew too rapidly, but we could estimate the nucleation rates of the 2D nuclei and the spiral growth rates. The bow-shaped growth spirals formed at the highest supersaturation showed a more curved outline than those for lower supersaturations, as mentioned above (Figure 4). A number of sector-shaped 2D nuclei immediately formed and coalesced on the initial surface, and new nuclei were deposited on the previous ones before a layer was completed (i.e., multi-nucleation growth [30]).

Table 1. Average advance rates of the various steps on the barite (001) surface.

BaSO$_4$ Concentration (µM)	Advance Rates of Steps on the (001) Surface (nm/s)				
	$<uv0>$ * "f" Step with Half-Layer	$<uv0>$ * "s" Step with Half-Layer	$<uv0>$ * Step with One Layer ("f" upper + "s" Lower Half-Layers)	[010] "f" Step with Half-Layer	[010] "s" Step with Half-Layer
20	0.06 ± 0.01	0.03 ± 0.01	0.03 ± 0.01	0.05 ± 0.01	0.02 ± 0.01
40	0.50 ± 0.03	0.08 ± 0.01	0.12 ± 0.01	0.51 ± 0.02	0.07 ± 0.01
60	0.62 ± 0.03	0.12 ± 0.01	0.21 ± 0.01	0.62 ± 0.03	0.11 ± 0.01
80	0.73 ± 0.05	0.16 ± 0.01	-**	-**	-**

* Mainly the $<120>$ step, but also including the $<130>$, $<140>$, and other steps, as well as the slightly curved steps; ** These rates could not be estimated accurately, due to the rapid formation and growth of many 2D nuclei.

Table 2. Average spiral growth rates and step separations.

BaSO$_4$ Concentration (µM)	Face or Corner Advance Rates in the Growth Spirals			The Ratio of the Advance Rates in the Growth Spirals			Mean Step Separations * (nm)	
	[100] Corner (nm/s)	[010] Corner (nm/s)	(001) (nm/min)	(001)/[100]	(001)/[010]	[010]/[100]	[100]	[010]
20	(0.002) **	(0.006) **	(0.006) **					
40	0.04 ± 0.01	0.11 ± 0.01	0.11 ± 0.02	4.58×10^{-2}	1.67×10^{-2}	2.9	15.7	43.2
60	0.10 ± 0.01	0.25 ± 0.01	0.22 ± 0.05	3.67×10^{-2}	1.47×10^{-2}	2.5	19.6	49.1
80	0.13 ± 0.01	0.32 ± 0.02	0.26 ± 0.05	3.33×10^{-2}	1.35×10^{-2}	2.4	21.6	53.2
100	0.19 ± 0.02	0.46 ± 0.04	0.33 ± 0.09	2.89×10^{-2}	1.20×10^{-2}	2.4	24.9	60.2

* When every parallel step on the growth spirals is a one-layer step; ** These rates may be somewhat rough estimates, because they are calculated from only two data points and represent extremely slow spiral growth.

Table 3. Nucleation rates and average growth rates of the sector-shaped 2D nuclei.

BaSO$_4$ Concentration (µM)	Advance Rates of the Steps (nm/s)		Growth Rates toward the [001] Direction (nm/min)	Nucleation Rates (N/(µm^2·min))
	[120] Step	Curved Step		
40	0.08 ± 0.01	0.50 ± 0.02	0.012 ± 0.001	0.029 ± 0.007
60	0.12 ± 0.01	0.60 ± 0.02	0.022 ± 0.002	0.030 ± 0.008
80	0.16 ± 0.01	0.73 ± 0.05	0.041 ± 0.002	0.14 ± 0.08
100	-*	-*	0.21 ± 0.06	2.2 ± 0.3

* These rates could not be estimated accurately, due to the rapid formation and growth of many 2D nuclei.

Table 4. Growth rates of spiral growth and 2D growth on the barite (001) surface.

BaSO$_4$ Concentration (μM)	Spiral Growth Rate * ($\times 10^{-9}$ mol/(m^2·s))	2D Growth Rate ** ($\times 10^{-9}$ mol/(m^2·s))	Total Growth Rate ($\times 10^{-9}$ mol/(m^2·s))
40	1.3 ± 0.5	3.3 ± 0.8	5 ± 1
60	3.1 ± 1.2	3.7 ± 1.0	7 ± 2
80	4.0 ± 1.7	7.1 ± 4.1	11 ± 6
100	5.7 ± 2.4	20 ± 1	26 ± 4

* Using a = 0.8884 nm, b = 0.5456 nm, c = 0.7157 nm, density = 4.50 g/cm^3 [31], rate data in Table 2, and mean growth spirals density = 2 ($\pm$1) $\times$ 10^{12} m^{-2} [31]; ** Using Equations (2) and (3), rate data in Table 3 as well as the above unit cell parameter and density [31].

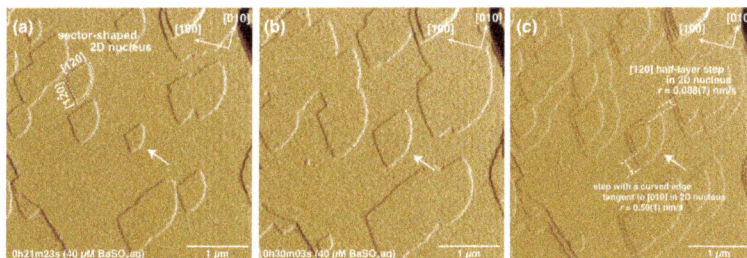

Figure 5. (a) and (b) Sequential CMAFM deflection images of a (001) surface in a 40 μM BaSO$_4$ solution at 30 °C after 21 min and 30 min, respectively; (c) a comparison of (a) and (b), which are overlapped. The lateral spreading process and step advance rates of the 2D nuclei are clearly demonstrated. See Figure 6 for the growth of a 2D nucleus that is indicated by the white arrows.

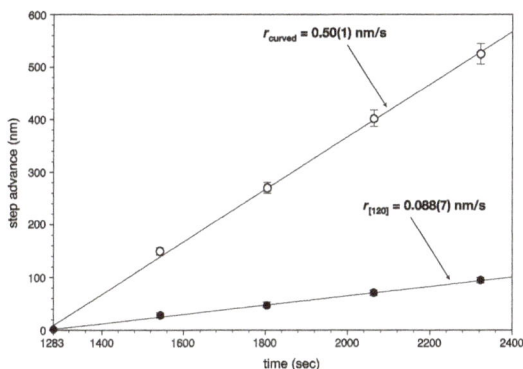

Figure 6. Changes in the [120] (solid circulars) and curved (open circulars) step advance distances in the 2D nucleus shown in Figure 5 as a function of time. We began measuring the step advance distance 21 min (Figure 5a) after the experiment began. The solid and open circular marks at 1803 s indicate the advance distances of the two steps in the 2D nucleus indicated by a white arrow in Figure 5b.

4. Discussion

4.1. Advance and Retreat Behavior of the Steps

Previous AFM studies on barite growth and dissolution reported that the <uv0> and [010] half-layer steps on the barite (001) surface have two different advance rates or retreat rates {e.g., step retreat (dissolution) [4,15,16,24,32]; step advance (growth) [17,18,21]}. However, the relationship between the anisotropic advance and retreat behaviors of the half-layer steps during barite growth and dissolution is not entirely understood or delineated. Our AFM observations revealed that the [010] "fast retreat" and "slow retreat" half-layer steps during dissolution in water shifted to "slow advance" and "fast advance" steps, respectively, during growth in the supersaturated solution (Figures 1 and 2). In contrast, the [120] "fast retreat" and "slow retreat" half-layer steps during dissolution showed "fast advance" and "slow advance" behaviors, respectively, during growth (Figure 3). The results of this study and our previous AFM studies on barite dissolution [15,16,24] lead to a schematic model showing the relationships between the advance and retreat behaviors of the half-layer steps, the reduction and growth of the etch pits, and the growth of the sector-shaped 2D islands during barite growth and dissolution on the barite (001) surface (Figure 7). What is the cause of the differences in the advance behaviors of the two half-layer steps? A possible cause is that the [120] and [010] steps have the opposite termination (obtuse or acute), in which the acute steps advance more slowly than the obtuse steps [33]. Another important result is that the rate-limiting reaction for Ba attachment and detachment involves escape of the ion from the inner-sphere adsorbed species [34]. Stack et al. [34,35] demonstrated that the rate-limiting reaction of Ba attachment to a step is the change from inner-sphere adsorbed species involving only one bond to a surface sulfate to the bidentate species where the departing Ba makes two bonds to two surface sulfates. The state of inner-sphere adsorbed species may affect the step advancing rate.

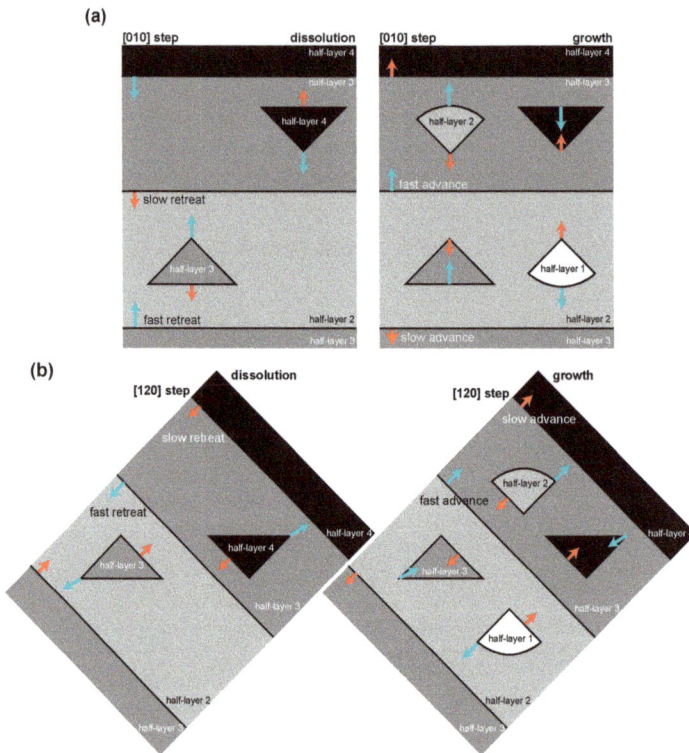

Figure 7. Schematic diagrams showing the relationships between the advance and retreat behaviors of the half-layer [010] (**a**) and [120] (**b**) steps, the reduction and growth of the etch pits, and the growth of sector-shaped 2D islands during growth and dissolution on the barite (001) surface. The uppermost and lowermost half-layers show half-layer 1 (white) and half-layer 4 (black), respectively. The red and blue arrows show the directions and relative rates (slow and fast, respectively) of the advance and retreat of the steps.

The advance rates of the <*uv*0> half-layer steps and the [120] and curved steps in the 2D islands were roughly proportional to the supersaturation index (*SI*) (Figure 8):

$$v_{<uv0>\text{f-step}} \text{ (nm/s)} = 0.25\ SI - 0.18\ (r = 0.985) \tag{2}$$

$$v_{<uv0>\text{s-step}} \text{ (nm/s)} = 0.047\ SI - 0.026\ (r = 0.989), \tag{3}$$

where $v_{<uv0>\text{f-step}}$ and $v_{<uv0>\text{s-step}}$ are the advance rates of the <*uv*0> "f" and "s" steps, respectively, and *r* is the correlation coefficient. The direct proportionality between the advance rates of the individual steps and the *SI* may be expressed by the lateral growth law using Brice's model, which states that an energy barrier (ΔG_I) is required to transfer an atom from the growth phase to the crystal, and moving the first growth-phase atom into the crystal is exactly equivalent to moving one step on the growth face [36,37]:

$$f_1\ (= v_{\text{step}}) = A \cdot \Delta \mu / kT = A \cdot SI, \tag{4}$$

where f_1 is the rate of sideways growth, v_{step} is the step advance rate, and $A = sv\ \exp(-\Delta G_I / RT)$. In this second equation, *s* is the interlayer spacing for the particular face, *v* is the vibration frequency of the particles, and *R* is the gas constant. Equation (4) is equivalent to the equation for the advance rate of a step described by Burton et al. [38]. Another expression, where the step advance rates are proportional to $S^{1/2}(S - 1)$, has been presented by Zhang and Nancollas [39,40]. However, the step advance rates in this study were not proportional to $S^{1/2}(S - 1)$.

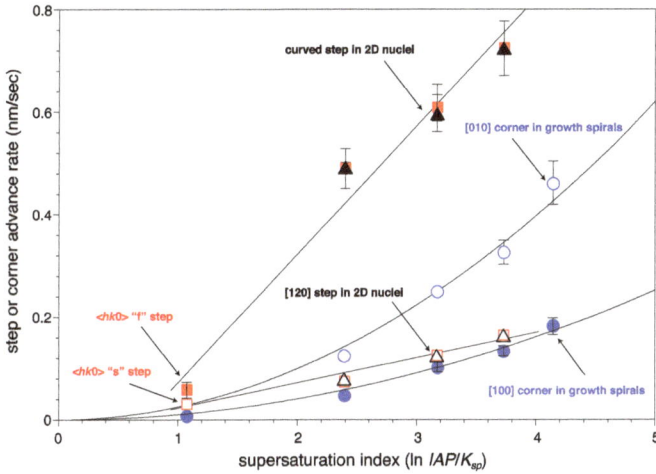

Figure 8. Changes in the step or corner advance rates of the initial steps, 2D nuclei, and growth spirals as a function of the supersaturation index (*SI*). The initial half-layer fast ("f") and slow ("s") steps were indicated by red solid and open squares, respectively. The [120] and curved steps in 2D nuclei were shown as open and solid triangles, respectively. The [010] and [100] corners of the growth spirals were indicated by blue open and solid circles, respectively. The lateral growth rates of the growth spirals were proportional to SI^2, while the advance rates of the initial steps and the two steps in the 2D nuclei were proportional to the *SI*.

4.2. Growth Spiral Formation and Development

The development of growth spirals formed from screw dislocations was somewhat complicated. The lateral growth rates of the growth spirals were proportional to SI^2 rather than *SI*, while the advance rates of the initial steps on the surface were proportional to *SI* (Figure 8):

$$f_{[010]} \text{ (nm/s)} = 0.025 \ SI^2 \ (r = 0.992) \tag{5}$$

$$f_{[100]} \text{ (nm/s)} = 0.010 \ SI^2 \ (r = 0.986), \tag{6}$$

where $f_{[010]}$ and $f_{[100]}$ are the advance rates of the [010] and [100] corners in the growth spirals, respectively. According to Burton et al. [38], a better derivation for Equation (4) yields

$$f_1 = v_{step} = A \ SI \ \tanh(\lambda/2x_s), \tag{7}$$

where λ is the step separation in the parallel step sequence and x_s is the mean displacement of the adsorbed molecules ($2x_s$ is the catchment area of a step for adsorbed molecules [30]). Equation (7) reduces to Equation (4) when $\lambda \gg 2x_s$ (namely, single steps or a step sequence where the step spacing is sufficiently wider than the catchment area of every step) and when $\lambda \ll 2x_s$, the equation predicts that

$$f_1 = v_{step} = A \ \lambda \ SI/2x_s. \tag{8}$$

Equation (8) indicates that the advance rate of every step in a parallel step sequence is proportional to the spacing between the parallel steps when the step spacing is sufficiently shorter than the catchment area of every step, generally in higher supersaturations [30]. The slopes (p) of the [100] and [010] directions of the growth spirals ($R_{sp}/f_{[100]}$ and $R_{sp}/f_{[010]}$, respectively, where R_{sp} is the vertical growth rate of the growth spirals) in this study tended to gradually decrease with SI (Table 3). This result indicates that the mean step separations (λ) of parallel steps on the growth spirals were proportional to SI because the growth spirals showed mainly one-layer step sequences and approximately regular step spacings at each supersaturation (Table 3, Figure 4d). Thus, Equation (8) can be represented as:

$$f_1 = v_{step} = A_2 \ SI^2, \tag{9}$$

where A_2 is a constant. Indeed, the lateral growth rates of the growth spirals in our AFM experiments were directly proportional to the step separations (Figure 9) and proportional to SI^2 (Figure 8, Equations (5) and (6)).

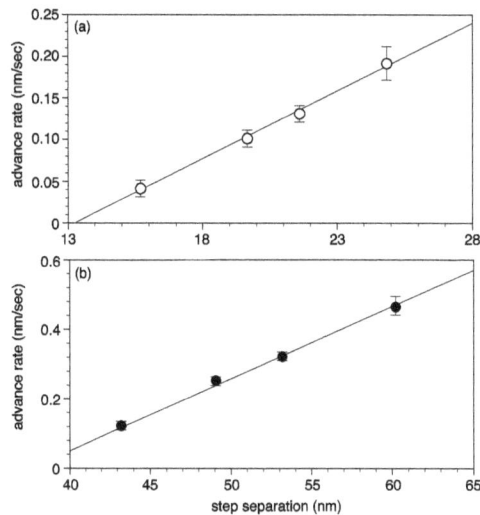

Figure 9. Changes in the lateral advance rates of parallel steps on the growth spirals as a function of the mean step separations of those that were proportional to SI. (**a**) Toward the [100] direction; (**b**) toward the [010] direction.

Pina et al. [17] explained the inhibition of the spiral growth through the structural control and anisotropy of the growth. On the (001) surface, alternate $BaSO_4$ layers are related by a 2_1 screw axis so that the anisotropy and the shape of the nuclei are reversed in each growth layer. The growth of the first $BaSO_4$ layer around the dislocation is restricted to one sector and cannot continue around the spiral because of the very slow growth in the opposite direction. In the next layer that forms around the dislocation, the directions are reversed but the rapid growth along is almost completely prevented by the very slow growth in the underlying layer. This alternation of fast and slow directions continues for subsequent layers and so growth around the screw dislocation is limited to an ever-tightening spiral. Similar results were also shown in anhydrite (100) and cerestite (001) growth [17,41].

The growth rate normal to the surface by spiral growth from a screw dislocation (R_{sp}) has been described by some well-known classical growth models (e.g., [37,38,42,43]):

$$R_{sp} = pv_{step} = dv_{step}/\lambda = dA \ SI \ \tanh(\lambda/2x_s)/\lambda, \tag{10}$$

where d is the height of a step. For $\lambda \gg 2x_s$, Equation (10) predicts that

$$R_{sp} = dA \ SI/\lambda. \tag{11}$$

Here, the step separation (λ) of parallel steps on the growth spirals depends on the critical radius (ρ_c) of a central island [30,36]:

$$\lambda = 4\pi\rho_c \tag{12}$$

$$\rho_c = \kappa V/\Delta\mu = \gamma V/kTS, \tag{13}$$

where κ is the step edge free energy and V is the molar volume. From Equations (12) and (13),

$$1/\lambda \propto SI, \tag{14}$$

therefore

$$R_{sp} = A_3 \ SI^2, \tag{15}$$

where A_3 is a constant. On the other hand, for $\lambda \ll 2x_s$, Equation (10) can be written in the form

$$R_{sp} = dA \ SI/2x_s = A_4 \ SI, \tag{16}$$

where A_4 is a constant.

The growth rates (R_{sp}) of the growth spirals formed on the barite (001) surface appeared to be proportional to SI^2 (Figure 10), similar to the lateral growth rates ($f_{[010]}$ and $f_{[100]}$) (Figure 8). Thus,

$$R_{sp} \ (nm/min) = 0.019 \ SI^2 \ (r = 0.995). \tag{17}$$

However, the aforementioned dependence of the advance rate of every step on the growth spirals on the step separations or a proportional relationship between the lateral growth rates of the growth spirals and SI^2 predicts that the step separations would be significantly shorter than the catchment area of every step ($\lambda \ll 2x_s$). Hence, at least for $SI = 2.4–4.1$ at 30 °C, where the lateral growth rates of growth spirals clearly depended on the step separations (Figure 9), the growth rates (R_{sp}) of the growth spirals likely followed the growth law described in Equation (16) (Figure 10):

$$R_{sp} \ (nm/min) = 0.124 \ SI - 0.187 \ (r = 0.993). \tag{18}$$

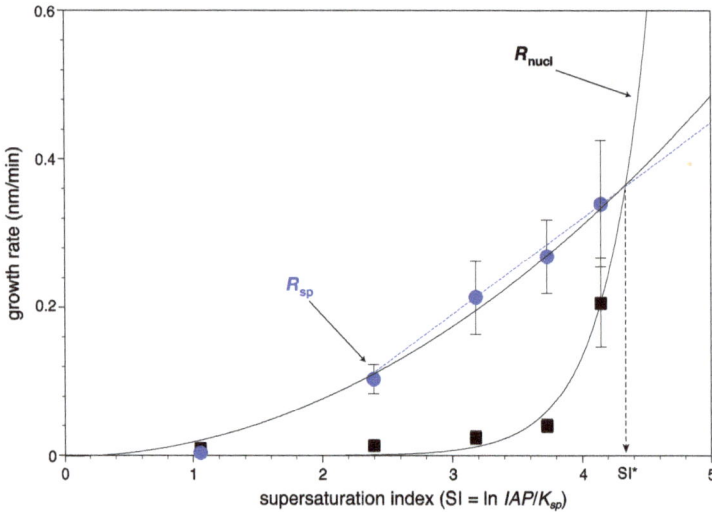

Figure 10. Changes in the growth rates toward the [001] direction of the growth spirals (R_{sp}) (blue solid circles) and 2D nuclei (R_{nucl}) (solid squares) as a function of the supersaturation index (*SI*). The growth rates (R_{sp}) of the growth spirals were likely proportional to SI^2 for lower supersaturations (a solid line) and to *SI* for higher supersaturations (a blue dotted line). In contrast, the growth rates (R_{nucl}) of the 2D nuclei increased sharply at higher supersaturations ($SI \geq 3.8$) and followed an exponential function (see text). It is expected that the main growth mechanism of the (001) face changes from a spiral growth to a 2D nucleation growth mechanism at an *SI** of approximately 4.3.

4.3. Formation and Growth of 2D Nuclei

The nucleation frequency and growth rate of the 2D nuclei on the singular interface have also been described by some growth models (e.g., [30,36,37,44]). According to their models, two limiting cases can be considered. The first case is the one where the nuclei rapidly grow laterally, namely, single nucleation growth, and the second case is one where the nuclei, once formed, spread slowly, such that new nuclei form on the old ones before layer growth is complete, namely, multi-nucleation growth. Here, our AFM observations revealed that the formation and growth of the 2D nuclei on the barite (001) surface followed the second limiting case (Figure 1). The nucleation frequency (J_L) of the 2D nuclei in the second limiting case [37] is:

$$J_L = A_5 \exp(-\Delta G^*/kT), \tag{19}$$

where A_5 is a constant and ΔG^* is the free energy of the nucleus:

$$\Delta G^* = \pi \rho_c \kappa. \tag{20}$$

Using Equations (13) and (20), Equation (19) can be written in the form

$$J_L = A_5 \exp(-B/SI), \tag{21}$$

where $B = \pi V \kappa^2$. Fitting our AFM data to Equation (21), we obtain a relationship between the formation frequency of 2D nuclei on the barite (001) surface and the *SI* (Figure 11):

$$J_L \ (N/(\mu m^2 \cdot min)) = 1.23 \times 10^{11} \exp(-1.03 \times 10^2/SI) \ (r = 0.999). \tag{22}$$

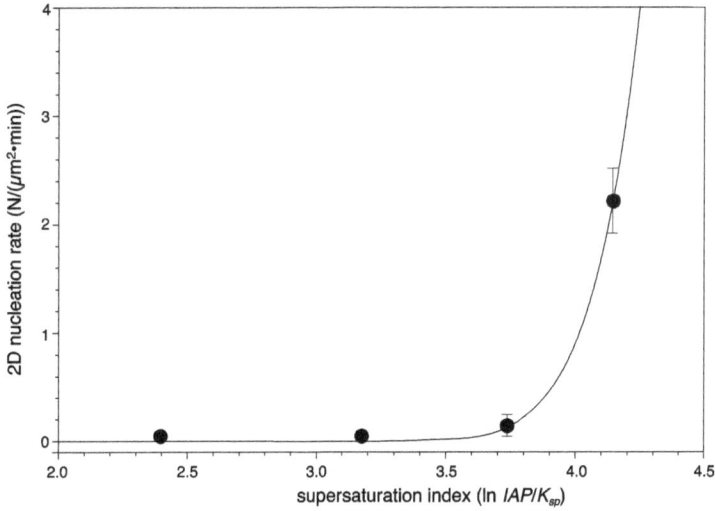

Figure 11. Changes in the 2D nucleation rates as a function of the supersaturation index (*SI*). The 2D nucleation rates increased sharply at higher supersaturations (*SI* ≥ 3.8) and followed an exponential function, similar to the growth rates (R_{nucl}) of the 2D nuclei.

Similar results showing that there is a sharp increase in the nucleation rate above a certain high supersaturation level have been reported in previous AFM studies of the barite growth at different temperatures [4,21]. The supersaturation index (*SI*) above which there is a sharp increase in the nucleation rate was approximately 4.1 at 30 °C in this study and 4.2 at 22 °C in a previous study [4].

The vertical growth rate (R_{nucl}) of 2D nucleation growth in the first limiting case of Brice's model is represented as

$$R_{nucl} = d\, J_S, \tag{23}$$

where d is the step height of the nucleus and J_S is the formation frequency of the 2D nuclei in the first limiting case [36]. In the second limiting case, the growth rate is the cube root of the volume deposited in unit time:

$$R_{nucl} = (d\pi f_1^2 J_L)^{1/3}. \tag{24}$$

Using Equations (4) and (21), Equation (24) can be written in the form

$$R_{nucl} = A_6\, SI^{2/3}\, \exp(-B/SI), \tag{25}$$

where A_6 is a constant for a given material. Again, fitting our AFM data to Equation (25), we predict that the growth rates (R_{nucl}) of 2D nucleation growth on the barite (001) surface followed the growth law of Equation (25) (Figure 10):

$$R_{nucl} = 4.89 \times 10^4\, SI^{2/3}\, \exp\left(-5.51 \times 10^1 /SI\right)\, (r = 0.993). \tag{26}$$

The lateral spreading rates of the sector-shaped 2D nuclei appeared to be proportional to the supersaturation index (*SI*), similar to the advance rates of the initial steps (Figure 8). The advance rates of the curved steps of the 2D nuclei were much higher than those of the [120] steps, although the difference in the rates gradually decreased with supersaturation (approximately 6.3 times at *SI* = 2.4 to 4.6 times at *SI* = 3.7) and was smaller than that (approximately 10 times) at a lower temperature (22 °C) in previous studies [4,21].

4.4. Implication of Microscopic Growth Kinetic Data for Macroscopic Features

Macroscopic barite growth rates reported in the previous studies showed 10^7–10^8 mol/(m^2·s) at a range of $3 < SI < 5$ at room temperature (e.g., [21,45]). The macroscopic growth rates are approximately an order of magnitude higher than the growth rates on the (001) surface that were estimated from our AFM data (Table 4). A possible cause of the difference is the growth of other surfaces that are faster than the growth of the (001) surface. Godinho and Stack [45] showed that growth rates of the {010} and {210} surfaces were significantly and slightly higher than that of the {001} surface, respectively. Another possible cause is the difference in the dislocation density (or growth spirals density). Boshbach [21] showed that the observed dislocation density on the (001) surface was up to 10^{13} m^{-2} and led to a macroscopic growth rate of about 10^{-8} mol/(m^2·s). Using the dislocation density instead of our growth spirals density data (2×10^{12} m^{-2}), we can also estimate growth rates of the (001) surface of about 10^{-8} mol/(m^2·s).

It is important to understand the process by which minerals that are naturally grown in solutions form to transfer microscopic growth features to macroscopic features [12]. Several studies on barite crystals in hydrothermal fluid or cold seep have reported that they increase in size and change their morphologies from platy to rectangular, prismatic or dendritic with the degree of supersaturation [3,46–48]. Kowacz and Putnis [13] have demonstrated that the nanoscale morphology of the growth features on the barite (001) surface in a supersaturated solution with 0.03 M KCl was reflected in the morphology of the bulk crystals (~10 μm in size) precipitated from the solution of the corresponding composition. The pseudo-hexagonal shape of the (001) face and elongation of the bulk crystals correlate with the hexagonal shape of the growth spirals and the very narrow layer spacing on the growth spirals that corresponds to its preferential growth in the direction normal to the (001) surface with little lateral spreading. Using four empirical kinetic equations (Equations (2), (5), (17) or (18) and (26)), we can estimate that for $SI < 4$, where the growth of the (001) face was mainly controlled by a spiral growth mechanism, the ratio (R_{sp}/f_l) of the vertical growth rates to the lateral advance rates of the growth spirals was one to two orders of magnitude higher than that (R_{nucl}/v_{step}) of the 2D nuclei. This result was consistent with the results reported by Kowacz and Putnis [13].

The 2D nuclei on the barite (001) surface were randomly distributed independent of the microtopography, similar to the nucleation under different conditions in previous AFM studies on barite growth [4,13,18]. Using two growth kinetic equations (Equations (18) and (26)), we can predict a critical supersaturation ($SI^* \approx 4.3$) at which the main growth mechanism of the (001) face would change from a spiral growth mechanism to a 2D nucleation growth mechanism (Figure 10). Therefore, over the critical supersaturation SI^*($\approx$4.3) the development of larger crystals with a wider singular (001) surface {e.g., tabular crystals bounded by {210} (closely related to the [120] and [1$\bar{2}$0] steps, with much slower advance rates than the curved step of 2D nuclei) precipitated from sea water or (low temperature) hydrothermal solutions [3,47]} may be predicted, due to the rapid nucleation and faster lateral spreading of the 2D nuclei. Interestingly, Pina et al. [49] reported that the transitional supersaturation values where 2D nuclei were observed for the first time were approximately $SI = 1.9$ for barite and $SI = 0.8$ for selestite, predicting the decrease of the critical supersaturation SI^* with increasing celestite composition in the (Ba,Sr)SO$_4$ solid solution.

The formation and growth of barite from aqueous supersaturated solutions are a well-known problem in industrial oil and gas production systems. Recent experimental studies on the coating of barite from supersaturated solutions in the absence of chemical scale inhibitors on a stainless steel surface have revealed the oriented crystallization where the (001) face mainly developed [50,51]. Our kinetic data regarding the vertical growth of the barite (001) face are certainly helpful in estimating the thickness of the barite film formed under such conditions. For instance, Equations (18) and (26), which predict the sharp increase of the growth rate of the barite (001) face over a critical supersaturation, estimate that the vertical growth rate of the (001) face rises more than two-fold when the SI increases from 4.0 to 4.5, while it increases only by 1.2-fold when the SI increases from 3.5 to 4.0. This explains why barite crystals deposited on a quartz crystal microbalance (QCM) surface at the highest *SI* (4.32) were considerably larger than those at the lower *SI* [52].

5. Conclusions

Our in situ AFM study carefully examined the microscopic growth behavior on the barite (001) surface at 30 °C and revealed the mechanisms and kinetics of barite growth across a range of supersaturation conditions. Regarding lateral growth on the barite (001) surface, the advance rates of the parallel steps with extremely short step spacings on the growth spirals were proportional to the square of the supersaturation index (SI^2), while those of the initial steps and the two steps of the 2D nuclei (namely, single steps or steps with wider step spacings) were proportional to *SI*. On the other hand, the growth perpendicular to the (001) surface was mainly controlled by a spiral growth mechanism, where the growth rates (R_{sp}) of the growth spirals were proportional to SI^2 for lower supersaturations and to *SI* for higher supersaturations. However, over a critical supersaturation ($SI \approx 3.8$) the nucleation rates (J_L) and growth rates (R_{nucl}) of the 2D nuclei formed on the (001) surface increased sharply and followed exponential functions. These microscopic and kinetic data on barite growth help us to predict and improve the growth processes and rate laws.

Acknowledgments: The authors thank K. Ishida, Y. Nakamuta, and S. Uehara of Kyushu University for their helpful suggestions. We also thank the three anonymous reviewers for their thorough reviews that improved the quality of the study. This study was supported in part by a Grant-in-Aid for Scientific Research (Y. Kuwahara, No. 26400518) from the Japan Society for the Promotion of Science.

Author Contributions: Yoshihiro Kuwahara and Wen Liu conceived and designed the experiments; Yoshihiro Kuwahara, Wen Liu, and Keisuke Otsuka performed the experiments; all authors analyzed the data; Yoshihiro Kuwahara, Wen Liu, and Masato Makio contributed reagents/materials/analysis tools; Yoshihiro Kuwahara wrote the paper.

Conflicts of Interest: The authors declare no conflict of interest.

References

1. Gaines, R.V.; Skinner, H.C.W.; Foord, E.E.; Mason, B.; Rosenzweig, A. *Dana's New Mineralogy*, 8th ed.; John Wiley & Sons: New York, NY, USA, 1997; pp. 571–582.

2. Hanor, J.S. Barite-Celestine geochemistry and environments of formation. In *Sulfate Minerals–Crystallography, Geochemistry, and Environmental Significance. Reviews in Mineralogy and Geochemistry*; Alpers, C.N., Jambor, J.L., Nordstrom, D.K., Eds.; Mineralogical Society of America: Chantilly, VA, USA, 2000; Volume 40, pp. 193–275.

3. Griffith, E.M.; Paytan, A. Barite in the ocean—Occurrence, geochemistry and palaeoceanographic applications. *Sedimentology* **2012**, *59*, 1817–1835. [CrossRef]

4. Bosbach, D.; Hall, C.; Putnis, A. Mineral precipitation and dissolution in aqueous solution: In-situ microscopic observations on barite (001) with atomic force microscopy. *Chem. Geol.* **1998**, *151*, 143–160. [CrossRef]

5. Putnis, C.V.; Kowacz, M.; Putnis, A. The mechanism and kinetics of DTPA-promoted dissolution of barite. *Appl. Geochem.* **2008**, *23*, 2778–2788. [CrossRef]

6. Wang, K.S.; Resch, R.; Koel, B.E.; Shuler, P.J.; Tang, Y.; Chen, H.J. Study of the dissolution of the barium sulfate (001) surface with hydrochloric acid by atomic force microscopy. *J. Colloid Interface Sci.* **1999**, *219*, 212–215. [CrossRef] [PubMed]

7. Wang, K.S.; Resch, R.; Dunn, K.; Shuler, P.; Tang, Y.; Koel, B.E.; Yen, T.F. Dissolution of the barite (001) surface by the chelating agent DTPA as studied with non-contact atomic force microscopy. *Colloids Surf.* **1999**, *160*, 217–227. [CrossRef]

8. Risthaus, P.; Bosbach, D.; Becker, U.; Putnis, A. Barite scale formation and dissolution at high ionic strength studied with atomic force microscopy. *Colloids Surf.* **2001**, *191*, 201–214. [CrossRef]

9. Becker, U.; Biswas, S.; Kendall, T.; Risthaus, P.; Putnis, C.V.; Pina, C.M. Interactions between mineral surfaces and dissolved species: From monovalent ions to complex organic molecules. *Am. J. Sci.* **2005**, *305*, 791–825. [CrossRef]

10. Brandt, F.; Curti, E.; Klinkenberg, M.; Rozov, K.; Bosbach, D. Replacement of barite by a (Ba,Ra)SO$_4$ solid solution at close-to-equilibrium conditions: A combined experimental and theoretical study. *Geochim. Cosmochim. Acta* **2015**, *155*, 1–15. [CrossRef]

11. Culti, E.; Fujiwara, K.; Iijima, K.; Tits, J.; Cuesta, C.; Kitamura, A.; Glaus, M.A.; Müller, W. Radium uptake during barite recrystallization at 23 ± 2 °C as a function of solution composition: An experimental [133]Ba and [226]Ra tracer study. *Geochim. Cosmochim. Acta* **2010**, *74*, 3553–3570.

12. De Yoreo, J.J.; Dove, P.M. Shaping crystals with biomolecules. *Science* **2004**, *306*, 1301–1302. [CrossRef] [PubMed]

13. Kowacz, M.; Putnis, A. The effect of specific background electrolytes on water structure and solute hydration: Consequences for crystal dissolution and growth. *Geochim. Cosmochim. Acta* **2008**, *72*, 4476–4487. [CrossRef]

14. Bosbach, D.; Charlet, L.; Bickmore, B.; Hochella, M.F., Jr. The dissolution of hectorite: In-situ, real-time observations using atomic force microscopy. *Am. Mineral.* **2000**, *85*, 1209–1216. [CrossRef]

15. Kuwahara, Y. In situ Atomic Force Microscopy study of dissolution of the barite (001) surface in water at 30 °C. *Geochim. Cosmochim. Acta* **2011**, *75*, 41–51. [CrossRef]

16. Kuwahara, Y. In situ hot-stage AFM study of the dissolution of the barite (001) surface in water at 30–55 °C. *Am. Mineral.* **2012**, *97*, 1564–1573. [CrossRef]

17. Pina, C.M.; Becker, U.; Risthaus, P.; Bosbach, D.; Putnis, A. Molecular-scale mechanisms of crystal growth in barite. *Nature* **1998**, *395*, 483–486.

18. Pina, C.M.; Bosbach, D.; Prieto, M.; Putnis, A. Microtopography of the barite (001) face during growth: AFM observations and PBC theory. *J. Cryst. Growth* **1998**, *187*, 119–125. [CrossRef]

19. Higgins, S.R.; Bosbach, D.; Eggleston, C.M.; Knauss, K.G. Kink dynamics and step growth on barium sulfate (001): A hydrothermal scanning probe microscopy study. *J. Phys. Chem. B* **2000**, *104*, 6978–6982. [CrossRef]

20. Becker, U.; Risthaus, P.; Bosbach, D.; Putnis, A. Selective attachment of monovalent background electrolyte ions and growth inhibitors to polar steps on sulfates as studied by molecular simulations and AFM observations. *Mol. Simul.* **2002**, *28*, 607–632. [CrossRef]

21. Bosbach, D. Linking molecular-scale barite precipitation mechanisms with macroscopic crystal growth rates. In *Water-Rock Interactions, Ore Deposits, and Environmental Geochemistry: A Tribute to David A. Crerar*; Hellmann, R., Wood, S.A., Eds.; The Geochemical Society: St. Louis, MO, USA, 2002; Special Publication No. 7; pp. 97–110.

22. Kuwahara, Y.; Ishida, K.; Uehara, S.; Kita, I.; Nakamuta, Y.; Hayashi, T.; Fujii, R. Cool-stage AFM, a new AFM method for in situ observations of mineral growth and dissolution at reduced temperature: Investigation of the responsiveness and accuracy of the cooling system and a prelim experiment on barite growth. *Clay Sci.* **2012**, *16*, 111–119.

23. Ruiz-Agudo, C.; Putnis, C.V.; Putnis, A. The effect of a copolymer inhibitor on baryte precipitation. *Mineral. Mag.* **2014**, *78*, 1423–1430. [CrossRef]

24. Kuwahara, Y.; Makio, M. In situ AFM study on barite (001) surface dissolution in NaCl solutions at 30 °C. *Appl. Geochem.* **2014**, *51*, 246–254. [CrossRef]

25. Parkhurst, D.L.; Appelo, C.A.J. *User's Guide to PHREEQC (Version 2)—A Computer Program for Speciation, Batch-Reaction, One-Dimensional Transport, and Inverse Geochemical Calculations*; U.S. Geological Survey; U.S. Department of the Interior: Washington, DC, USA, 1999.

26. Henry Teng, H.; Dove, P.M.; De Yoreo, J.J. Kinetics of calcite growth: Surface processes and relationships to macroscopic rate laws. *Geochim. Cosmochim. Acta* **2000**, *64*, 2255–2266. [CrossRef]

27. Kowacz, M.; Putnis, C.V.; Putnis, A. The effect of cation: Anion ratio in solution on the mechanism of barite growth at constant supersaturation: Role of the desolvation process on the growth kinetics. *Geochim. Cosmochim. Acta* **2007**, *71*, 5168–5179. [CrossRef]

28. Zhang, F.; Yan, C.; Henry Teng, H.; Roden, E.E.; Xu, H. In situ AFM observations of Ca-Mg carbonate crystallization catalyzed by dissolved sulfide: Implications for sedimentary dolomite formation. *Geochim. Cosmochim. Acta* **2013**, *105*, 44–55. [CrossRef]

29. Kuwahara, Y. In situ observations of muscovite dissolution under alkaline conditions at 25–50 °C by AFM with an air/fluid heater system. *Am. Mineral.* **2008**, *93*, 1209–1216. [CrossRef]

30. Kuroda, T. *Crystal Lives (Kesshou-ha-ikiteiru): The Mechanism of Its Growth and Transformation of Morphology*; Science-sha: Tokyo, Japan, 1984. (In Japanese)

31. Hill, R.J. A further refinement of the barite structure. *Can. Mineral.* **1977**, *15*, 522–526.

32. Higgins, S.R.; Jordan, G.; Eggleston, C.M. Dissolution kinetics of the barium sulfate (001) surface by hydrothermal atomic force microscopy. *Langmuir* **1998**, *14*, 4967–4971. [CrossRef]

33. Bracco, J.N.; Gooijer, Y.; Higgins, S.R. Hydrothermal atomic force microscopy observantions of barite step growth rates as a function of the aqueous barium-to-sulfate ratio. *Geochim. Cosmochim. Acta* **2016**, *183*, 1–13. [CrossRef]

34. Stack, A.G.; Raiteri, P.; Gale, J.D. Accurate rates of the complex mechanisms for growth and dissolution of minerals using a combination of rare-event theories. *J. Am. Chem. Soc.* **2012**, *134*, 11–14. [CrossRef] [PubMed]

35. Stack, A.G. Molecular dynamics simulations of solvation and kink site formation at the {001} barite-water interface. *J. Phys. Chem. C* **2009**, *113*, 2104–2110. [CrossRef]

36. Brice, J.C. The Growth of Crystals from Liquids. In *Selected Topics in Solid State Physics*; Wohlfarth, E.P., Ed.; North-Holland Publishing Company: Amsterdam, The Netherlands, 1973.

37. Brice, J.C. *Crystal Growth Processes*; Blackie Halsted Press: Glasgow/London, UK, 1986.

38. Burton, W.K.; Cabrera, N.; Frank, F.C. The growth of crystals and the equilibrium structure of their surfaces. *Philos. Trans. R. Soc. Lond. A* **1951**, *243*, 299–358. [CrossRef]

39. Zhang, J.; Nancollas, G.H. Kink densities along a crystal surface step at low temperatures and under nonequilibrium conditions. *J. Cryst. Growth* **1990**, *106*, 181–190. [CrossRef]

40. Zhang, J.; Nancollas, G.H. Kink density and rate of step movement during growth and dissolution of an AB crystal in a nonstoichiometric solution. *J. Colloid Interface Sci.* **1998**, *200*, 131–145. [CrossRef]

41. Morales, J.; Astilleros, J.M.; Fernández-Díaz, L. Nanoscopic characteristics of anhydrite (001) growth. *Cryst. Growth Des.* **2012**, *12*, 414–421. [CrossRef]

42. Chernov, A.A. The Spiral Growth of Crystals. *Sov. Phys. Uspekhi* **1961**, *4*, 116–148. [CrossRef]

43. Chernov, A.A. *Modern Crystallography III. Crystal Growth*; Springer Series in Solid-State Science; Springer: Berlin/Heidelberg, Germany, 1984.

44. Walton, A.G. Nucleation in Liquids and Solutions. In *Nucleation*; Zettlemoyer, A.C., Ed.; Marcel Dekker Inc.: New York, NY, USA, 1969; pp. 225–307.

45. Godinho, J.R.A.; Stack, A.G. Growth kinetics and morphology of barite crystals derived from face-specific growth rates. *Cryst. Growth Des.* **2015**, *15*, 2064–2071. [CrossRef]

46. Shikazono, N. Precipitation mechanisms of barite in sulfate-sulfide deposits in back-arc basins. *Geochim. Cosmochim. Acta* **1994**, *58*, 2203–2213. [CrossRef]

47. Paytan, A.; Mearon, S.; Cobb, K.; Kastner, M. Origin of marine barite deposits: Sr and S isotope characterization. *Geology* **2002**, *30*, 747–750. [CrossRef]

48. Ray, D.; Kota, D.; Das, P.; Surya Prakash, L.; Khedekar, V.D.; Paropkari, A.L.; Mudholkar, A.V. Microtexture and distribution of minerals in hydrothermal barite-silica chimney from the Franklin seamount, SW Pacific: Constraints on the mode of formation. *Acta Geol. Sin.* **2014**, *88*, 213–225.

49. Pina, C.M.; Enders, M.; Putnis, A. The composition of solid solutions crystallising from aqueous solutions: The influence of supersaturation and growth mechanisms. *Chem. Geol.* **2000**, *168*, 198–210. [CrossRef]

50. Dinamani, M.; Kamath, P.V.; Seshadri, R. Electrochemical deposition of $BaSO_4$ coatings on stainless steel substrates. *Chem. Mater.* **2001**, *13*, 3981–3985. [CrossRef]

51. Mavredaki, E.; Neville, A.; Sorbie, K.S. Assessment of barium sulphate formation and inhibition at surfaces with synchrotron X-ray diffraction (SXRD). *Appl. Surf. Sci.* **2011**, *257*, 4264–4271. [CrossRef]

52. Mavredaki, E.; Neville, A.; Sorbie, K.S. Initial stages of barium sulfate formation at surfaces in the presence of inhibitors. *Cryst. Growth Des.* **2011**, *11*, 4751–4758. [CrossRef]

minerals

MDPI

Communication

Desiccator Volume: A Vital Yet Ignored Parameter in CaCO$_3$ Crystallization by the Ammonium Carbonate Diffusion Method

Joe Harris [1] and Stephan E. Wolf [1,2,*]

[1] Department of Materials Science and Engineering, Institute of Glass and Ceramics (WW3), Friedrich-Alexander University Erlangen-Nürnberg (FAU), Martensstrasse 5, Erlangen 91058, Germany; joe.harris@fau.de

[2] Interdisciplinary Center for Functional Particle Systems (FPS), Friedrich-Alexander University Erlangen-Nürnberg (FAU), Haberstrasse 9a, Erlangen 91058, Germany

* Correspondence: stephan.e.wolf@fau.de; Tel.: +49-9131-85-27565

Received: 30 May 2017; Accepted: 14 July 2017; Published: 19 July 2017

Abstract: Employing the widely used ammonium carbonate diffusion method, we demonstrate that altering an extrinsic parameter—desiccator size—which is rarely detailed in publications, can alter the route of crystallization. Hexagonally packed assemblies of spherical magnesium-calcium carbonate particles or spherulitic aragonitic particles can be selectively prepared from the same initial reaction solution by simply changing the internal volume of the desiccator, thereby changing the rate of carbonate addition and consequently precursor formation. This demonstrates that it is not merely the quantity of an additive which can control particle morphogenesis and phase selectivity, but control of other often ignored parameters are vital to ensure adequate reproducibility.

Keywords: crystal growth; amorphous calcium carbonate; colloidal crystals

1. Introduction

Nature precisely controls crystallization pathways to form functional crystalline materials with non-equilibrium morphologies that feature properties which often exceed those of their artificial counterparts [1–3]. As a consequence, significant research efforts have focused upon the synthesis and characterization of ordered structures of assembled crystallites which mimic the architectures found in nature [4,5]. Typically, organic and inorganic additives are employed to guide polymorph selection and crystallite assembly. These additives are selected based on their ability to favor formation of amorphous phases, interact with growing crystal faces, or assemble into architectures which guide crystal formation. In nature, numerous organic molecules are found in the crystallization milieu, a quantity which reflects the complexity of the mineralization processes performed by the biomineralizing organism [6]. Extraction, isolation, and identification of these molecules is a complex and time intensive process and quite often insufficient material can be extracted to assess the role the molecule takes during crystallization. Nevertheless, crystallization in the presence of such additives is an intensively studied topic, particularly in the case of the most prevalent biomineral calcium carbonate. Work often focuses on how relative quantities of such additives affect crystallization pathways and promote the formation of amorphous phases. Biomineralizing organisms however, do not merely regulate the quantities of additives within the crystallization milieu; they also control the kinetics of the process by mediating the rate at which ions are added. The effect of reaction kinetics on formation and morphogenesis of crystals in the presence of such additives is a much less discussed and well understood process than additive quantity, and is a parameter that is often omitted from studies.

Herein, we demonstrate that alteration of the rate at which carbonate concentration in the reaction solution is increased changes the polymorph selectivity and morphogenesis of calcium carbonates.

To do so, we employed the well-known ammonium carbonate diffusion method in which vessels containing calcium ion containing solutions are sealed in desiccator with vials of ammonium carbonate powder. Decomposition of the powder into carbon dioxide and ammonia, and subsequent absorption of these gaseous molecules into the solution increases carbonate ion concentration and maintains an alkaline pH of the solution. In this study, to vary the reaction kinetics, the size of the desiccator was simply altered. By decreasing the volume of the desiccator, the rates at which both the atmosphere in the desiccator becomes saturated with gas and the carbonate concentration in the solution increases are both increased and the kinetics of carbonate precipitation are influenced. To demonstrate how reaction kinetics can easily influence the effect of an additive on the crystallization system, we selected the magnesium ion as a simple additive. Previous work has shown that magnesium ions interacts with calcium carbonate during formation and stabilizes amorphous phases [7].

2. Materials and Methods

2.1. Materials

Double deionized water (Milli-Q) was used throughout all experiments. Ammonium carbonate (Carl Roth, Karlsruhe, Germany), calcium chloride dihydrate (Merck, Darmstadt, Germany), magnesium chloride hexahydrate (Sigma Aldrich, St. Louis, MO, USA), and hydrochloric acid (Carl Roth) were used without further purification. Glassware was first cleansed with deionized water, immersed in 1 M HCl for 5 min and then rinsed with deionized water before air drying.

2.2. Carbonate Precipitation

In a typical experiment, calcium carbonate precipitation was performed in a desiccator of 21, 11.5 or 3 dm^3 internal volume at 25 °C, for 24 to 72 h. A 9 cm diameter petri dish with a height of 1.2 cm was used as a crystallization vessel. An aqueous 25 mL reaction solution of composition 10 mM $CaCl_2 \cdot 2H_2O$ and 100 mM $MgCl_2 \cdot 6H_2O$ was poured into the petri dish which was then covered with cling film that was punctured with one needle hole 0.9 mm in diameter in the center. One glass vial (10 mL) containing 3 g of freshly crushed ammonium carbonate powder was covered with parafilm punctured with a single hole and placed at the bottom of the desiccator. The petri dish was then placed onto a plate above the glass vial and the desiccator was sealed, effectively starting the reaction. After the desired time, the precipitates were recovered from the air–water interface by use of a glass slide with careful attention paid to prevent aggregation of the particles. The particles were then rinsed with ethanol and air dried. In addition, particle assemblies were also examined by light microscopy whilst still at the air–water interface. Control measurements in the absence of magnesium were performed as above but with the omission of magnesium ions from the reaction solution, but with 200 mM NaCl in the solution to maintain the ionic strength.

2.3. Characterization

Optical microscopy was performed on an Olympus BX51 polarized optical microscope equipped with an Olympus XC50 digital camera (Olympus, Tokyo, Japan). Scanning electron microscopy was performed using a Quanta 200 scanning electron microscope (FEI, Brno-Černovice, Czech Republic) equipped with an energy dispersive X-ray analysis detector on gold sputtered samples mounted on SEM stubs using carbon pads. Powder X-ray diffraction patterns were recorded on a Kristalloflex D500 diffractometer (Siemens, Mannheim, Germany) using monochromatic Cu Kα radiation, (λ = 1.54184 Å) at a scan rate of 1 min^{-1} between 10° and 70°. Infrared spectra were obtained between 600 and 4000 cm^{-1} on a Nicolet iS10 infrared spectrometer equipped with an attenuated total reflection (ATR) accessory (Thermofisher Scientific, Waltham, MA, USA).

2.4. Carbonate Dissolution Quantification

The temporal evolution of the reaction solution pH as a result of ammonium carbonate dissolution was measured using a 905 Titrando titration device equipped with a micro pH probe (Metrohm, Herisau, Switzerland) and recorded on tiamo 2.4 software. Prior to each assay, the probe was calibrated against three standardized buffers of pH 7, 9, and 10. For each desiccator system, a 9 cm diameter petri dish with a height of 1.2 cm was used as a crystallization vessel. An aqueous 25 mL reaction solution of composition 10 mM $CaCl_2 \cdot 2H_2O$ and 100 mM $MgCl_2 \cdot 6H_2O$ was poured into the petri dish which was then covered with cling film that was punctured with two needle holes in the centre, one to allow ammonium carbonate influx and the second to allow the pH probe to be placed in solution. One glass vial (10 mL) containing 3 g of freshly crushed ammonium carbonate powder was covered with parafilm punctured with a single hole and placed at the bottom of the desiccator. The petri dish was then placed onto a plate above the glass vial and the desiccator was sealed, effectively starting the reaction. The solution pH was monitored for 3 h with pH measurements recorded every 5 s. Carbonate content was then calculated from a calibration curve constructed from pH measurements of 10 mL solutions of composition 10 mM $CaCl_2 \cdot 2H_2O$ and 100 mM $MgCl_2 \cdot 6H_2O$ with known quantities of ammonium carbonate.

3. Results

The temporal dissolution of ammonium carbonate into the reaction solution for each of the three desiccator volumes was studied by monitoring the evolution of solution pH against time (Figure S1). Dissolution of ammonia gas into water results in an increase in solution pH. By monitoring the change in solution pH the rate of ammonium carbonate dissolution (Figure 1) could be calculated by use of a calibration curve constructed from pH measurements of solutions of composition 10 mM $CaCl_2 \cdot 2H_2O$ and 100 mM $MgCl_2 \cdot 6H_2O$ and known quantities of ammonium carbonate. The rates of ammonium carbonate dissolution were 0.045, 0.080, and 0.115 mM·min^{-1} for the large, medium, and small desiccators respectively. An increase in internal volume of the desiccator decreased the rate at which ammonium carbonate dissolution into the solution occurred. The ammonium carbonate dissolution rates are lower than the 0.6 mmol·min^{-1} reported previously [8], in which a chamber of lower internal volume was used and the reaction solution and ammonium carbonate were uncovered. These observed differences in dissolution rates highlight the role of desiccator size on determining the rate of carbonate addition into solution.

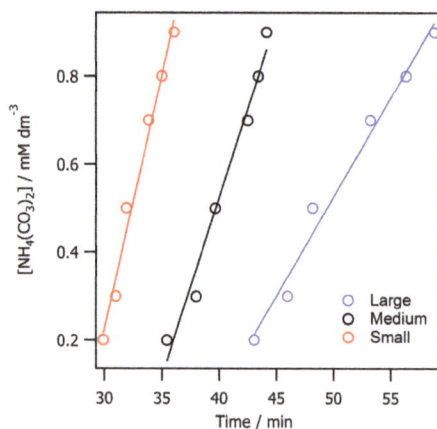

Figure 1. Ammonium carbonate dissolution rates. Temporal evolution of ammonium carbonate concentration for each of the three desiccators; rates of dissolution are 0.045, 0.080, and 0.115 mM·min^{-1} for the large, medium, and small desiccators respectively.

To demonstrate how desiccator size can highly impact $CaCO_3$ saturation levels in the reaction solution, and potentially $CaCO_3$ crystallization pathways, the formation and morphogenesis of calcium carbonate particles precipitated at the air-water interface were investigated. Scanning electron microscopy (SEM) studies of the precipitates removed from solution after 72 h revealed that alteration of the size of the desiccator resulted in the formation of precipitates with differing morphologies and states of crystallinity. In the largest desiccator (21 dm^3 internal volume), roughly spherical polycrystalline grains composed of radially arranged fibres typical of spherulitic growth were observed (Figure 2a). The powder X-ray diffraction (PXRD) pattern of the dried precipitate showed reflections at 2θ = 26.3°, 27.3°, 31.2°, 33.3°, 36.2°, 37.5°, 38.0°, 38.6°, 41.4°, 43.0°, 44.0°, 48.5°, 50.4°, 52.6°, and 53.0° corresponding to the miller indices {111}, {021}, {002}, {012}, {200}, {031}, {112}, {130}, {211}, {132}, {113}, and {231} of aragonite respectively (Figure 3a). The high concentration of magnesium ions in the crystallization solution is known to favour the formation of aragonite instead of the more thermodynamically stable polymorph calcite. This is because Mg^{2+} is incorporated into the calcite lattice but not into the lattice of aragonite, this therefore increases the solubility of calcite and decreases its growth rate at the expense of aragonite [9]. When the volume of the desiccator was reduced to 11.5 dm^3 (herein referred to as medium desiccator), SEM studies on particles extracted from the air-water interface after 72 h revealed that aggregates of particles of no defined shape were formed, and that these aggregates were several hundred micrometers in diameter (Figure 2b). The PXRD pattern (Figure 3a) of these precipitates was devoid of any peaks, and the infrared (IR) spectra (Figure 3b) showed bands at 859 (v_2), 1082 (v_1), 1394 (v_3), and 1633 cm^{-1} (water) and a broad absorbance centered at 3300 cm^{-1} (OH stretch) indicating that particles are highly-hydrated magnesium–substituted amorphous calcium carbonate. Absorbance at 747 and 714 cm^{-1} characteristic of the v_4 bands of crystalline calcium carbonate polymorphs vaterite and calcite respectively were not observed. Further reduction of the volume of the desiccator to 3 dm^3 (herein referred to as small desiccator), resulted in the formation of hexagonally-packed aggregates of spherical particles at the air–water interface after 72 h (Figure 2c). PXRD patterns (Figure 3a) of the dried aggregates contained no peaks, and IR spectra of the aggregates showed bands at 860 (v_2), 1087 (v_1), 1375 (v_3), 1631 (water), and 3300 cm^{-1} (OH stretch) again indicating that the particles are highly-hydrated magnesium-substituted amorphous calcium carbonate. The presence of magnesium in the particles is confirmed by a magnesium peak in the energy dispersive X-ray analysis spectra recorded from the aggregates (Figure S2). This demonstrates the importance of reaction kinetics for not only polymorph selection, as the large and small desiccators contained different polymorphs, but also in morphogenesis as the small and large desiccators contained particles of the same polymorph, but differing morphologies. The IR bands from the precipitate formed in the medium desiccator are at higher wavenumbers than the corresponding bands of the precipitates formed in the small desiccator. This shift is indicative of a higher magnesium content of the precipitates produced in the small desiccator [7]. SEM studies performed on samples from control experiments where magnesium ions where omitted from the reaction solution resulted in the formation of a mixture of rhombohedral calcite and quasi-spherical polycrystalline vaterite at the air–water interface after 72 h (Figure 2d), regardless of the size of desiccator used. This demonstrates that in the absence of Mg the change in desiccator size, and concomitant change in $CaCO_3$ saturation levels, were not sufficient to promote formation of different $CaCO_3$ polymorphs and morphologies after the 72 h of reaction, and that a strong $CaCO_3$ modulating additive such as Mg is required in addition to drive the formation of different structures.

Figure 2. Particles precipitated at the air-water interface after 72 h. (**a**) SEM image of aragonite particles precipitated in the large desiccator; (**b**) SEM image of highly hydrated magnesium-substituted amorphous calcium carbonate precipitated in the medium desiccator; (**c**) SEM image of hexagonally-packed spherical highly hydrated magnesium-substituted amorphous calcium carbonate particles precipitated in the small desiccator; (**d**) Optical microscopy image of representative calcite and vaterite particles precipitated in the large desiccator after 72 h.

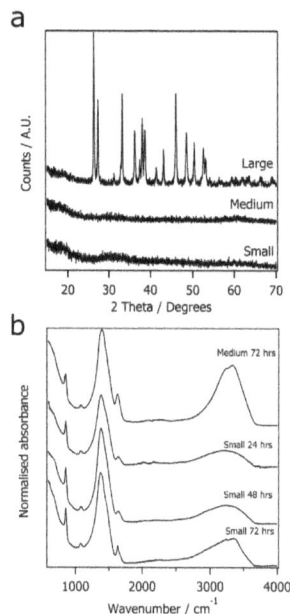

Figure 3. PXRD patterns and IR spectra of particles formed at air-water interface. (**a**) PXRD patterns of precipitates removed from the air-water interface after 72 h from reaction vessels in the small, medium, and large desiccators; (**b**) IR spectra of precipitates removed from the small and medium desiccator at varying time points.

Further studies were undertaken to probe the temporal evolution of structures in the small desiccator system. After 24 h a mixture of micrometre scale spherical particles (1.01 ± 0.24 μm diameter) and aggregates of no defined morphology were present at the air water interface (Figure 4a,d). IR spectra exhibited bands at 852 (v_2), 1098 (v_1), 1398 (v_3), and 1665 cm^{-1} and a broad absorbance centred at 3300 cm^{-1} (Figure 3b) concomitant with hydrated and magnesium-substituted amorphous calcium carbonate. Extension of the reaction time to 48 h resulted in the growth of the spherical particles to an average diameter of 5.78 ± 1.06 μm diameter (Figure 4b,d), and dissolution of the misshapen aggregates so that the most prevalent structure at the air–water interface were the spherical particles. The IR spectra featured bands at 852 (v_2), 1100 (v_1), 1397 (v_3), and 1659 cm^{-1} and a broad absorbance centered at 3300 cm^{-1} (Figure 3b). Further extension of the reaction time to 72 h resulted in increased growth of the spherical particles to 9.45 ± 1.07 μm (Figure 4c,d). The increase in mean particle size resulted in aggregation of the particles at the air-water interface due to the increased surface area occupied by the colloids. Packing of the particles was observed to be largely hexagonal, however, some defects in the colloidal crystal could be observed due to particle movement resulting from incomplete coverage of the air-water interface by the particles. Nevertheless, this demonstrates that colloidal crystals of calcium carbonate can be generated in the absence of an organic additive by control over precipitation rates.

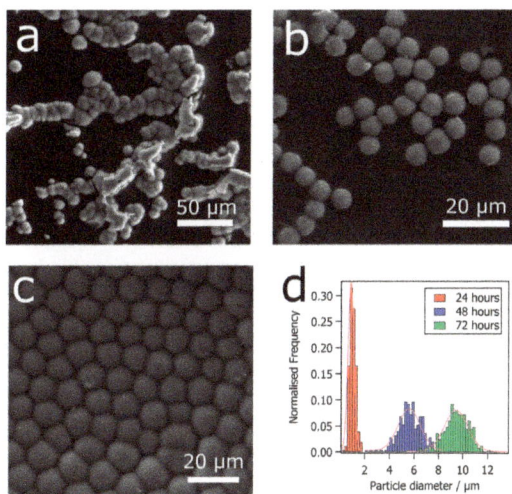

Figure 4. Growth and aggregation of spherical magnesium-calcium carbonate in the small desiccator. (**a**,**b**) SEM images of particles formed at the air-water interface in the small desiccator after 24 and 48 h, respectively; (**c**) Optical microscopy image of spherical particles at the air water interface after 72 h; (**d**) A histogram of particle sizes measured from optical microscopy images after 24, 48, and 72 h.

In conclusion, this report demonstrates that hexagonally packed arrays of spherical magnesium-doped amorphous calcium carbonate particles or spherulitic aragonite particles can be selectively synthesized from the same magnesium and calcium bearing solution by altering the size of the desiccator used. This shows that use of a simple additive (Mg ions) in combination with precise reaction kinetics is sufficient to generate higher-order structures and indicates even more elaborate structures can be generated by the combination of more complex additives and control over reaction kinetics. Moreover, this report highlights how the endemic problem of reproducibility can develop between laboratories, even for simple crystallization procedures. Often, vital experimental parameters are simply omitted from reports either through complacency or due to a lack of perception of their importance. Here, we have altered the size of the desiccator used, an extrinsic experimental parameter which is rarely disclosed in reports, yet one that is shown to be an essential variable which can drastically change crystallization by altering the reaction kinetics. Other variables associated with the ammonium carbonate diffusion method have been identified and characterized [9], however, the free volume of the reaction chamber remains a parameter which is still neglected. This report therefore demonstrates that an established protocol is required that can be used as an assay to compare crystallization experiments between laboratories. This is essential for systems in which difficult to acquire or expensive additives are used as it will increase reproducibility, thereby increasing research output.

Supplementary Materials: The following are available online at www.mdpi.com/2075-163X/7/7/122/s1, Figure S1: Temporal evolution of solution pH in the large, medium, and small desiccators; Figure S2: Energy dispersive X-ray analysis of magnesium doped amorphous calcium carbonate particles formed at the air-water interface in the small desiccator after 72 h.

Acknowledgments: Stephan E. Wolf acknowledges financial support by an Emmy Noether starting grant issued by the German Research Foundation (DFG, No. WO1712/3-1). Joe Harris and Stephan E. Wolf acknowledge financial support from the Cluster of Excellence 315 "Engineering of Advanced Materials—Hierarchical Structure Formation for Functional Devices" funded by the German Research Foundation. Eva Springer is thanked for assistance with scanning electron microscopy imaging.

Author Contributions: Joe Harris and Stephan E. Wolf conceived and designed the experiments; Joe Harris performed the experiments and analysed the data; Joe Harris and Stephan E. Wolf wrote the paper.

Conflicts of Interest: The authors declare no conflict of interest.

References

1. Ma, Y.; Aichmayer, B.; Paris, O.; Fratzl, P.; Meibom, A.; Metzler, R.A.; Politi, Y.; Addadi, L.; Gilbert, P.U.P.A.; Weiner, S. The grinding tip of the sea urchin tooth exhibits exquisite control over calcite crystal orientation and Mg distribution. *Proc. Natl. Acad. Sci. USA* **2009**, *106*, 6048–6053. [CrossRef] [PubMed]

2. Hovden, R.; Wolf, S.E.; Holtz, M.E.; Marin, F.; Muller, D.A.; Estroff, L.A. Nanoscale assembly processes revealed in the nacroprismatic transition zone of Pinna nobilis mollusc shells. *Nat. Commun.* **2015**, *6*, 10097. [CrossRef] [PubMed]

3. Weaver, J.C.; Milliron, G.W.; Miserez, A.; Evans-Lutterodt, K.; Herrera, S.; Gallana, I.; Mershon, W.J.; Swanson, B.; Zavattieri, P.; DiMasi, E.; et al. The stomatopod dactyl club: A formidable damage-tolerant biological hammer. *Science* **2012**, *336*, 1275–1280. [CrossRef] [PubMed]

4. Lee, K.; Wagermaier, W.; Masic, A.; Kommareddy, K.P.; Bennet, M.; Manjubala, I.; Lee, S.-W.; Park, S.B.; Cölfen, H.; Fratzl, P. Self-assembly of amorphous calcium carbonate microlens arrays. *Nat. Commun.* **2012**, *3*, 725. [CrossRef] [PubMed]

5. Harris, J.; Mey, I.; Hajir, M.; Mondeshki, M.; Wolf, S.E. Pseudomorphic transformation of amorphous calcium carbonate films follows spherulitic growth mechanisms and can give rise to crystal lattice tilting. *CrystEngComm* **2015**, *17*, 6831–6837. [CrossRef]

6. Marie, B.; Joubert, C.; Tayalé, A.; Zanella-Cléon, I.; Belliard, C.; Piquemal, D.; Cochennec-Laureau, N.; Marin, F.; Gueguen, Y.; Montagnani, C. Different secretory repertoires control the biomineralization processes of prism and nacre deposition of the pearl oyster shell. *Proc. Natl. Acad. Sci. USA* **2012**, *109*, 20986–20991. [CrossRef] [PubMed]

7. Loste, E.; Wilson, R.M.; Seshadri, R.; Meldrum, F.C. The role of magnesium in stabilising amorphous calcium carbonate and controlling calcite morphologies. *J. Cryst. Growth* **2003**, *254*, 206–218. [CrossRef]

8. Ihli, J.; Bots, P.; Kulak, A.; Benning, L.G.; Meldrum, F.C. Elucidating mechanisms of diffusion-based calcium carbonate synthesis leads to controlled mesocrystal formation. *Adv. Funct. Mater.* **2013**, *23*, 1965–1973. [CrossRef]

9. Davis, K.J.; Dove, P.M.; De Yoreo, J.J. The role of Mg^{2+} as an impurity in calcite growth. *Science* **2000**, *290*, 1134–1138. [CrossRef] [PubMed]

minerals

MDPI

Article

Crystal Chemistry and Stability of Hydrated Rare-Earth Phosphates Formed at Room Temperature

Asumi Ochiai and Satoshi Utsunomiya *

Department of Chemistry, Kyushu University, 744 Motooka, Nishi-ku, Fukuoka 819-0395, Japan;
a.ochia@chem.kyushu-univ.jp
* Correspondence: utsunomiya.satoshi.998@m.kyushu-u.ac.jp; Tel.: +81-92-802-4168

Academic Editor: Denis Gebauer
Received: 29 April 2017; Accepted: 17 May 2017; Published: 19 May 2017

Abstract: In order to understand the crystal chemical properties of hydrous rare-earth (RE) phosphates, $REPO_{4,hyd}$, that form at ambient temperature, we have synthesized $REPO_{4,hyd}$ through the interaction of aqueous RE elements (REEs) with aqueous P at room temperature at pH < 6, where the precipitation of RE hydroxides does not occur, and performed rigorous solid characterization. The second experiment was designed identically except for using hydroxyapatite (HAP) crystals as the P source at pH constrained by the dissolved P. Hydrated RE phosphate that precipitated at pH 3 after 3 days was classified into three groups: $LREPO_{4,hyd}$ (La → Gd) containing each REE from La-Gd, $MREPO_{4,hyd}$ (Tb → Ho), and $HREPO_{4,hyd}$ (Er → Lu). The latter two groups included increasing fractions of an amorphous component with increasing ionic radius, which was associated with non-coordinated water. $RE_{all}PO_{4,hyd}$ that contains all lanthanides except Pm transformed to rhabdophane structure over 30 days of aging. In the experiments using HAP, light REEs were preferentially distributed into nano-crystals, which can potentially constrain initial RE distributions in aqueous phase. Consequently, the mineralogical properties of hydrous RE phosphates forming at ambient temperature depend on the aging, the pH of the solution, and the average ionic radii of REE, similarly to the well-crystalline RE phosphates.

Keywords: rare earth elements; hydrated rare earth phosphate; apatite; nanoparticle; TEM

1. Introduction

Rare-earth (RE) elements (REEs), such as the lanthanides (La-Lu), are important rare metals in geological resources, and commonly used for technological application owing to their fluorescent, catalytic, and magnetic properties [1]. REEs are also potential fission products and used as surrogates for trivalent actinides in various experimental studies in nuclear chemistry [2–4]. Among a variety of REE-bearing minerals, RE phosphate is one of the important REE hosts that have further applications to phosphors, though post-heat treatment is required [1].

Crystalline RE phosphates occur in four different structures: monazite (light $REPO_4$; $P2_1/n$), rhabdophane (light $REPO_4 \cdot nH_2O$; $P6_222$ or $C2$), xenotime (heavy $REPO_4$; $I4_1/amd$), and churchite (heavy $REPO_4 \cdot nH_2O$; $I2/a$) [5–8]. In the crust (granite, metamorphic and/or metasomatic rocks, and pegmatite), monazite and xenotime are the stable phases for light REE and heavy REE, respectively, while rhabdophane and churchite occurrences are almost exclusively encountered in soil [9–13].

The structure depends on physico-chemical parameters such as pH, reaction time, and temperature [14–16]. Ionic radii also influence the crystal structure and morphology. Although the previous studies have reported on the mineralogical properties of all four types of RE phosphates and their application to accommodate foreign cations [17,18], these studies were mainly concerned with the well-crystalline phase formed at elevated temperatures at acidic pH ~1, e.g., [16,19]. The crystal chemical property of the hydrous RE phosphates that formed at ambient temperature has not been explored in detail. In addition, recent studies reported a novel synthesis of RE phosphate nanocrystals in P-free solution using microorganisms that release P from cell interiors at pH 3–5 at room temperature [20,21]; however, the electron diffraction pattern revealed vague patterns with diffused diffraction maxima owing to the presence of amorphous matter, which prevents the full identification of these nanoparticles. Hence, the first half of the present study reports a systematic characterization of hydrated RE phosphates, $REPO_{4,hyd}$, that formed at ambient temperature to understand their crystal-chemical properties. The formation of RE hydroxides formation was suppressed by using the pH condition of <6 in the present experiment.

In geological media, RE phosphates occur primarily as an accessary mineral such as monazite and xenotime, and the secondary precipitates of RE phosphates such as rhabdophane frequently play a key role as an indicative of various geochemical signatures such as for estimating paleo-environment [22,23] and migration of actinides such as Pu in the Oklo natural fission reactor [17,18,24–29]. The secondary RE phosphates commonly occur in weathered profiles in association with original phosphate minerals, such as apatite, which acts as a source of phosphorus (P) and provides unique reaction interfaces for the formation of secondary phosphates such as epitaxial growth and pseudomorphism [30]; however, the effects of the substrates has not been evaluated for the formation of hydrous RE phosphate. Thus, the second half of the present study demonstrates the time-course of the formation process of hydrated RE phosphates, $REPO_{4,hyd}$, over the HAP crystal at ambient temperatures at varying pHs constrained by dissolved phosphate to elucidate the evolution of secondary RE phosphate at the interfaces and the REE fractionation between solution and solid.

2. Materials and Methods

2.1. $REPO_{4,hyd}$ Precipitation from Aqueous Trivalent REE and Phosphate in Solution

In the present study, RE_i refers to phases containing individual La-Lu elements but excludes Pm, and RE_{all} refers to phases containing all La-Lu elements except for Pm. To simplify the discussion, Sc and Y were not included in the present experiment. A series of hydrated RE phosphates, $RE_iPO_{4,hyd}$ (where RE = La-Lu except for Pm) were synthesized from aqueous solutions. The reactants used as starting materials were $La(NO_3)_3 \cdot 6H_2O$ (99.9%), $Ce(NO_3)_3 \cdot 6H_2O$ (98%), $Pr(NO_3)_3 \cdot xH_2O$ (99.5%), $Nd(NO_3)_3 \cdot 6H_2O$ (99.5%), $Sm(NO_3)_3 \cdot 6H_2O$ (99.5%), $Gd(NO_3)_3 \cdot 6H_2O$ (99.5%), $Dy(NO_3)_3 \cdot 6H_2O$ (99.5%), $Ho(NO_3)_3 \cdot xH_2O$ (99.5%), $Er(NO_3)_3 \cdot xH_2O$ (99.5%), and $Yb(NO_3)_3 \cdot xH_2O$ (99.9%) supplied by Wako Pure Chemical (Osaka, Japan), $Eu(NO_3)_3 \cdot 6H_2O$ (99.9%) and $Tb(NO_3)_3 \cdot 6H_2O$ (99.9%) supplied by Strem Chemicals (Newburyport, MA, USA), $Tm(NO_3)_3 \cdot xH_2O$ (99.9%) supplied by Alfa Aesar (Ward Hill, MA, USA), and $Lu(NO_3)_3 \cdot 4H_2O$ (99.95%) supplied by Kanto Chemical (Tokyo, Japan). A 0.1 mol·L^{-1} RE^{3+} solution was prepared for each REE by dissolving the reactants in ultrapure water (Milli-Q$^{®}$, Merck Millipore, Billerica, MA, USA). The pH of each solution was adjusted to pH 3.0 ± 0.1 with NaOH and HNO_3 using a pH meter (Toko TXP-999i, Tokyo, Japan) with an electrode (Toko PCE108CW-SR). The margin of error in the pH measurements was within 0.1 pH units. In addition, 10 mL of a 0.1 mol·L^{-1} NaH_2PO_4 (Wako Pure Chemical) aqueous solution adjusted to pH 9.0 ± 0.1 was added to 10 mL of a 0.1 mol·L^{-1} RE solution with continuous stirring, and the pH was adjusted to 3.0 ± 0.1. The solutions were shaken by a rotary shaker for 3 days at room temperature. The suspensions were filtered with an Omnipore membrane filter with a pore size of 0.1 μm (Merck Millipore JVWP04700, Billerica, MA, USA) and rinsed with ultrapure water. The precipitates were dried in air at room temperature. $RE_{all}PO_{4,hyd}$ was synthesized using the same procedure as $RE_iPO_{4,hyd}$ and a solution with a total RE concentration of 0.1 mol·L^{-1} (RE = La-Lu except Pm), in which the concentration of each REE was set to 7.1 mmol·L^{-1}.

To evaluate the effects of crystal aging, $RE_iPO_{4,hyd}$ (RE = La, Tb, Dy, Ho, and Yb) and $RE_{all}PO_{4,hyd}$ were reacted for an extended duration ranging from 3 to 30 days. The effect of pH was examined in the case of $LaPO_{4,hyd}$ formation by varying the pH from 1 to 5. $LaPO_{4,hyd}$ was synthesized at pH 1 by pouring 10 mL of 0.1 mol·L^{-1} NaH_2PO_4 into 10 mL of 0.1 mol·L^{-1} $La(NO_3)_3$. The pH of the NaH_2PO_4 and $La(NO_3)_3$ solutions were adjusted beforehand to 9.0 ± 0.1 and 1.1, respectively, to account for the number of protons that would be released to the solution during the reaction. For synthesis at pH 5, 10 mL of 0.1 mol·L^{-1} NaH$_2$PO$_4$ was poured into 10 mL of 0.1 mol·L^{-1} La(NO$_3$)$_3$, for which the pH values were adjusted beforehand to 12.2 and 5.3, respectively. After mixing the solutions, the pH was subsequently adjusted again to 5.0 ± 0.1. The same incubation and filtration procedures were performed as for the other RE$_i$PO$_{4,hyd}$ phases.

Additional experiments were performed to investigate the thermal stability of RE$_i$PO$_{4,hyd}$ formed at room temperature. RE$_i$PO$_{4,hyd}$ (RE = Tb, Dy, Ho, and Yb) and RE$_{all}$PO$_{4,hyd}$ synthesized from a solution at pH 3 for 3 days were annealed at 70, 120, 200, 300, 400, or 500 °C for 1 h in a muffle furnace (Yamato Scientific FM38, Tokyo, Japan).

2.2. REPO$_{4,hyd}$ Formation on HAP Crystals

RE^{3+} solutions at a concentration of 2.0 mmol·L^{-1} (RE = La, Tb, Yb) were prepared by dissolving RE(NO$_3$)$_3$·nH$_2$O in Milli-Q® water, which was adjusted to a pH of 5 using HNO$_3$. 0.2 g of synthetic HAP in the form of Ca$_5$(PO$_4$)$_3$OH powder (Wako Pure Chemical) was added to 100 mL of the RE^{3+} solutions at room temperature. The HAP grains are rod-shaped, with approximate dimensions of 1.5 μm × 300 nm and a specific surface area of 6.31 m^2·g^{-1} [30]. The *S*/*V* ratio (where *S* is the apatite surface area and *V* is the volume of solution) was 1.3 × 10^2 cm^{-1} in these experiments. The suspensions of HAP powders were gently agitated using a magnetic stirrer for 1, 3, 9, and 24 h and 3 and 10 days. The solution pH was measured for each duration. The precipitate was collected by centrifugation at 3000 rpm for 10 min at 25 °C, washed with ultrapure water, and dried in air at room temperature. The supernatants remaining after centrifugation were filtered under reduced pressure with an Omnipore membrane filter with a pore size of 0.025 μm (Merck Millipore VSWP04700), which was small enough to separate the precipitates from the supernatant. RE$_{all}$PO$_{4,hyd}$ was also synthesized on HAP through the same procedure using a solution with a total RE concentration of 2.0 mmol·L^{-1} (RE = La-Lu except Pm), in which the concentration of each REE was 0.14 mmol·L^{-1}.

2.3. Analytical Methods

The crystal structures of RE$_i$PO$_{4,hyd}$ (RE = La-Lu) and RE$_{all}$PO$_{4,hyd}$ prepared from solution at room temperature were determined using powder X-ray diffraction analysis (XRD, Rigaku MultiFlex, Tokyo, Japan or Rigaku SmartLab). The analysis was conducted using Cu *K*α radiation (λ = 1.5418 Å) at 40 kV and 40 mA for the MultiFlex instrument, and at 40 kV and 30 mA for the SmartLab instrument in a scan range of 2θ = 10°–63° with a scan speed of 2° per min. A non-reflective silicon (111) holder was used to hold the specimens. The REPO$_{4,hyd}$ that formed on HAP crystals was analyzed using the SmartLab XRD in a scan range of 10°–50° using a SiO$_2$ glass plate as a specimen holder. Each spectrum was smoothed with a modified Savitzky–Golay filter and the linear background was subtracted using JADE 7 software (Materials Data Incorporated, Livermore, CA, USA). Subsequently, peak separation and fitting were conducted using the same software to calculate the crystallite size from the Scherrer equation:

$$D = \frac{K\lambda}{\beta \cos\theta} \tag{1}$$

where *D*, *K*, λ, β, and θ refer to the crystallite size (Å), the Scherrer constant (in this study *K* = 0.9), the wavelength of X-rays (1.5418 Å), the full-width at half maximum of the peak (rad), and the Bragg angle at the peak (rad), respectively.

RE$_i$PO$_{4,hyd}$ (RE = La-Lu) and RE$_{all}$PO$_{4,hyd}$ prepared from solution at pH 3 for 3 days and 30 days were analyzed by Fourier transform infrared spectroscopy (FT-IR) using KBr disks (1 mg sample/100 mg KBr, 10 mm in diameter). Spectra were recorded at a resolution of 4 cm^{-1} by summation of 32 scans.

Thermogravimetry–differential thermal analysis (TG–DTA) was performed on RE$_i$PO$_{4,hyd}$ (RE = La-Lu) and RE$_{all}$PO$_{4,hyd}$ prepared from solution at pH 3 for a duration of 3 days. Furthermore, RE$_i$PO$_{4,hyd}$ (RE = La, Tb, Yb) and RE$_{all}$PO$_{4,hyd}$ were reacted at pH 3 for 30 days and were analyzed to investigate the effect of aging. Each specimen was heated to 500 °C at a heating rate of 10 °C/min and then held for 10 min at 500 °C under a nitrogen gas flow of 250 mL/min using α-alumina as a reference material (Hitachi TG/DTA7300, Tokyo, Japan).

The supernatant after filtration was analyzed to obtain the concentrations of RE (RE = La-Lu), Ca, and P using inductively coupled plasma atomic emission spectroscopy (ICP-AES, PerkinElmer Optima 530 DV, Waltham, MA, USA).

High-resolution transmission electron microscopy (HRTEM), selected-area electron diffraction (SAED), high-angle annular dark-field scanning transmission electron microscopy (HAADF-STEM), and energy dispersive X-ray spectroscopy (EDS) were performed for all precipitate samples obtained in the syntheses using a JEOL JEM-ARM200F (Tokyo, Japan) with an accelerating voltage of 200 kV. The spherical aberration coefficient C_s is ~0 mm. The contrast in HAADF-STEM mode correlates with the mass and thickness of the target material [31]. TEM specimens were prepared by placing drops of a suspension of the synthesized precipitates onto a holey carbon mesh supported by a Cu grid.

3. Results

3.1. XRD Analysis

XRD patterns for $RE_iPO_{4,hyd}$ (RE = La-Lu except Pm) and $RE_{all}PO_{4,hyd}$ reacted at pH 3 for 3 days are shown in Figure 1a. The peak positions of $RE_iPO_{4,hyd}$ (RE = La-Gd except Pm) correspond well, with a slight peak shift, to the structure of rhabdophane [8]. The patterns for $RE_iPO_{4,hyd}$ (RE = Tb-Lu) and $RE_{all}PO_{4,hyd}$ could not be identified due to the broadened peaks and lack of matching structures. Although a few of the $RE_iPO_{4,hyd}$ (RE = Er-Lu) peaks appear to be similar to those of xenotime [5], the patterns lack a (200) peak and, thus, the structures were not determined convincingly. Furthermore, the patterns for $RE_iPO_{4,hyd}$ (RE = Tb-Ho) display intermediate profiles between those of $RE_iPO_{4,hyd}$ (RE = La-Gd except Pm) and $RE_iPO_{4,hyd}$ (RE = Er-Lu). Based on the structural characteristics indicated by XRD analysis in this study, the structure of the $RE_iPO_{4,hyd}$ precipitates are classified as $LREPO_{4,hyd}$ for phases containing RE = La-Gd, $MREPO_{4,hyd}$ containing RE = Tb-Ho, and $HREPO_{4,hyd}$ containing RE = Er-Lu, respectively. In $LREPO_{4,hyd}$, which has a rhabdophane structure, the coordination numbers (CNs) of RE^{3+} ions are one third eight-fold and two-thirds nine-fold [8], resulting in an average CN of 8.7. RE^{3+} in $MREPO_{4,hyd}$ and $HREPO_{4,hyd}$ can be considered to have eight-fold coordination because the CN of xenotime and churchite is eight-fold. Based on the CNs of all possible $RE_iPO_{4,hyd}$ phases, the average ionic radius of $RE_{all}PO_{4,hyd}$ containing all RE (RE = La-Lu except Pm) is calculated to be 1.07 Å, which lies just between that of Gd (CN = 8.7, 1.09 Å) and of Tb (CN = 8, 1.04 Å) [32].

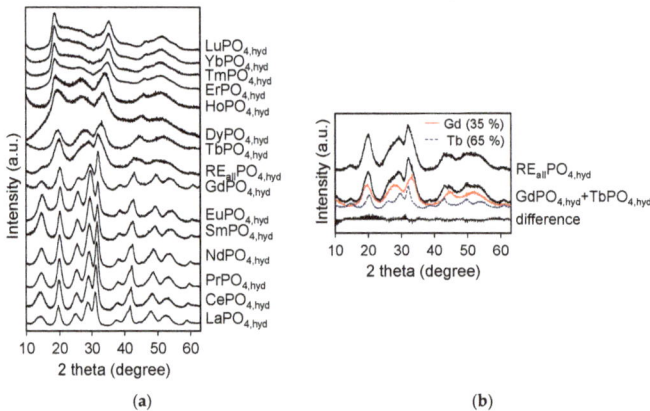

(a)

(b)

Figure 1. *Cont.*

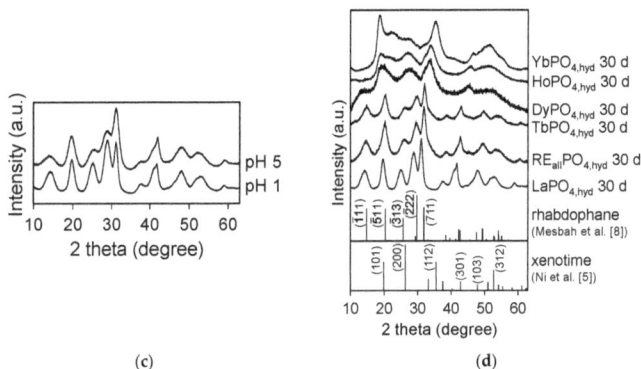

(c) (d)

Figure 1. (a) XRD patterns of $RE_iPO_{4,hyd}$ (RE =La-Lu except Pm) and $RE_{all}PO_{4,hyd}$ reacted at pH 3 for 3 days at room temperature. The patterns are arranged in order of increasing effective ionic radius from top to bottom; (b) comparison between XRD pattern of $RE_{all}PO_{4,hyd}$ and a calculated linear combination of the patterns of $GdPO_{4,hyd}$ + $TbPO_{4,hyd}$, which indicate proportions of 35% $GdPO_{4,hyd}$ and 65% $TbPO_{4,hyd}$ in the pattern based on a least-squares fitting method. The difference in the profiles between $RE_{all}PO_{4,hyd}$ and the calculation is also shown; (c) XRD patterns of $LaPO_{4,hyd}$ reacted at pH 1 and 5 for 3 days; (d) XRD patterns of $RE_iPO_{4,hyd}$ (RE = La, Tb, Dy, Ho, Yb) and $RE_{all}PO_{4,hyd}$ reacted at pH 3 for 30 days, accompanied by XRD patterns of rhabdophane-(Sm) and xenotime-(Lu) with Miller indexes from [5,8].

The XRD pattern of $RE_{all}PO_{4,hyd}$ correlates to a combination of the $GdPO_{4,hyd}$ and $TbPO_{4,hyd}$ patterns. A linear combination of these two profiles indicates volume fractions of 35% $GdPO_{4,hyd}$ structure and 65% $TbPO_{4,hyd}$ structure (Figure 1b).

The effect of pH on $LaPO_{4,hyd}$ precipitates are shown in Figure 1a,c. The full width at half maximum (FWHM) of each peak becomes smaller as the pH lowers.

In the aging experiments, there was little difference in the XRD profiles between 3 and 30 days for $RE_iPO_{4,hyd}$ (RE = La, Dy, Ho, and Yb); however, the structures of $RE_{all}PO_{4,hyd}$ and $TbPO_{4,hyd}$ changed significantly to become rhabdophane-structured after 30 days of aging (Figure 1a,d). In the $DyPO_{4,hyd}$ pattern, a small peak corresponding to the $(\bar{1}11)$ plane of rhabdophane appeared after 30 days of reaction, although the other peaks remained unchanged. The crystallite sizes of $LaPO_{4,hyd}$ calculated on the basis of the XRD patterns in Figure 1 using Scherrer equation are clearly a function of pH and the aging period (Figure 2a,b) and are summarized in Table 1.

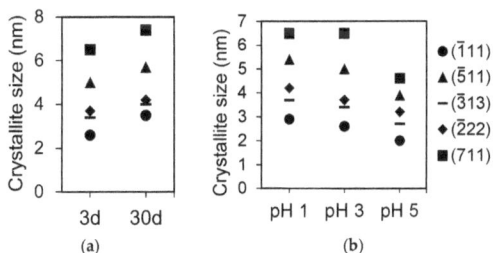

(a) (b)

Figure 2. Crystallite sizes of $LaPO_{4,hyd}$ calculated based on the Scherrer equation using JADE software. (a) $LaPO_{4,hyd}$ formed at pH 3 after 3 or 30 days; (b) $LaPO_{4,hyd}$ formed at pH values of 1, 3, and 5 after 3 days.

The calculated lattice volume of $LREPO_{4,hyd}$ reacted at a pH of 3 for 3 days are plotted as a function of the effective ionic radius in Figure 3a by fitting the $(\bar{1}11)$, $(\bar{5}11)$, $(\bar{3}13)$, $(\bar{2}22)$, and (711) peaks to the peaks of the rhabdophane structure using the JADE software. The calculated lattice parameters are summarized in Table 2. The calculated lattice volume displays a positive linear correlation with the effective ionic radius, indicating that the crystal lattice shrinks uniformly as the ionic radius decreases (Figure 3a). The lattice volumes calculated in the present experiment are larger than that of well-crystalline rhabdophane structure reported in [8], most likely owing to the presence of more water molecules in the structure of $LREPO_{4,hyd}$.

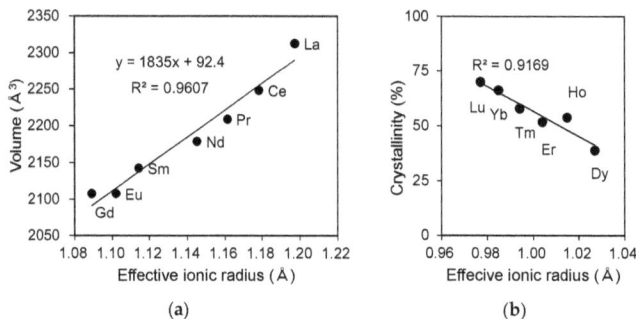

(a)

(b)

Figure 3. (a) Lattice volume of $LREPO_{4,hyd}$ (RE = La-Gd) calculated using JADE software, accompanied by a linear regression line; (b) crystallinity of $RE_iPO_{4,hyd}$ (RE = Dy-Lu) determined from XRD patterns calculated using JADE. These values indicate the intensity percentage of a crystalline peak I_c, which is obtained by subtracting the intensity of the amorphous halo I_a appearing at a 2-theta value of 20–30 degrees as: Crystallinity (%) = $I_c/(I_c + I_a)$.

Table 1. Crystallite size of $LaPO_{4,hyd}$ samples formed under four different conditions. The calculation was performed for five major peaks in the XRD pattern using the JADE software based on the Scherrer equation with a Scherrer constant K of 0.9.

Miller Index	$LaPO_{4,hyd}$ Crystallite Size (nm)			
	pH 3, 3 days	pH 3, 30 days	pH 1, 3 days	pH 5, 3 days
(11)	2.6	3.5	2.9	2.0
(11)	5.0	5.7	5.4	3.9
(13)	3.4	4.0	3.7	2.7
(22)	3.7	4.2	4.2	3.2
(711)	6.5	7.4	6.5	4.6

Table 2. Lattice parameters a, b, c and β, and lattice volume for $RE_iPO_{4,hyd}$ (RE = La-Gd) calculated based on the peaks in the XRD patterns using the JADE software.

REE	Effective Ionic Radius (Å) [1]	Lattice Parameter (Å)				Lattice Volume (Å³)
		a	b	c	β	
La	1.197	29.17	7.19	12.29	116.3	2313
Ce	1.178	28.50	7.22	12.13	115.7	2249
Pr	1.161	28.61	7.06	12.14	115.8	2209
Nd	1.145	28.48	7.04	12.07	115.8	2179
Sm	1.114	28.34	6.99	12.01	115.8	2142
Eu	1.102	27.93	7.02	11.91	115.5	2108
Gd	1.089	27.89	7.06	11.85	115.5	2108

[1] The effective ionic radius $r = 1/3\ r_{CN8} + 2/3\ r_{CN9}$.

The crystallinity of $RE_iPO_{4,hyd}$ (RE = Dy-Lu) was calculated based on the profile fitting of amorphous halos in the XRD patterns (Figure S2, Table 3) and is plotted in Figure 3b as a function of effective ionic radius. The results indicate that $DyPO_{4,hyd}$ has the largest amorphous volume and that the amorphous fraction linearly decreases towards $LuPO_{4,hyd}$.

The XRD patterns of $RE_iPO_{4,hyd}$ (RE = Tb, Ho, Dy, and Yb) and $RE_{all}PO_{4,hyd}$ obtained from the annealing experiments are summarized in Figure 4. $TbPO_{4,hyd}$ transformed to a monazite structure at 400 °C with small peaks occurring at a 2θ value around 30°, as indicated by the circle. $DyPO_{4,hyd}$ appeared to transform to a xenotime structure at 400 °C, as indicated by the appearance of small peaks at a 2θ value around 25.8°. $HoPO_{4,hyd}$ and $YbPO_{4,hyd}$ also transformed to a xenotime structure at around 300 and 200 °C, respectively. $RE_{all}PO_{4,hyd}$ appeared to retain its initial structure up to 200 °C and then transformed to a rhabdophane structure, which was retained up to 500 °C. In contrast, $TbPO_{4,hyd}$ did not become rhabdophane-structured but directly transformed to a monazite structure when annealed in air. The thermal behavior of $TbPO_{4,hyd}$ is different from that observed in the aging experiments, in which the structure simply changed to that of rhabdophane.

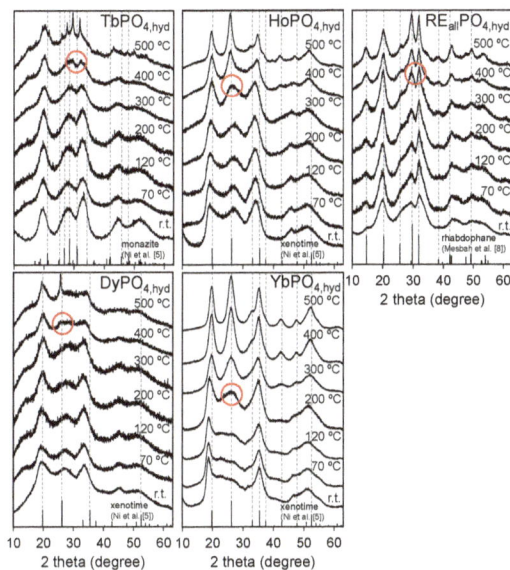

Figure 4. Transition in XRD patterns of $RE_iPO_{4,hyd}$ (RE = Tb, Dy, Ho, Yb) and $RE_{all}PO_{4,hyd}$ formed at pH 3 for 3 days at room temperature during annealing up to 500 °C. The patterns of each $REPO_{4,hyd}$ are accompanied by the spectrum of the ideal RE phosphate minerals; rhabdophane, monazite, and xenotime, based on [5,8], where r.t. represents room temperature. A circle indicates peaks exhibiting structural transformation.

Table 3. The results of quantification of the crystallinity of $RE_iPO_{4,hyd}$ (RE = Dy-Lu) calculated using the JADE software.

REE	Effective Ionic Radius (Å) [1]	Crystallinity (%)
Dy	1.027	38.9
Ho	1.015	53.9
Er	1.004	51.8
Tm	0.994	57.9
Yb	0.985	66.2
Lu	0.977	70.0

[1] The effective ionic radius from [32].

3.2. FT-IR Analysis

Infrared spectra for $RE_iPO_{4,hyd}$ (RE = La-Lu except Pm) and $RE_{all}PO_{4,hyd}$ reacted at pH 3 for 3 days are shown in Figure 5a. Two bands at 520–650 cm^{-1} (labeled 1 and 2) correspond to the asymmetric P–O bending vibration, ν_3, while the bands at 1000–1100 cm^{-1} (labeled 3 and 4) correspond to the antisymmetric P–O stretching vibration, ν_4 [33–35]. Bands 3 and 4 are not clearly separated for $LREPO_{4,hyd}$. The symmetric stretching mode ν_1, which typically appears as a small absorption band at ~966–985 cm^{-1} [36] and the bending vibration ν_2 were not observed, possibly because those bands were superimposed in the ν_3 and ν_4 regions for the hydrated RE phosphates [35]. The wavenumbers of band 4 in the ν_3 region plotted as a function of the effective ionic radius (Figure 5b and Table 4) exhibit a linear decrease with increasing ionic radius for $LREPO_{4,hyd}$ and $HREPO_{4,hyd}$, while $MREPO_{4,hyd}$ and $RE_{all}PO_{4,hyd}$ did not exhibit such trends. The broad bands at ~1600 cm^{-1} (band 5) and at ~3000–3500 cm^{-1} (band 6) are derived from coordinated H_2O [34]. Nitrate impurities derived from the initial REE reagents produce the band at ~1380 cm^{-1}.

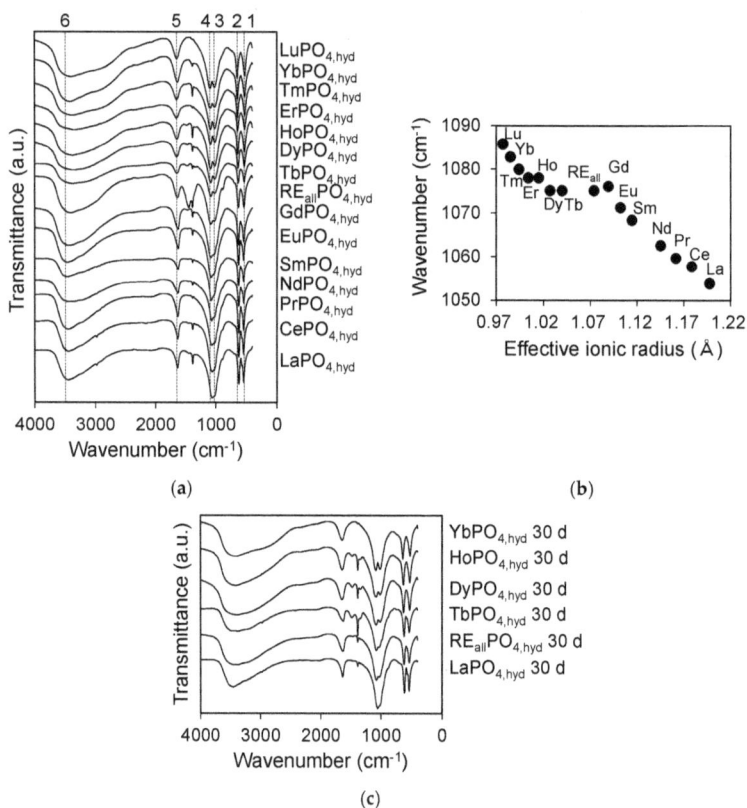

(a)

(b)

(c)

Figure 5. (a) IR spectra of $RE_iPO_{4,hyd}$ (RE = La-Lu) and $RE_{all}PO_{4,hyd}$ reacted at pH 3 for 3 days. The bands labeled 1–2, 3–4, 5, and 6 correspond to the P–O bending, P–O antisymmetric stretching, O–H bending, and O–H stretching vibrations, respectively; (b) a diagram showing the wavenumber of the P–O antisymmetric stretching vibration (labeled 4 in Figure 5a) of $RE_iPO_{4,hyd}$ (RE = La-Lu) and $RE_{all}PO_{4,hyd}$ as a function of the effective ionic radius; (c) IR spectra of $RE_iPO_{4,hyd}$ (RE = La, Tb, Dy, Ho, Yb) and $RE_{all}PO_{4,hyd}$ reacted at pH 3 for 30 days.

Table 4. List of the wavenumber of band 4 that corresponds to P–O antisymmetric stretching vibration, ν_4 in the FT-IR spectra obtained for RE_iPO_4 (RE = La-Lu) and $RE_{all}PO_{4,hyd}$ from FT-IR spectra.

REE	Effective Ionic Radius (Å) [1]	Wavenumber (cm^{-1})
La	1.197	1053.9
Ce	1.178	1057.8
Pr	1.161	1059.7
Nd	1.145	1062.6
Sm	1.114	1068.4
Eu	1.102	1071.3
Gd	1.089	1076.1
RE_{all}	1.074	1075.1
Tb	1.040	1075.1
Dy	1.027	1075.1
Ho	1.015	1078.0
Er	1.004	1078.0
Tm	0.994	1079.9
Yb	0.985	1082.8
Lu	0.977	1085.7

[1] Effective ionic radius: $r_{La-Gd} = 1/3 \, r_{CN8} + 2/3 \, r_{CN9}$, $r_{Tb-Lu} = r_{CN8}$, and r_{all} is the average of r_{La-Lu}.

FT-IR spectrum of the aged $TbPO_{4,hyd}$ and $RE_{all}PO_{4,hyd}$ appeared to be similar to that of $LaPO_{4,hyd}$, especially at bands 3 and 4 (Figure 5c). These FT-IR results concur with the XRD analyses, i.e., they indicate a change in the structures of $RE_{all}PO_{4,hyd}$ and $TbPO_{4,hyd}$ to a rhabdophane structure after 30 days of aging in solution.

3.3. TG-DTA Analysis

Derivatives of the TG (DTG) and DTA spectra for $RE_iPO_{4,hyd}$ (RE = La-Lu) and $RE_{all}PO_{4,hyd}$ reacted at a pH of 3 for 3 days are shown in Figure 6. $LREPO_{4,hyd}$ and $HREPO_{4,hyd}$ display a two-step dehydration, which is indicated by the presence of a DTG upper peak with an endothermic DTA peak. Assuming that the material remaining after heating to 500 °C entirely comprise anhydrous RE phosphates and the weight loss is derived solely from the dissociation of H_2O in $REPO_{4,hyd}$, the amount of dissociated water was calculated for the total heating duration (Figure 7a) and for each dehydration step (Figure 7b,c), and compiled in Table 5. The starting temperature of the second dehydration step was determined by selecting the local minimum of the DTG spectrum or the peak of the DTA spectrum between the first and second dehydration steps. A diagram of the dehydration extent in the first step in $HREPO_{4,hyd}$ indicates that the amount of water loss increases as a function of ionic radius (Figure 7b). The amount of water loss in the second dehydration step also appears to exhibit a linear decrease as the ionic radii decrease in $LREPO_{4,hyd}$ and $HREPO_{4,hyd}$ (Figure 7c). $MREPO_{4,hyd}$ and $RE_{all}PO_{4,hyd}$ each exhibited only one dehydration peak. For these phases, a DTG upper peak and a DTA exothermic peak are present, indicating that crystallization occurred during dehydration.

Table 5. Summary of dehydration or crystallization temperature and the amount of dissociated water n from the $RE_iPO_{4,hyd}$ (RE = La-Lu) and $RE_{all}PO_{4,hyd}$ during the TG-DTA analysis heated up to 500 °C. The value n corresponds to the number of H_2O molecule per unit formula.

REE	Effective Ionic Radius (Å) [1]	Second Dehydration Temperature (°C)	Crystallization Temperature (°C)	Dissociated Water n		
				Total	First	Second
La	1.197	165	-	1.95	1.30	0.66
Ce	1.178	147	-	1.71	1.03	0.68
Pr	1.161	143	-	1.86	1.14	0.72
Nd	1.145	137	-	1.76	1.00	0.76
Sm	1.114	127	-	1.54	0.80	0.73
Eu	1.102	117	-	1.58	0.75	0.83
Gd	1.089	133	-	1.96	1.13	0.83

Table 5. *Cont.*

REE	Effective Ionic Radius (Å) [1]	Second Dehydration Temperature (°C)	Crystallization Temperature (°C)	Dissociated Water n		
				Total	First	Second
RE$_{all}$	1.074	-	220	2.22	-	-
Tb	1.040	-	300	2.24	-	-
Dy	1.027	-	326	2.36	-	-
Ho	1.015	-	300	2.43	-	-
Er	1.004	178	-	2.64	1.77	0.86
Tm	0.994	169	-	2.63	1.66	0.97
Yb	0.985	150	-	2.56	1.54	1.02
Lu	0.977	160	-	2.39	1.40	0.99

[1] Effective ionic radius: $r_{La-Gd} = 1/3\, r_{CN8} + 2/3\, r_{CN9}$, $r_{Tb-Lu} = r_{CN8}$, and r_{all} is the average of r_{La-Lu}.

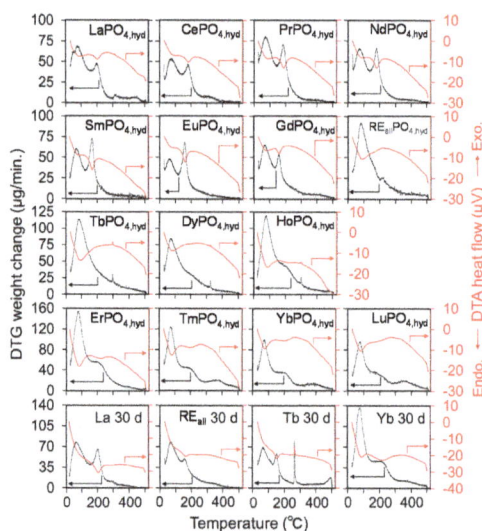

Figure 6. Differential thermogravimetric and thermal spectra of RE$_i$PO$_{4,hyd}$ (RE = La-Lu) and RE$_{all}$PO$_{4,hyd}$ reacted at pH 3 for 3 days. RE$_i$PO$_{4,hyd}$ (RE = La, Tb, Yb) and RE$_{all}$PO$_{4,hyd}$ reacted for 30 days are also shown. The black dotted line indicates the DTG curve and the red solid line represents the DTA heat flow.

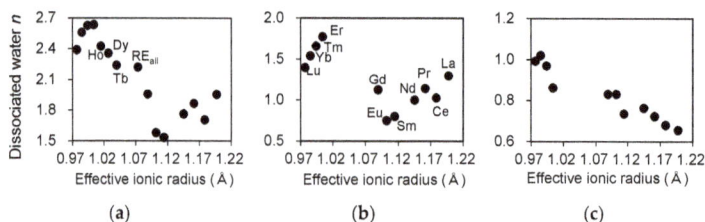

Figure 7. (**a**) The number of water molecules per formula unit in REPO$_{4,hyd}$ plotted as a function of REE ionic radius determined based on the TG-DTA results (Figure 6); the amounts of water dissociated during (**b**) the first dehydration step and (**c**) the second dehydration step calculated for LREPO$_{4,hyd}$ and HREPO$_{4,hyd}$.

In addition, the DTG and DTA spectra of RE$_i$PO$_{4,hyd}$ (RE = La, Tb, and Yb) and RE$_{all}$PO$_{4,hyd}$ after 30 days of reaction time are shown in Figure 6 for comparison with samples measured after

3 days of reaction time. Although $LaPO_{4,hyd}$ and $YbPO_{4,hyd}$ did not appear significantly different from those samples that had reacted for 3 days, additional dehydration peaks appeared in the spectra of $TbPO_{4,hyd}$ and $RE_{all}PO_{4,hyd}$ reacted for 30 days.

3.4. ICP-AES Analysis

In experiments of HAP dissolution in a solution containing all REEs, the concentrations of RE (RE = La-Lu except Pm), Ca, and P were measured for each duration using ICP-AES. The results are summarized in Table 6 and the total REE concentrations are plotted in Figure 8a as a function of reaction time. The P concentration is under the detection limit for all durations. The Ca concentration increased linearly for periods from 3 to 24 h, whereas the total concentration of all REEs decreased inversely with the Ca concentration. The reaction did not reach equilibrium within the duration of this dissolution experiment (Figure 8a); however, more than 80% of the total initial REE concentration was removed from solution after 240 h.

Figure 8b shows the transition in the patterns of REEs in solution reacted with HAP. The four curves at La-Nd, Sm-Gd, Gd-Ho, and Er-Lu exhibit the lanthanide tetrad effect [37]. Furthermore, more heavy REEs remained in solution compared with light REEs, and this trend becomes more pronounced with reaction time. The possible concentrations of free RE^{3+}, $RECO_3^+$, and $REOH^{2+}$ at these experimental conditions were calculated using Geochemist's Workbench using their formation constants [38], and input data of $P_{CO2} = 10^{-3.5}$ bar [39] and phosphate free solution. The calculation results clearly demonstrate that the concentration of free RE^{3+} decreases as the ionic radius decreases because the amount of other REE complexes such as $RECO_3^+$ and $REOH^{2+}$ increase with decreasing ionic radius (Figure 8c,d).

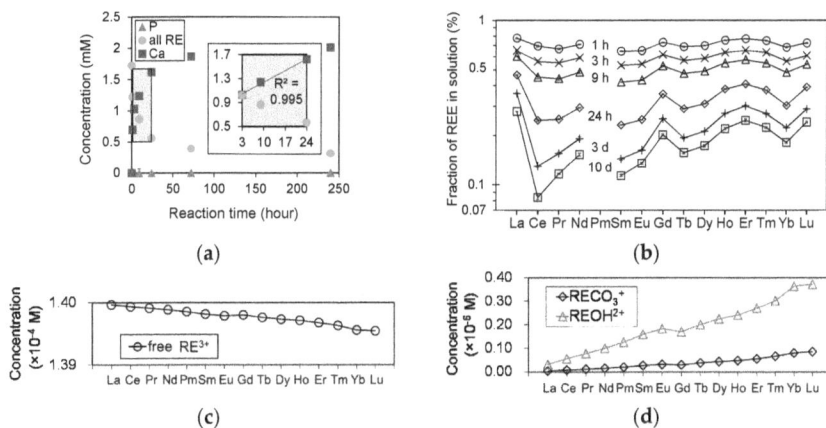

Figure 8. (a) Time-dependent concentrations of P (filled green triangles), Ca (filled purple squares), and all RE (filled light green circles) in solution during the experiment with HAP in contact with a RE^{3+} solution; (b) development of REE patterns in the solution as a function of time. The concentrations were normalized to the initial concentration; (c) thermodynamic calculation of aqueous RE^{3+} species in solution at a pH of 5 under the conditions of the present study. Concentration of free (non-coordinated) RE^{3+} calculated by subtracting the concentration of $RECO_3^+$ and $REOH^{2+}$ from the initial concentration of RE^{3+} (0.14 mM of each RE^{3+}); (d) the concentrations of $RECO_3^+$ and $REOH^{2+}$ obtained by multiplying the initial concentrations of RE^{3+} (0.14 mM of each La-Lu), the concentration of CO_3^{2-} or OH^-, and the formation constant $_{CO3}\beta_1$ or $_{OH}\beta_1$ [38]. The concentrations of CO_3^{2-} and OH^- were calculated using Geochemist's Workbench with the thermodynamic database *thermo.dat* and input data of pH 5 and $P_{CO2} = 10^{-3.5}$ bar [39].

Table 6. The solution composition determined by ICP-AES at each reaction time during the formation of $RE_{all}PO_{4,hyd}$, in which HAP crystals were in contact with the solution containing all REEs.

Reaction Time	Concentration (mmol·dm^{-3})																
	La	Ce	Pr	Nd	Sm	Eu	Gd	Tb	Dy	Ho	Er	Tm	Yb	Lu	Ca	P [1]	pH
0 h	0.12	0.11	0.13	0.12	0.12	0.13	0.11	0.13	0.11	0.13	0.13	0.13	0.13	0.13	udl[1]	udl	5.06
1 h	0.10	0.08	0.09	0.09	0.08	0.08	0.08	0.09	0.08	0.10	0.10	0.10	0.09	0.10	0.70	udl	5.16
3 h	0.08	0.06	0.07	0.07	0.06	0.07	0.06	0.07	0.07	0.08	0.08	0.08	0.08	0.08	1.03	udl	5.53
9 h	0.07	0.05	0.06	0.06	0.05	0.06	0.06	0.06	0.06	0.07	0.07	0.07	0.06	0.07	1.23	udl	5.55
24 h	0.06	0.03	0.03	0.04	0.03	0.03	0.04	0.04	0.03	0.05	0.05	0.05	0.04	0.05	1.62	udl	5.63
3 days	0.04	0.01	0.02	0.02	0.02	0.02	0.03	0.02	0.02	0.03	0.04	0.03	0.03	0.04	1.86	udl	5.66
10 days	0.03	0.01	0.01	0.02	0.01	0.02	0.02	0.02	0.02	0.03	0.03	0.03	0.02	0.03	2.01	udl	-[2]

[1] udl: Under detection limits (0.05 ppb for Ca and 15 ppb for P). [2] The pH after 10 days was not available.

3.5. TEM Analysis

The grain shapes of $REPO_{4,hyd}$ gradually change with ionic radius. HRTEM images show that $RE_iPO_{4,hyd}$ (RE = La, Nd, Tb) and $RE_{all}PO_{4,hyd}$ form rod-shaped particles of ~50 nm in length. $TmPO_{4,hyd}$ and $LuPO_{4,hyd}$ are spherically shaped with diameters of 10 nm (Figure 9). $DyPO_{4,hyd}$ and $HoPO_{4,hyd}$ precipitates do not exhibit any specific shape due to their amorphous characteristics. The SAED patterns of $LREPO_{4,hyd}$ and $RE_{all}PO_{4,hyd}$ display the same ring pattern as that of the rhabdophane structure. Although the SAED patterns of other $REPO_{4,hyd}$ are indistinct due to the large amorphous fractions, the $RE_iPO_{4,hyd}$ (RE = Tb-Lu) structure can be distinguished from the rhabdophane structure by the lack of an innermost diffraction ring. In EDS analyses, most analytical points from the $RE_{all}PO_4$ crystals contain uniform collective peaks at 4–8 keV corresponding to the L lines of REEs (Figure 10a). However, a few points display non-uniform spectra, as shown in Figure 10b,c, in which enhanced peaks for Er and Nd occur, respectively.

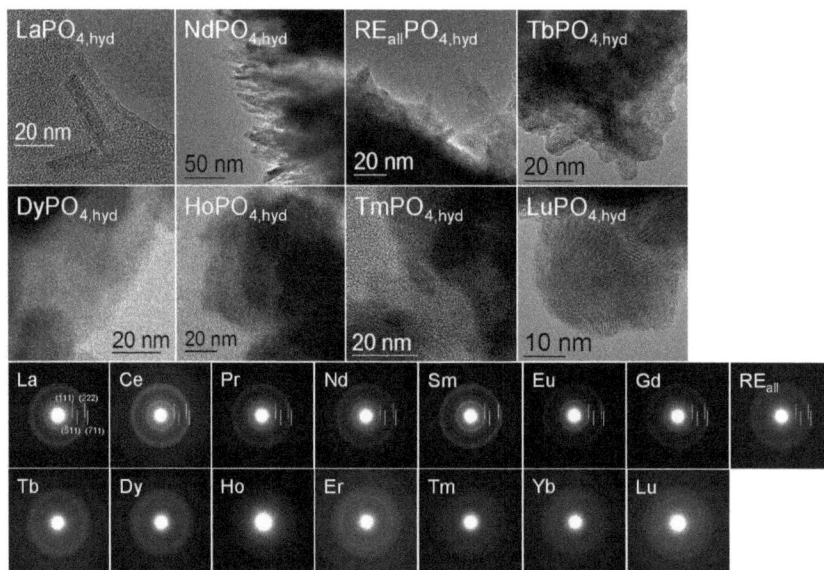

Figure 9. Top panels: Bright-field TEM image of $RE_iPO_{4,hyd}$ (RE = La, Nd, Tb, Dy, Ho, Tm, Lu) and $RE_{all}PO_{4,hyd}$ formed at pH 3 for 3 days. Bottom panels: Selected-area electron diffraction pattern of $RE_iPO_{4,hyd}$ (RE = La-Lu) and $RE_{all}PO_{4,hyd}$ formed at pH 3 for 3 days. White lines with mirror indices are shown for rhabdophane structure.

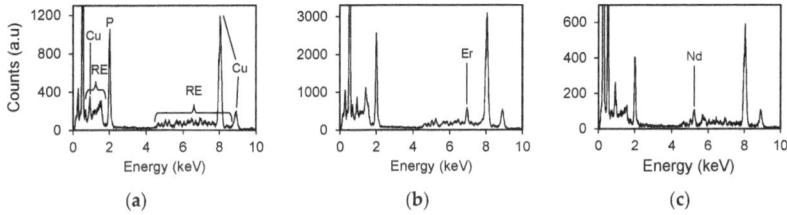

Figure 10. STEM-EDX spectra of $RE_{all}PO_{4,hyd}$ at (**a**) a typical point; (**b**) an Er-rich point; (**c**) a Nd-rich point. The Cu peak originates from the Cu TEM grid.

$RE_iPO_{4,hyd}$ (RE = La, Tb, and Yb) and $RE_{all}PO_{4,hyd}$ formed on HAP crystals appear to differ slightly from those formed in mixed solutions. For example, $LaPO_4$ precipitated as fibrous crystals on HAP (Figure 11a–c), in contrast to the rod-shaped crystals precipitated from the mixed solutions. $RE_{all}PO_{4,hyd}$ precipitates that formed on HAP exhibit shorter fibrous shaped crystals than the $LaPO_{4,hyd}$ precipitates on HAP (Figure 11d–i). The crystal shapes were not significantly affected by the reaction time. $YbPO_{4,hyd}$ precipitated at the rim of HAP crystals with a non-fibrous morphology (Figure 11j–k), reflecting the rapid precipitation of $YbPO_{4,hyd}$ and consumption of phosphate upon the release of P from the surface of HAP crystals.

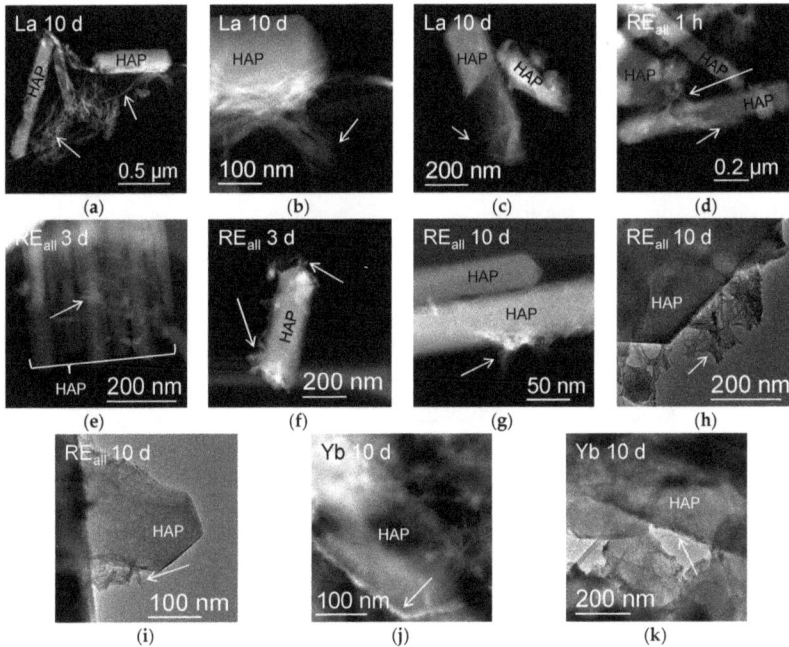

Figure 11. (**a–c**) HAADF-STEM images of $LaPO_{4,hyd}$ during HAP dissolution in contact with a La solution for 10 days. The fibrous nano-crystals indicated by the arrows correspond to $LaPO_{4,hyd}$ precipitation. (**d**) HAADF-STEM image of $RE_{all}PO_{4,hyd}$ formed during HAP dissolution after 1 h; (**e,f**) 3 days; (**g**) 10 days. (**h,i**) Bright-Field TEM images of the reaction products in all REE solutions after 10 days. The shorter fibrous nano-crystals indicated by the arrows correspond to $RE_{all}PO_{4,hyd}$ precipitation. (**j,k**) HAADF-STEM images of $YbPO_{4,hyd}$ precipitated on HAP during HAP dissolution in contact with a Yb^{3+} solution for 10 days. $YbPO_{4,hyd}$ nano-particles occur at the rim of HAP.

4. Discussion

4.1. Properties of REPO$_{4,hyd}$ Formed at Room Temperature

The REPO$_{4,hyd}$ that formed at room temperature in this study and its characteristics are summarized schematically in Figure 12. The LREPO$_{4,hyd}$ that formed at room temperature exhibited a rhabdophane structure similar to [40]. In this study, monazite did not form an primary product from solution at room temperature, although Hikichi and Hukuo [41] and Hikichi et al. [14] reported that monazite formed from aqueous solutions at 50 °C and pH < 1 within 28 days and at pH 5 after 900 days. Previous studies also reported that monazite was formed at 90 °C from solutions at pH < 1 [40,42], proceeded by the rhabdophane structure losing hydrated waters as a function of the aging time [40]. The differences in pH and the elevated temperatures may account for the absence of monazite in the present experiments at room temperature. This study considers mildly-acidic pH ranges relevant to natural environments without forming RE hydroxides, and the results show that monazite is not likely to form at room-temperature conditions.

Hikichi et al. [14] reported that a churchite structure in DyPO$_4$, YPO$_4$, and YbPO$_4$ formed at a pH of 3 at 20 °C; however, Hikichi et al. [43] did not identify a churchite structure for YPO$_4$ at a pH of 3.7 at 50 °C. The present study also did not detect a churchite structure at a pH of 3 at room temperature. The HREPO$_{4,hyd}$ structure did not match the structure of any RE phosphate minerals, and only a few peaks in the XRD pattern corresponding to the anhydrous tetragonal structure of the xenotime structure appeared in the HREPO$_{4,hyd}$ patterns. Such characteristics in the XRD pattern were also recognized in a previous study by Lucas et al. [40] reporting the synthesis of YPO$_4 \cdot n$H$_2$O at 50 °C. As described in the results section, the structure of these phases cannot be identified conclusively. However, it is clear that the HREPO$_{4,hyd}$ structure involves not only amorphous features but also some periodic features. In addition, the structural periodicity is very limited, most likely less than 10 nm as a first approximation based on the XRD peak broadness. Indeed, HRTEM images revealed aggregates of particles as small as ~10 nm. In some cases, although the structure could not be clearly seen initially, electron-beam irradiation resulted in the formation of xenotime structure (Figure 9). Such transformation prevented the acquisition of HRTEM images of HREPO$_{4,hyd}$ particles with their original structure. The evidence implies that the original structure is closely related to very small, nanometer-sized particles with a tetragonal structure similar to that of xenotime.

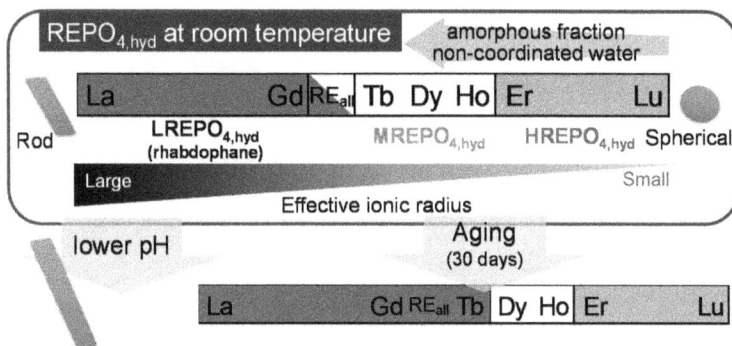

Figure 12. Schematic illustration summarizing the properties of initially formed hydrated RE phosphates.

The presence of a structured component in HREPO$_{4,hyd}$ is also supported by the gradual decrease in wavenumber for band 4, corresponding to the antisymmetric stretching vibration, as a function of the REE ionic radius in the FT-IR spectrum. This feature indicates lattice contraction of a configured structure (Figure 5b). There is evidence that the lattices of monazite, rhabdophane, xenotime, and churchite contract and appear as a shift in the bands derived from the P–O stretching vibration mode [34,36,44,45]. The lattice shrinks as it incorporates smaller REEs, and the P–O distance becomes slightly shorter, resulting in greater wavenumbers for this vibration. It is noted that $HREPO_{4,hyd}$ is partially composed of amorphous matter in addition to material with a configured structure, as shown in Figure 3b, and only the structured portion contributes to the shift in the wavenumber. The fact that there was no correlation between the effective ionic radius and the antisymmetric stretching vibration wavenumber in the $MREPO_{4,hyd}$ sample may indicate that no configured structure is present in $MREPO_{4,hyd}$. Although Lucas et al. [40] detected a structural transformation to churchite after 1 h of reaction at 50 °C in solution, no such structural transition was observed in the present study in aging experiments for $HREPO_{4,hyd}$ at room temperature. This indicates that the $HREPO_{4,hyd}$ structure is stable at room temperature and that structural transformation is slow on the time-scale of the present study.

The XRD analysis revealed that the structure of $RE_{all}PO_{4,hyd}$ comprises a combination of the $GdPO_{4,hyd}$ and $TbPO_{4,hyd}$ structures (Figure 1b); however, the both structures in $RE_{all}PO_{4,hyd}$ are not composed of the end member compositions of $GdPO_{4,hyd}$ and $TbPO_{4,hyd}$. Rather they are composed of all REE at almost same composition as that in the starting reagents as evidenced by the EDX analysis (Figure 10a).

In the $LaPO_{4,hyd}$ precipitation experiments, the crystallite size increased with increasing reaction time from 3 days to 30 days (Figure 1a,d and Figure 2a). This ripening behavior in crystallites was also seen for other REEs, as evidenced by the smaller full-width at half-maximum of diffraction peaks after aging (Figure 1a,d). The crystallite size also increased as pH decreased (Figure 1a,c and Figure 2b). This can be explained by the saturation ratio Ω:

$$\Omega = \left(\frac{IAP_0}{K_{s0}} \right)^{\frac{1}{\eta}} \tag{2}$$

where IAP_0, K_{s0}, and η refer to the ion activity product ($mol^2 \cdot dm^{-6}$), the solubility product ($mol^2 \cdot dm^{-6}$), and the number of ions per formula unit, respectively [46]. Using the parameters $K_{s0} = 10^{-25.7}$ [47] and $\eta = 2$ for $LaPO_4$, the value of Ω at pH 1 was calculated in the present study to be four orders of magnitude smaller than that at pH 5 due to the smaller amount of PO_4^{3-} (Table 7). Stumm [46] recorded that Ω values greater than 1 exponentially induce faster formation of smaller nuclei; hence, lower pH produces larger crystal sizes.

Table 7. Saturation ratio, Ω of the reacting solution with respect to $LaPO_{4,hyd}$. Activities of RE^{3+} and PO_4^{3-} were calculated using Geochemist's Workbench with the thermodynamic database *minteq*. The input data are 0.05 mM RE^{3+}, 0.05 mM PO_4^{3-}, 0.1 M Na^+, and 0.5 M NO_3^-.

pH	Activity ($mol \cdot dm^{-3}$)		Ion Activity Product IAP_0	Saturation Ratio Ω
	aRE^{3+}	aPO_4^{3-}		
1	2.06×10^{-3}	8.80×10^{-21}	1.81×10^{-23}	3.01×10^1
3	1.29×10^{-3}	4.95×10^{-16}	6.37×10^{-19}	5.65×10^3
5	1.26×10^{-3}	5.16×10^{-12}	6.53×10^{-15}	5.72×10^5

4.2. Thermal Stability of RE Phosphates Formed at Room Temperature

Because secondary RE phosphate minerals that form during rock weathering in surface environments are frequently subjected to subsequent thermal events, the thermal stability of $REPO_{4,hyd}$ that formed at room temperature was evaluated. In the present study, the structures of $REPO_{4,hyd}$ were classified into three categories: $LREPO_{4,hyd}$ corresponds to the rhabdophane structure, whereas the structures of both $MREPO_{4,hyd}$ and $HREPO_{4,hyd}$ could not be conclusively identified. The thermal stability of rhabdophane has been thoroughly investigated in previous studies and the rhabdophane structure is known to be stable up to ~500 °C–800 °C [15,48,49], ~750 °C [50]. The present study focused on the thermal stability of the unidentified structures, $MREPO_{4,hyd}$ and $HREPO_{4,hyd}$, which were stable in air up to 100–300 °C on the time-scale of the experiments. The thermal stability of these two types of phosphates was comparable to the stability of $YPO_4 \cdot 2H_2O$, in which the churchite structure transforms to the xenotime structure by annealing at ~100 °C for 73 h and at ~200 °C within 1 h [50,51]. Non-coordinated water seems to be unrelated to the crystal structure, based on a comparison between the TG-DTA charts and the XRD patterns of heated samples (Figures 4 and 6). Although Mesbah et al. [52] reported that two-step dehydration processes of well-crystalline rhabdophane in detail, the present study demonstrated that structural transitions occurred during the second dehydration step (after the local minimum in the DTG spectrum) for $LREPO_{4,hyd}$ and $HREPO_{4,hyd}$, or during crystallization (at the DTA peak of exothermic) for $MREPO_{4,hyd}$ and $RE_{all}PO_{4,hyd}$. As noted above, the initial structure can survive up to a certain temperature, at least 120 °C. The $TbPO_{4,hyd}$ structure does not possess a perfect rhabdophane structure, but persists up to 300 °C under dry conditions, although rhabdophane forms after 30 days of aging in solution (Figure 1d). Thus, $REPO_{4,hyd}$ formed at room temperature is more stable under dry conditions than wet conditions. Therefore, when the average ionic radii are larger than that of Tb (1.04 Å), the initial structure of $REPO_{4,hyd}$ is not stable and will transform to the rhabdophane structure by aging under wet conditions at room temperature.

4.3. Formation of $REPO_{4,hyd}$ on Apatite at Room Temperature

In $REPO_{4,hyd}$ formation experiments using apatite as a source of P, the dissolution rate of HAP in a RE^{3+} solution was determined to be 7.3×10^{-9} mol·m^{-2}·min^{-1} based on the rate of Ca release from HAP for a period of 3–24 h, as shown in Figure 8a. The pH of the solution remained within 5.5–5.6 during the period, after an initial increase in pH from ~5.1 in the first 3 h. The dissolution rate calculated in the present study is comparable with the dissolution rate determined by a previous study in pure water; 10^{-7}–10^{-9} mol·m^{-2}·min^{-1} at pHs ranging from 2–7 [53], indicating that the presence of soluble REEs in solution does not significantly affect the dissolution rate of HAP.

The solution pH adjacent to the HAP surface could be slightly higher than that of the bulk solution. However, RE hydroxide was not characterized in the present study, indicating that the local pH at the HAP surface did not exceed 8.

RE phosphates precipitated at the surface of apatite crystals through dissolution–precipitation mechanisms without the involvement of further mechanisms, such as epitaxial growth and pseudomorphism, which were previously reported in experiments on apatite interaction with aqueous Pb [30]. In general, epitaxial growth and pseudomorphism only occur when the crystallographic characteristics, such as the lattice parameter of the secondary precipitate, are closely related to those of the substrate and when a difference in solubility between the original and secondary mineral acts as the driving force [54–56]. The crystallographic parameters of the RE phosphate precipitates that were characterized in the present study ($LREPO_{4,hyd}$ (rhabdophane), $MREPO_{4,hyd}$, and $HREPO_{4,hyd}$) are not related to those of HAP, preventing the occurrence of those two specific mechanisms. Thus, dissolution–precipitation on apatite surfaces without the lattice relationship is the only mechanism expected for the initial formation of hydrous RE phosphates at ambient temperature in nature. Indeed, rhabdophane crystals associated with apatite were found to be crystallographically unrelated to apatite in natural environments [23], whereas the dissolution–precipitation mechanisms at high temperature >300 °C were reported to exhibit the topotaxial lattice relationship between apatite and monazite [57,58].

5. Implications

In the present study, only the rhabdophane structure was conclusively identified, which was constrained by the average ionic radius calculated according to the proportion of each constituent REE. This finding can be applied to predict the initial structure of hydrated RE phosphates formed in natural surface environments. However, natural systems involve more complicated species, such as other ions and organic matter. Organic ligands can produce smaller particles [59]. Organic matter, in particular, has the potential to control the stability of the rhabdophane structure, as some previous studies have reported that the rhabdophane structure was stable even for HREE (Yb, Lu) when organic matter was added to the starting reagent [60–62] or mechanical treatment was used to introduce it to the churchite structure [49,63]. A natural occurrence of rhabdophane-Y was reported by Takai and Uehara [64]; however, the mineral they discovered contains a wide variety of other LREE including ~20 wt % La_2O_3 and 15 wt % Nd_2O_3 in addition to ~15 wt % Y_2O_3. Thus, their report is consistent with the results obtained in the present experiments, which conclude that the structure is determined by the average ionic radius for phases that include multiple REEs.

Aqueous REEs frequently become immobilized in phosphate minerals by the consumption of apatite or they are simply bound to aqueous P during alteration processes [27,64]. Under typical environmental conditions, the hydrated RE phosphates observed in the present study are expected to initially form at nanoscale sizes. As natural water is not supersaturated with regard to RE phosphates [65] when compared with the composition of solutions used in this study, actual RE phosphates may slowly precipitate as larger crystals locally near apatite grains in natural aquifers than those observed in the present study.

The observed development of REE patterns during the formation and aging of $RE_{all}PO_{4,hyd}$ suggests that REE fractionation during the formation of nanoscale $RE_{all}PO_{4,hyd}$ is plausible. Therefore, the REE pattern for rhabdophane is not identical to that of the reacting solution. This is of particular importance when mineral REE patterns are used to estimate certain geochemical conditions of the ambient environment, such as oxidation states, compared with REE patterns of bedrock [22]. If specific phosphate minerals such as rhabdophane are dominant in a weathering profile, the chemical composition of the RE phosphate may constrain the bulk REE composition of the weathered profile, as demonstrated by a previous field study [66]. In the present study, only REE fractionation into the rhabdophane structure was examined. Further experiments are needed in order to evaluate REE fractionation into other structures such as $MREPO_{4,hyd}$ and $HREPO_{4,hyd}$ by varying the average ionic radius over all REEs.

Previous studies highlighted the importance of biomineralization of RE phosphates in constraining the mobility of REEs in the environment [20,21,67]. In these studies, biogenic Ce phosphate was characterized as having a needle-shaped rhabdophane structure [20], while Yb phosphate precipitated as amorphous nanoparticles ~50 nm in size [21]. The structure and morphology of these phases are very similar to those of $LREPO_{4,hyd}$ and $HREPO_{4,hyd}$ observed in the present study. Hydrous RE phosphate precipitation as a result of microbial activity was also reported in a weathering profile of granite [67]. Given that the majority of P in extracellular substances occurs as orthophosphate [68], the crystallographic nature of biogenic RE phosphates may be mainly controlled by the combined effects of the inorganic reaction processes elucidated in the present study and the effects of organically-bound P species.

Supplementary Materials: The following are available online at www.mdpi.com/2075-163X/7/5/84/s1, Figure S1: XRD patterns of $RE_iPO_{4,hyd}$ (RE = (a) La, (b) Tb, (c) Yb) and (d) $RE_{all}PO_{4,hyd}$ formed on HAP, Figure S2: Results of deconvolution of the XRD pattern obtained for $REPO_{4,hyd}$.

Acknowledgments: This study is partially supported by Japan Science and Technology Agency (JST) Initiatives for Atomic Energy Basic and Generic Strategic Research and by a Grant-in-Aid for Scientific Research (KAKENHI) from the Japan Society for the Promotion of Science (16K12585, 16H04634). The authors would like to thank the emeritus professor, Takashi Murakami, of the University of Tokyo for his detailed comments on the early version of this manuscript, and Hiroyuki Shiotsu for helpful discussions during the course of this study. The authors are grateful to Midori Watanabe for her assistance on XRD, TG-DTA, and FT-IR analyses at the Center of Advanced Instrumental Analysis, Kyushu University. The findings and conclusions of the authors of this paper do not necessarily state or reflect those of the JST.

Author Contributions: Satoshi Utsunomiya conceived and designed the experiments; Asumi Ochiai performed the experiments and analyzed the data; and Asumi Ochiai and Satoshi Utsunomiya wrote the paper.

Conflicts of Interest: The authors declare no conflict of interest.

References

1. Gai, S.; Li, C.; Yang, P.; Lin, J. Recent progress in rare earth micro/nanocrystals: Soft chemical synthesis, luminescent properties, and biomedical applications. *Chem. Rev.* **2014**, *114*, 2343–2389. [CrossRef] [PubMed]
2. Chapman, N.A.; Smellie, J.A.T. Special issue—Natural analogs to the conditions around a final repository for high-level radioactive-waste—Introduction and summary of the workshop. *Chem. Geol.* **1986**, *55*, 167–173. [CrossRef]
3. Choppin, G.R. Comparison of the solution chemistry of the actinides and lanthanides. *J. Less Common Met.* **1983**, *93*, 323–330. [CrossRef]
4. Choppin, G.R. Comparative solution chemistry of the 4f and 5f elements. *J. Alloys Compd.* **1995**, *223*, 174–179. [CrossRef]
5. Ni, Y.; Hughes, J.M.; Mariano, A.N. Crystal chemistry of the monazite and xenotime structures. *Am. Mineral.* **1995**, *80*, 21–26. [CrossRef]
6. Mooney, R.C.L. X-ray diffraction study of cerous phosphate and related crystals. I. Hexagonal modification. *Acta Crystallogr.* **1950**, *3*, 337–340. [CrossRef]
7. Kohlmann, M.; Sowa, H.; Reithmayer, K.; Schulz, H. Structure of a $Y_{1-x}(Gd,Dy,Er)_xPO_4 \cdot 2H_2O$ microcrystal using synchrotron radiation. *Acta Crystallogr. C* **1994**, *50*, 1651–1652. [CrossRef]
8. Mesbah, A.; Clavier, N.; Elkaim, E.; Gausse, C.; Kacem, I.B.; Szenknect, S.; Dacheux, N. Monoclinic form of the rhabdophane compounds: $REEPO_4 \cdot 0.667H_2O$. *Cryst. Growth Des.* **2014**, *14*, 5090–5098. [CrossRef]
9. Braun, J.-J.; Pagel, M.; Herbillon, A.; Rosin, C. Mobilization and redistribution of REEs and thorium in a syenitic lateritic profile: A mass balance study. *Geochim. Cosmochim. Acta* **1993**, *57*, 4419–4434. [CrossRef]
10. Braun, J.-J.; Viers, J.; Dupré, B.; Polve, M.; Ndam, J.; Muller, J.-P. Solid/liquid REE fractionation in the lateritic system of Goyoum, East Cameroon: The implication for the present dynamics of the soil covers of the humid tropical regions. *Geochim. Cosmochim. Acta* **1998**, *62*, 273–299. [CrossRef]
11. Stille, P.; Pierret, M.-C.; Steinmann, M.; Chabaux, F.; Boutin, R.; Aubert, D.; Pourcelot, L.; Morvan, G. Impact of atmospheric deposition, biogeochemical cycling and water-mineral interaction on REE fractionation in acidic surface soils and soil water (the Strengbach case). *Chem. Geol.* **2009**, *264*, 173–186. [CrossRef]
12. Sanematsu, K.; Kon, Y.; Imai, A.; Watanabe, K.; Watanabe, Y. Geochemical and mineralogical characteristics of ion-adsorption type REE mineralization in Phuket, Thailand. *Miner. Depos.* **2013**, *48*, 437–451. [CrossRef]
13. Berger, A.; Janots, E.; Gnos, E.; Frei, R.; Bernier, F. Rare earth element mineralogy and geochemistry in a laterite profile from Madagascar. *Appl. Geochem.* **2014**, *41*, 218–228. [CrossRef]
14. Hikichi, Y.; Hukuo, K.; Shiokawa, J. Synthesis of rare earth orthophosphates. *Bull. Chem. Soc. Jpn.* **1978**, *51*, 3645–3646. [CrossRef]
15. Hikichi, Y.; Murayama, K.; Ohsato, H.; Nomura, T. Thermal changes of rare earth phosphate minerals. *J. Mineral. Soc. Jpn.* **1989**, *19*, 117–126. [CrossRef]
16. Fang, Y.P.; Xu, A.W.; Song, R.Q.; Zhang, H.X.; You, L.P.; Yu, J.C.; Liu, H.Q. Systematic synthesis and characterization of single-crystal lanthanide orthophosphate nanowires. *J. Am. Chem. Soc.* **2003**, *125*, 16025–16034. [CrossRef] [PubMed]
17. Dacheux, N.; Clavier, N.; Podor, R. Monazite as promising long-term radioactive waste matrix: Benefits of high-structural flexibility and chemical durability. *Am. Mineral.* **2013**, *98*, 833–847. [CrossRef]
18. Deschanels, X.; Seydoux-Guillaume, A.M.; Magnin, V.; Mesbah, A.; Tribet, M.; Moloney, M.P.; Serruys, Y.; Peuget, S. Swelling induced by alpha decay in monazite and zirconolite ceramics: A XRD and TEM comparative study. *J. Nucl. Mater.* **2014**, *448*, 184–194. [CrossRef]
19. Yan, R.; Sun, X.; Wang, X.; Peng, Q.; Li, Y. Crystal structures, anisotropic growth, and optical properties: Controlled synthesis of Lanthanide orthophosphate one-dimensional nanomaterials. *Chem. Eur. J.* **2005**, *11*, 2183–2195. [CrossRef] [PubMed]
20. Jiang, M.; Ohnuki, T.; Kozai, N.; Tanaka, K.; Suzuki, Y.; Sakamoto, F.; Kamiishi, E.; Utsunomiya, S. Biological nano-mineralization of Ce phosphate by Saccharomyces cerevisiae. *Chem. Geol.* **2010**, *277*, 61–69. [CrossRef]
21. Jiang, M.; Ohnuki, T.; Tanaka, K.; Kozai, N.; Kamiishi, E.; Utsunomiya, S. Post-adsorption process of Yb phosphate nano-particle formation by Saccharomyces cerevisiae. *Geochim. Cosmochim. Acta* **2012**, *93*, 30–46. [CrossRef]

22. Murakami, T.; Utsunomiya, S.; Imazu, Y.; Prasad, N. Direct evidence of late Archean to early Proterozoic anoxic atmosphere from a product of 2.5 Ga old weathering. *Earth Planet. Sci. Lett.* **2001**, *184*, 523–528. [CrossRef]

23. Ichimura, K.; Murakami, T. Formation of rare earth phosphate minerals in 2.45-Ga paleosol. *J. Mineral. Petrol. Sci.* **2009**, *104*, 86–91. [CrossRef]

24. Forster, H.J. The chemical composition of REE-Y-Th-U-rich accessory minerals in peraluminous granites of the Erzgebirge-Fichtelgebirge region, Germany. Part II: Xenotime. *Am. Mineral.* **1998**, *83*, 1302–1315. [CrossRef]

25. Bingen, B.; Demaiffe, D.; Hertogen, J. Redistribution of rare earth elements, thorium, and uranium over accessory minerals in the course of amphibolite to granulite facies metamorphism: The role of apatite and monazite in orthogneisses from southwestern Norway. *Geochim. Cosmochim. Acta* **1996**, *60*, 1341–1354. [CrossRef]

26. Ohnuki, T.; Kozai, N.; Samadfam, M.; Yasuda, R.; Kamiya, T.; Sakai, T.; Murakami, T. Analysis of uranium distribution in rocks by μ-PIXE. *Nucl. Inst. Methods Phys. Res. B* **2001**, *181*, 586–592. [CrossRef]

27. Stille, P.; Gauthier-Lafaye, F.; Jensen, K.A.; Salah, S.; Bracke, G.; Ewing, R.C.; Louvat, D.; Million, D. REE mobility in groundwater proximate to the natural fission reactor at Bangombé (Gabon). *Chem. Geol.* **2003**, *198*, 289–304. [CrossRef]

28. Horie, K.; Hidaka, H.; Gauthier-Lafaye, F. Isotopic evidence for trapped fissiogenic REE and nucleogenic Pu in apatite and Pb evolution at the Oklo natural reactor. *Geochim. Cosmochim. Acta* **2004**, *68*, 115–125. [CrossRef]

29. Hidaka, H.; Kikuchi, M. SHRIMP in situ isotopic analyses of REE, Pb and U in micro-minerals bearing fission products in the Oklo and Bangombe natural reactors: A review of a natural analogue study for the migration of fission products. *Precambr. Res.* **2010**, *183*, 158–165. [CrossRef]

30. Kamiishi, E.; Utsunomiya, S. Nano-scale reaction processes at the interface between apatite and aqueous lead. *Chem. Geol.* **2013**, *340*, 121–130. [CrossRef]

31. Utsunomiya, S.; Ewing, R.C. Application of high-angle annular dark field scanning transmission electron microscopy, scanning transmission electron microscopy-energy dispersive X-ray spectrometry, and energy-filtered transmission electron microscopy to the characterization of nanoparticles in the environment. *Environ. Sci. Technol.* **2003**, *37*, 786–791. [PubMed]

32. Shannon, R.D. Revised effective ionic radii and systematic studies of interatomic distances in halides and chalcogenides. *Acta Crystallogr. A* **1976**, *32*, 751–767. [CrossRef]

33. Hezel, A.; Ross, S.D. Forbidden transitions in the infra-red spectra of tetrahedral anions—III. Spectra-structure correlations in perchlorates, sulphates and phosphates of the formula MXO_4. *Spectrochim. Acta* **1966**, *22*, 1949–1961. [CrossRef]

34. Assaaoudi, H.; Ennaciri, A.; Rulmont, A. Vibrational spectra of hydrated rare earth orthophosphates. *Vib. Spectrosc.* **2001**, *25*, 81–90. [CrossRef]

35. Heuser, J.; Bukaemskiy, A.A.; Neumeier, S.; Neumann, A.; Bosbach, D. Raman and infrared spectroscopy of monazite-type ceramics used for nuclear waste conditioning. *Prog. Nucl. Energy* **2014**, *72*, 149–155. [CrossRef]

36. Kijkowska, R.; Cholewka, E.; Duszak, B. X-ray diffraction and Ir-absorption characteristics of lanthanide orthophosphates obtained by crystallization from phosphoric acid solution. *J. Mater. Sci.* **2003**, *38*, 223–228. [CrossRef]

37. Masuda, A.; Ikeuchi, Y. Lanthanide tetrad effect observed in marine environment. *Geochem. J.* **1979**, *13*, 19–22. [CrossRef]

38. Lee, J.H.; Byrne, R.H. Examination of comparative rare earth element complexation behavior using linear free-energy relationships. *Geochim. Cosmochim. Acta* **1992**, *56*, 1127–1137. [CrossRef]

39. Langmuir, D. *Aqueous Environmental Geochemistry*; Prentice-Hall: Upper Saddle River, NJ, USA, 1997; p. 602.

40. Lucas, S.; Champion, E.; Breigiroux, D.; Bernache-Assollant, D.; Audubert, F. Rare earth phosphate powders $RePO_4 \cdot nH_2O$ (Re = La, Ce or Y)—Part I. Synthesis and characterization. *J. Solid State Chem.* **2004**, *177*, 1302–1311. [CrossRef]

41. Hikichi, Y.; Hukuo, K. Synthesis of monazite in aqueous solution. *Nippon Kagaku Kaishi* **1975**, *8*, 1311–1314. [CrossRef]

42. Hukuo, K.; Hikichi, Y. Synthesis of hexagonal cerium phosphate from aqueous solution. *Nippon Kagaku Kaishi* **1975**, *4*, 622–626. [CrossRef]

43. Hikichi, Y.; Hukuo, K.; Shiokawa, J. Synthesis of xenotime (YPO$_4$) by precipitation from aqueous solution. *Nippon Kagaku Kaishi* **1978**, *2*, 186–189. [CrossRef]
44. Begun, G.M.; Beall, G.W.; Boatner, L.A.; Gregor, W.J. Raman spectra of the rare earth orthophosphates. *J. Raman Spectrosc.* **1981**, *11*, 273–278. [CrossRef]
45. Assaaoudi, H.; Ennaciri, A. Vibrational spectra and structure of rare earth orthophosphates, weinschenkite type. *Spectrochim. Acta A* **1997**, *53*, 895–902. [CrossRef]
46. Stumm, W. *Chemistry of the Solid-Water Interface. Processes at the Mineral-Water and Particle-Water Interface in Natural Systems*; John Wiley & Sons: New York, NY, USA, 1992; p. 428.
47. Cetiner, Z.S.; Wood, S.A.; Gammons, C.H. The aqueous geochemistry of the rare earth elements. Part XIV. The solubility of rare earth element phosphates from 23 to 150 °C. *Chem. Geol.* **2005**, *217*, 147–169. [CrossRef]
48. Utsunomiya, S.; Murakami, T.; Nakada, M.; Kasama, T. Iron oxidation state of a 2.45-Byr-old paleosol developed on mafic volcanics. *Geochim. Cosmochim. Acta* **2003**, *67*, 213–221. [CrossRef]
49. Hikichi, Y.; Sasaki, T.; Murayama, K.; Nomura, T. Mechanochemical changes of weinschenkite-type RPO$_4$·2H$_2$O (R = Dy, Y, Er, or Yb) by grinding and thermal reactions of the ground specimens. *J. Am. Ceram. Soc.* **1989**, *72*, 1073–1076. [CrossRef]
50. Lucas, S.; Champion, E.; Bernache-Assollant, D.; Leroy, G. Rare earth phosphate powders RePO$_4$·nH$_2$O (Re = La, Ce or Y)—Part II. Thermal behavior. *J. Solid State Chem.* **2004**, *177*, 1312–1320. [CrossRef]
51. Hikichi, Y.; Hukuo, K.; Shiokawa, J. Synthesis of monoclinic YPO$_4$·2H$_2$O and its thermal change. *Nippon Kagaku Kaishi* **1977**, *11*, 1634–1638. [CrossRef]
52. Mesbah, A.; Clavier, N.; Elkaim, E.; Szenknect, S.; Dacheux, N. In pursuit of the rhabdophane crystal structure: From the hydrated monoclinic LnPO$_4$·0.667H$_2$O to the hexagonal LnPO$_4$ (Ln = Nd, Sm, Gd, Eu and Dy). *J. Solid State Chem.* **2017**, *249*, 221–227. [CrossRef]
53. Valsami-Jones, E.; Ragnarsdottir, K.V.; Putnis, A.; Bosbach, D.; Kemp, A.J.; Cressey, G. The dissolution of apatite in the presence of aqueous metal cations at pH 2–7. *Chem. Geol.* **1998**, *151*, 215–233. [CrossRef]
54. Putnis, A. Mineral replacement reactions: from macroscopic observations to microscopic mechanisms. *Mineral. Mag.* **2002**, *66*, 689–708. [CrossRef]
55. Putnis, A.; Putnis, C.V. The mechanism of reequilibration of solids in the presence of a fluid phase. *J. Solid State Chem.* **2007**, *180*, 1783–1786. [CrossRef]
56. Xia, F.; Brugger, J.; Chen, G.; Ngothai, Y.; O'Neill, B.; Putnis, A.; Pring, A. Mechanism and kinetics of pseudomorphic mineral replacement reactions: A case study of the replacement of pentlandite by violarite. *Geochim. Cosmochim. Acta* **2009**, *73*, 1945–1969. [CrossRef]
57. Harlov, D.E.; Wirth, R.; Förster, H.-J. An experimental study of dissolution-reprecipitation in fluorapatite: Fluid infiltration and the formation of monazite. *Contrib. Mineral. Petrol.* **2005**, *150*, 268–286. [CrossRef]
58. Harlov, D.E. Apatite: A fingerprint for metasomatic processes. *Elements* **2015**, *11*, 171–176. [CrossRef]
59. Boakye, E.E.; Mogilevsky, P.; Hay, R.S. Synthesis of nanosized spherical rhabdophane particles. *J. Am. Ceram. Soc.* **2005**, *88*, 2740–2746.
60. Hikichi, Y.; Ota, T.; Hattori, T.; Imaeda, T. Synthesis and thermal reactions of rhabdophane-(Y). *Mineral. J.* **1996**, *18*, 87–96. [CrossRef]
61. Min, W.; Daimon, K.; Ota, T.; Matsubara, T.; Hikichi, Y. Synthesis and thermal reactions of rhabdophane-(Yb or Lu). *Mater. Res. Bull.* **2000**, *35*, 2199–2205. [CrossRef]
62. Ito, H.; Fujishiro, Y.; Sato, T.; Okuwaki, A. Preparation of lanthanide orthophosphates by homogeneous precipitation under hydrothermal conditions using lanthanide-EDTA chelates. *Br. Ceram. Trans.* **1995**, *94*, 146–150.
63. Hikichi, Y.; Yogi, K.; Ota, T. Preparation of rhabdophane-type RPO$_4$·nH$_2$O (R = Y or Er, n = 0.7–0.8) by pot-milling churchite-type RPO$_4$·2H$_2$O at 20–25 °C in air. *J. Alloys Compd.* **1995**, *224*, L1–L3. [CrossRef]
64. Takai, Y.; Uehara, S. Rhabdophane-(Y), YPO$_4$·H$_2$O, a new mineral in alkali olivine basalt from Hinodematsu, Genkai-cho, Saga Prefecture, Japan. *J. Mineral. Petrol. Sci.* **2012**, *107*, 110–113. [CrossRef]
65. Roncal-Herrero, T.; Rodríguez-Blanco, J.D.; Oelkers, E.H.; Benning, L.G. The direct precipitation of rhabdophane (REEPO$_4$·nH$_2$O) nano-rods from acidic aqueous solutions at 5–100 °C. *J. Nanopart. Res.* **2011**, *13*, 4049–4062. [CrossRef]

66. Aubert, D.; Stille, P.; Probst, A. REE fractionation during granite weathering and removal by waters and suspended loads: Sr and Nd isotopic evidence. *Geochim. Cosmochim. Acta* **2001**, *65*, 387–406. [CrossRef]
67. Taunton, A.E.; Welch, S.A.; Banfield, J.F. Microbial controls on phosphate and lanthanide distributions during granite weathering and soil formation. *Chem. Geol.* **2000**, *169*, 371–382. [CrossRef]
68. Masaki, S.; Shiotsu, H.; Ohnuki, T.; Sakamoto, F.; Utsunomiya, S. Effects of CeO_2 nanoparticles on microbial metabolism. *Chem. Geol.* **2015**, *391*, 33–41. [CrossRef]

Communication

A Micro-Comb Test System for In Situ Investigation of Infiltration and Crystallization Processes

Dominik Gruber [†], Stefan L. P. Wolf [†], Andra-Lisa M. Hoyt, Julian P. Konsek and Helmut Cölfen [*]

Department of Chemistry, Physical Chemistry, University of Konstanz, Universitätsstraße 10, 78457 Konstanz, Germany; dominik.gruber@uni-konstanz.de (D.G.); stefan.3.wolf@uni-konstanz.de (S.L.P.W.); andra-lisa.hoyt@uni-konstanz.de (A.-L.M.H.); Julian.Konsek@uni-konstanz.de (J.P.K.)
* Correspondence: Helmut.Coelfen@uni-konstanz.de; Tel.: +49-7531-88-4063
† These authors contributed equally to this work.

Received: 31 August 2017; Accepted: 3 October 2017; Published: 6 October 2017

Abstract: The investigation of mineralization and demineralization processes is important for the understanding of many phenomena in daily life. Many crystalline materials are exposed to decay processes, resulting in lesions, cracks, and cavities. Historical artifacts, for example, often composed of calcium carbonate ($CaCO_3$), are damaged by exposure to acid rain or temperature cycles. Another example for lesions in a crystalline material is dental caries, which lead to the loss of dental hard tissue, mainly composed of hydroxyapatite (HAp). The filling of such cavities and lesions, to avoid further mineral loss and enable or support the remineralization, is a major effort in both areas. Nevertheless, the investigation of the filling process of these materials into the cavities is difficult due to the non-transparency and crystallinity of the concerned materials. In order to address this problem, we present a transparent, inexpensive, and reusable test system for the investigation of infiltration and crystallization processes in situ, being able to deliver datasets that could potentially be used for quantitative evaluation of the infiltration process. This was achieved using a UV-lithography-based micro-comb test system (MCTS), combined with self-assembled monolayers (SAMs) to mimic the surface tension/wettability of different materials, like marble, sandstone, or human enamel. Moreover, the potential of this test system is illustrated by infiltration of a $CaCO_3$ crystallization solution and a hydroxyapatite precursor (HApP) into the MCTS.

Keywords: porous mineral; infiltration analysis; self-assembled monolayer; surface modification; calcium phosphate; calcium carbonate; dental hard tissue; artifact restoration; remineralization; biomaterials

---

1. Introduction

Crystallization processes in confined environments are essential for a wide range of fields, ranging from scaling in pipelines, to biomineralization, to the construction industry [1–12]. All of these processes are characterized by a hindered ion diffusion, locally higher supersaturation, and often heterogeneous crystallization. Therefore, crystallization is observed in areas where nucleation is unexpected under the applied conditions, like acute steps or pockets [13–15]. However, it is also commonly observed that crystallization is significantly slowed down in confinement or even completely inhibited [2,11,15–19].

In the case of concrete, calcium oxide links the already crystallized silicon dioxide to form a brittle but very hard material, like calcium silicate hydrate in cements [9]. In the early centuries such high-end materials were not available for architecture. Therefore, naturally-abundant materials like marble or sandstone were used to build monuments that are still impressive nowadays, like the Pantheon in Rome, the Acropolis in Athens, or the Cologne Cathedral.

Sandstones, like Pietra Serena (PS), are composed of a mixture of oxides and carbonates and are characterized by a relatively high porosity [20,21]. In addition, they are relatively soft and easy to handle, which is among the reasons why this material was often used in architecture. In comparison, marbles are materials containing varying degrees of carbonates, depending on the quarrying area, especially of calcium [22]. Carrara marble (CM) represents a class of rock with a carbonate content of over 80% and a generally lower porosity [20]. Due to the exquisite white color, CM was often used for statues and architecture, as well [22]. Over the centuries of their existence, historical monuments and artworks are exposed to many decay processes [22]. The porosity of the stones enables the infiltration of water, which can result in mechanical stress and, finally, in damage of the structural integrity of the stones due to thermal cycles and crystallization processes inside the pores. This also shows how important a better understanding of crystallization in confined environments is [23]. Additionally, chemical reactions—e.g., caused by acid rain—can lead to cracks and cavities, for example, due to the dissolution of $CaCO_3$. To preserve historical artifacts by preventing further damage and restoring the integrity, several restoration formulations have been developed [24–26].

Similar problems related to acid-induced decay processes and subsequent mineral loss are a major challenge in dentistry—called dental caries. It is the most common chronic disease around the world affecting more than 95% of concerned adults [27,28]. Caries result from bacteria, which form a biofilm on the teeth. These bacteria metabolize carbohydrates and produce acids as by-products [29]. The acidic metabolites demineralize the dental hard tissues—enamel and dentin—mainly composed of hydroxyapatite (HAp), resulting in the formation of lesions in the range of nanometers to hundreds of microns [30–32]. Due to the enormous number of diseased individuals, several different treatments for refilling the cavities have been established. The developed filling materials range from cements [33–35] over (nano)ceramics [36,37] to peptide-amorphous calcium phosphate composites [38,39]. Nevertheless, the mode of action of many fillings is not completely understood.

The investigation of the infiltration behavior of new filling materials in dental lesions, as well as in pores and cracks in historical artifacts, is a great challenge. Furthermore, the investigation of the underlying mechanism of the interactions with the substrates (stones and teeth) is crucial for the development and validation of the filling methods and materials. In both cases, the production of suitable tests systems (TSs) is complex and time consuming. The stone samples need to be aged artificially to create controlled and reproducible cavities and damage. This is usually achieved by multiple cycles of thermal shock or hot-cold cycling, purposeful infiltration of soluble salts, or treatment with acids [40–42]. For the dental TSs, the usage of artificial tooth lesions, made of bovine or human teeth, is the state of the art [39,43,44]. Therefore, the teeth need to be embedded in a polymer matrix, polished afterwards and etched with acids [39,44,45]. Both tooth and stone TSs, need to be replaced after each experiment, resulting in a high consumption of the TSs. Apart from the sophisticated preparation of the TSs, it is difficult to obtain an insight into the lesions due to the crystallinity of the TSs and non-transparency to visible light. The investigation of infiltration depth and crystallization processes of fillers with light microscopy or scattering techniques is hindered.

In this paper, we present the fabrication of a UV-lithography-based, transparent, inexpensive, and reusable micro-comb test system (MCTS) for the investigation of infiltration and crystallization processes in situ by using common methods, like optical light microscopy (LM). The lesions of the MCTSs were varied from 10 μm to 100 μm in width. In addition, the surface tension/wetting properties of PS and CM were imitated, as well as human enamel, by using a mixture of different thiols in self-assembled monolayers (SAMs). Furthermore, we investigated the infiltration and crystallization of $CaCO_3$ and HAp in the MCTSs using an infiltration setup.

2. Materials and Methods

2.1. Materials

Poly(acrylic acid) sodium salt (PAA, M_w = 8000 g/mol, 45 wt % in water, M_w = 15,000 g/mol, 35 wt % in water), calcium chloride dihydrate ($CaCl_2 \cdot 2H_2O$, ≥99%), 2-Amino-2-(hydroxylmethyl)-1,3-propanediol (Tris, ≥99,9%), 2-Propanol (iPrOH, p.a.), and 11-mercapto-1-undecanol (MUO, 97%) were purchased from Sigma Aldrich (Taufkirchen, Germany). Disodium hydrogen phosphate (Na_2HPO_4, ≥98%) and ethanol (EtOH, p.a.) were purchased from Carl Roth (Karlsruhe, Germany). Potassium chloride (KCl, 99.5%), sodium hydrogen carbonate ($NaHCO_3$, p.a.), and sodium hydroxide solution (NaOH, 0.1 N) were obtained from Merck Chemicals (Darmstadt, Germany). 1-Dodecanethiol (DDT, ≥98%) and poly(acrylic acid) (PAA, M_w = 2000 g/mol, 63 wt % in water) were purchased from Acros Organics (Geel, Belgium). The photoresist SU-8 3050 and photo developer mr-Dev 600 were purchased from Micro Resist Technology (Berlin, Germany). All chemicals were used without further purification. All experiments were carried out using double-deionized water (18.2 MΩ) using a Milli-Q Direct 8 machine from Merck Millipore (Darmstadt, Germany).

2.2. UV-Lithopgraphy of the Micro-Comb Test System (MCTS)

For the production of the MCTSs, 1 mL of photoresist was spin-coated onto a 2.5 × 2.5 cm^2 glass slide with a defined coating program (500 rpm for 10 s with a ramp-up of 100 rpm/s and then 30 s at 1400 rpm with a ramp-up of 300 rpm/s). The coated layer was heated for 45 min at 95 °C, covered with the lithography mask and exposed to UV-light (λ = 365 nm) for 7 min. Afterwards, the substrate was heated for 1 min at 65 °C and 5 min at 95 °C. The resin was placed in the photo developer and was ultrasonicated for 15 min. The structured resin was lifted off the glass, washed with iPrOH and dried with a nitrogen stream. To receive a straight infiltration border at the open end of the channels, the comb-teeth were cut to the same size using a microtome.

2.3. Surface Modification Using Self-Assembled Monolayers (SAMs)

All substrates (glass platelets and MCTSs) were rinsed with water and iPrOH, then coated with 5 nm chromium to facilitate the adhesion of the 20 nm gold layer. Both layers were deposited by thermal evaporation. The substrates were modified with freshly-prepared thiol solutions or mixtures of DDT and MUO (5 mM in EtOH each) by incubating for 12 h (Figure 1D). After washing the samples with iPrOH and drying with a nitrogen stream, contact angles where measured statically. A drop of 10 μL water was manually deposited on the surface. Drop deposition and spreading was recorded via video. For analysis, the video was replayed and stopped 5 s after drop deposition to measure the contact angle θ. The measurements where repeated 5 times to calculate the average and the standard deviation.

2.4. Synthesis and Infiltration of the CaCO3 Crystallization Solution

The solution was adapted from Wolf et al. [46]. For the $CaCO_3$ crystallization solution, a stock solution of 9.9 mM $CaCl_2$ and 10 μg/mL PAA (M_w = 2000 g/mol) was prepared and the pH was adjusted to 9.1. 1.04 mL $NaHCO_3$ (100 mM) were added dropwise to 10 mL of the stock solution while stirring, and the pH was adjusted to 9. The solution became slightly turbid. The freshly prepared $CaCO_3$ crystallization solution was infiltrated into the MCTS using the infiltration setup (Figure 1C).

2.5. Synthesis and Mineralization of the Hydroxyapatite Precursor (HApP)

The HApP was synthesized as described by Wang et al. [47]. In summary, 10 mL of a $CaCl_2$ solution (40 mM) were added slowly to 10 mL of a stirred PAA-sodium salt solution (400 mg/L). Afterwards, 10 mL Na_2HPO_4-solution (20 mM) was added to the turbid solution. The mixture was stirred for 10 min, then centrifuged for 10 min at 1000 rpm and washed with 30 mL water.

The washing steps were repeated three times in total. Prior to infiltration, the HApP was concentrated by centrifugation for 10 min at 1000 rpm.

The freshly prepared HApP was used for infiltration experiments. The infiltrated MCTS were incubated in a remineralization solution (1.5 mM $CaCl_2$, 0.9 mM Na_2HPO_4, 130 mM KCl, 60 mM Tris with different pH values, ranging from 7.4 to 11) as described by Kirkham et al. [45] for seven days. Prior to the investigation of the results using LM, PLM, SEM, and PXRD, the top glass platelet was removed, the sample was carefully cleaned with water to eliminate soluble salts, and air-dried.

2.6. Instruments

The substrates were spin-coated using an OPTIcoat SB20+ from SSE (ATM Group, Degotec GmbH, Singen, Germany). The lithography mask was manufactured by Compugraphics Jena GmbH (Jena, Germany). The UV-lithography was performed using a MJB3 Mask Aligner from Süss MicroTec AG (Garching, Germany) with a mercury vapor lamp. A Leica EMFC6 microtome (Wetzlar, Germany) was used for the fine cutting of the MCTS. For the thermal evaporation of Cr and Au, a Tectra Minicoater (Frankfurt a. M., Germany) was used. A Zeiss Stemi 2000-C (Jena, Germany) stereomicroscope with an AVC 535 Color CCD camera was used for contact angle measurements. All (polarized-)light microscopy ((P)LM) images were recorded on a Zeiss Imager M2Mm with a Zeiss AxioCam MRc3. Scanning electron microscopy (SEM) and energy-dispersive X-ray spectroscopy (EDX) were performed on a Hitachi TM 3000 instrument (Tokyo, Japan) coupled with a Quantax EDX detector from Bruker (Berlin, Germany). Powder X-ray diffraction (PXRD) measurements were performed with a D8 Discover X-Ray diffractometer (Cu, Kα) with a VANTEC-500 detector from Bruker (Karlsruhe, Germany).

3. Results and Discussion

3.1. The Micro-Comb Test System (MCTS) and the Surface Modification with Self-Assembled Monolayers (SAMs)

The investigation of the infiltration process is essential for the development of filling materials for dental applications and other restorative products as well. To facilitate this, a UV-lithography-based micro-comb test system (MCTS) was designed. The MCTSs are inexpensive, reusable, and transparent to visible light. The channels between the comb-teeth are varied from 10 to 100 μm in width and up to 10 mm in length to mimic the pores and lesions in the real systems. A gradient in the channel width was designed, as well, to investigate the influence of the lesion width on the infiltration behavior (Figure 1A). To ensure a uniform infiltration, the comb-teeth were cut to the same length using microtome cutting. It should be mentioned that this step is rather delicate due to the brittle epoxy resin and the high shearing forces perpendicular to the comb-teeth during the microtome cutting. It is desirable for the infiltration and mineralization experiments to produce MCTS with just a few microns in thickness to mime the confinement in real systems. The MCTS is placed into the newly-developed infiltration setup, shown in Figure 1B. To avoid spilling of the infiltration solution, the MCTS is closed and sealed with an upper and a bottom glass platelet ((2) and (6) in Figure 1B). This sandwich-like system is covered by a metal frame (1), which is fixed in place with four screws (4) to distribute the pressure evenly and seal the system completely. In the acrylic glass base plate (8), a reservoir (7) is provided for the infiltration solution in front of the MCTS channel entrance. The reservoir can be filled via the inlet (3) in the metal frame. Another advantage is that the reservoir enables an equal infiltration of all channels in the MCTS. The whole infiltration setup is transparent for visible light, which allows the analysis with LM and PLM in situ and in real time.

To determine the infiltration behavior of filling materials, the wettability of the substrate with the filler is a crucial factor. The MCTS is made of an epoxy resin, which results in different wetting properties than the highly hydrophilic calcium carbonate or phosphate surfaces. To overcome this problem, we modified the surface of the MCTSs as well as the upper and bottom glass platelets with self-assembled monolayers (SAMs). The formation of SAMs on gold surfaces using thiols is a well understood and robust system [48,49]. The structure of a SAM on a gold surface is shown in Figure 1C. The thiol groups bind with high affinity to the gold surface and stabilize the surface atoms [49]. The organic spacer, in this case a C_{11}-alkane chain, provides a defined thickness of the monolayer and acts as a physical protection barrier for the surface. The terminal functional groups determine the surface properties like the wettability of the SAM-covered surface. By using a mixture of different functionalized thiols, like 11-mercapto-1-undecanol (MUO) and 1-dodecanethiol (DDT), mixed SAMs are available, resulting in an average of the surface properties (Figure 1C, bottom). The wettability of a surface with a liquid phase correlates to the contact angle θ [50]. The usage of different mixtures of MUO and DDT for the formation of SAMs results in different contact angles, as shown in Figure 1D. A SAM made of pure MUO yields a highly hydrophilic surface ($\theta = 26° \pm 3°$), which is nearly similar to the wettability of PS ($\theta = 33° \pm 2°$) [40], whereas pure DDT results in a SAM with a hydrophobic surface ($\theta = 104° \pm 2°$). The mixture of different volume ratios of MUO and DDT leads to a variety of contact angles between 26° and 104°. A ratio of 70:30 (v/v) of MUO:DDT results in a SAM with an contact angle of $\theta = 58° \pm 3°$ which perfectly matches the wettability of healthy human enamel ($57° < \theta < 60°$) [50,51] and is also in good accordance with CM ($\theta = 50° \pm 4°$) [40]. Prior to infiltration experiments all MCTSs and glass platelets are functionalized with SAMs to yield a material with similar wettability to the material to be imitated.

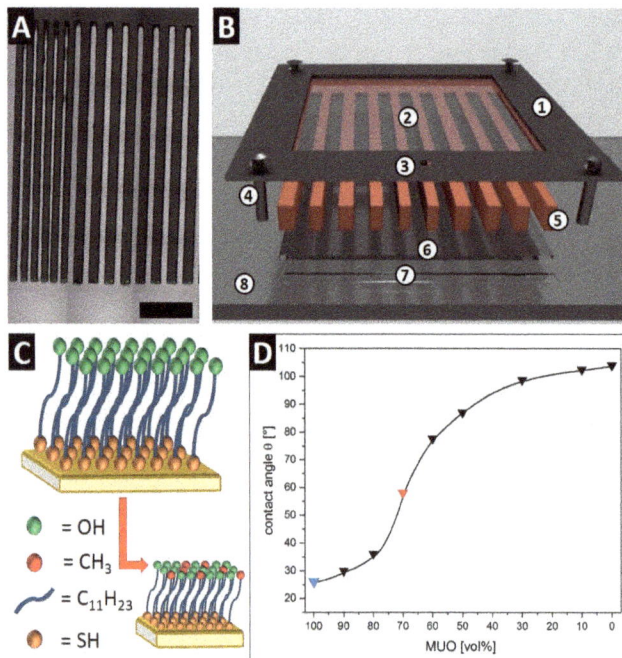

Figure 1. (**A**) LM image of a micro-comb test system (MCTS) (scale 1000 μm). (**B**) Schematic drawing of the infiltration setup with all compounds numbered: (1) metal frame; (2) top glass platelet; (3) inlet for solutions; (4) screws; (5) MCTS; (6) bottom glass platelet; (7) reservoir for infiltration solution; and (8) acrylic glass base plate. (**C**) Schematic image of a homogenous, hydroxyl-terminated self-assembled monolayer (SAM) on a gold surface (top) and a mixed SAM with 70% hydroxyl-terminated and 30% methyl-terminated thiols (bottom). (**D**) Contact angle measurements of SAM-modified, gold-coated substrates with different mixtures of MUO and DDT. The blue triangle symbolizes the pure MUO-SAM to mimic the wettability of PS. The red triangle symbolizes the thiol mixture 70/30 (v/v) to mimic the wettability of enamel and CM. The black line is the expected guideline for other mixtures.

3.2. Infiltration of a MUO-Modified MCTS with a CaCO₃ Crystallization Solution

To validate the newly-designed MCTS, the micro-comb and the glass platelets were modified, using a MUO-SAM ($\theta = 26° \pm 3°$) to mimic a realistic situation of a damaged PS sample ($\theta = 33° \pm 2°$), and infiltrated with a $CaCO_3$ crystallization solution. For the infiltration experiments, the newly-developed infiltration setup was used (Figure 1B), which enables the investigation in situ using LM. Figure 2A shows a LM image where small $CaCO_3$ particles are observable, which are transported into the channels by capillary forces after the crystallization solution was added to the dry comb channel system. The particles accumulate at the liquid-air boundary and form little $CaCO_3$ islands (Figure 2A, red circle) although the liquid meniscus is not concave as expected for aqueous solutions in a hydrophilic capillary, which indicates perturbations in our present system. The particles are also present as non-crystalline agglomerations on the channel walls (Figure 2A, blue square) suggesting that our present test system is not yet perfectly tight due to the brittle nature of the photo resist used to manufacture the comb system clamped in between the two glass plates. However, the PLM image of the same area at the same time in Figure 2B proves the crystallinity of the $CaCO_3$ particles as they show birefringence. Figure 2C,D are an overview LM (Figure 2C) and PLM (Figure 2D) image of the dried sample. These images show again the infiltrated, crystalline $CaCO_3$ particle islands inside the channels. The channels of the MCTS are not completely filled due to the low mineral content of the crystallization solution and just one application cycle. To increase the mineral content in the channels, either the mineral content in the precursor solution needs to be increased or a multiple infiltration process should be applied. A second strategy would be the decreasing of the channel size and volume. By optimizing the MCTS, more complex systems, like multi-component fillers, will be infiltrated. Nevertheless, the potential and the operating principle of the modified MCTS, combined with an infiltration setup, is demonstrated descriptively.

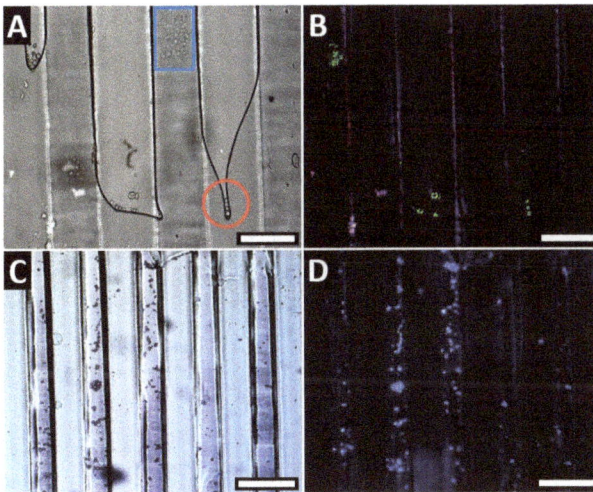

Figure 2. (**A**) In situ LM image during the infiltration of the $CaCO_3$ crystallization solution in a MUO-modified MCTS shows small $CaCO_3$ particles at the liquid-air boundary. The red circle highlights the resulting $CaCO_3$ islands and the blue square indicates non-crystalline aggregates on the channel walls (scale 50 μm); (**B**) PLM image of the area of (**A**) shows the crystallinity of the $CaCO_3$ particles in green (scale 50 μm); (**C**) LM image after drying in air shows the $CaCO_3$ particles inside the channels (scale 100 μm); and (**D**) PLM image of the area of (**C**) indicates the crystallinity of the particles in the channel system by their birefringence (scale 100 μm).

3.3. Infiltration and Crystallization of a Hydroxyapatite Precursor (HApP)

Dental caries was described as a second important system often affected by acid-caused mineral loss. The demineralization of dental hard tissue starts at the outer enamel layer, which is mainly composed of HAp. To mimic the surface wettability of human enamel, the MCTS and the glass platelets were modified using a thiol mixture of MUO and DDT in a 70:30 (v/v) ratio (Figure 1D, red triangle). To simulate real conditions, a hydroxyapatite precursor (HApP) was synthesized and used as a potential filling material. The synthesis of the HApP was described by Wang et al. [47]. We modified the synthesis by utilizing higher molecular weights of PAA (M_w = 8000 g/mol and 15,000 g/mol) to increase the complexation kinetics of the PAA [52]. The freshly-prepared HApP was infiltrated into a MCTS with a channel width of 50 μm using the infiltration setup shown in Figure 1B. The infiltrated MCTS was stored in a remineralization solution for seven days [45]. The LM image of an infiltrated MCTS shows the almost completely filled channels (dark areas in Figure 3A). The PLM image (Figure 3B) of the same area proves the crystallinity of the calcium phosphate (CaP) filling material. The SEM image in Figure 3C shows the successful infiltration and filling of several channels with a depth of at least 500 μm—the same dimension as deep human enamel lesions. The filling mineral has an unordered structure, made up of small particles, which fused together partially (Figure 3D). The PXRD of the mineralized HApP after incubation for seven days in the remineralization solution at pH 7.4 displays the crystallization to HAp (Figure 4). The broad reflexes indicate the presence of nano-crystalline hydroxyapatite, which was further confirmed by the calculation of the crystallite size of 26.3 nm utilizing the Scherrer equation [53,54].

At the moment the system is limited to small lesions and cavities due to the straightforward synthesis procedure, the used photo resin and the spin coating program In the future we aim to increase the diversity of the MCTS and the precursor solutions as well. Furthermore, we want to adapt the system for more complex systems, like a mixture of fillers.

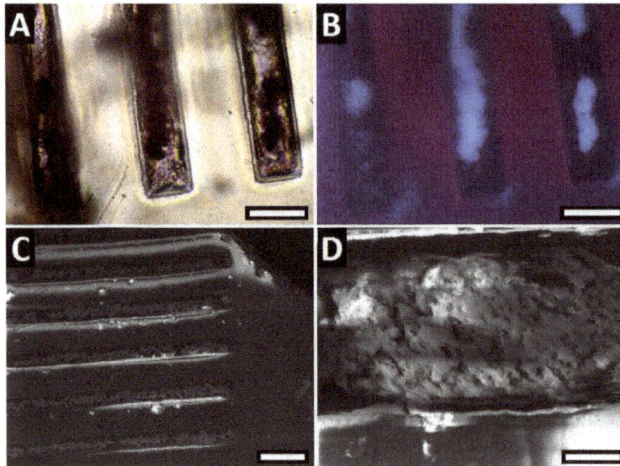

Figure 3. (**A**) LM image of an HApP-infiltrated MCTS, after an incubation time of seven days in the remineralization solution showing CaP-filled channels (dark area, scale 50 μm); (**B**) PLM image of (**A**) shows the crystallinity of the CaP-filling material (blue) in the channels (scale 50 μm); (**C**) SEM overview image of several CaP-filled MCTS channels (scale 100 μm); and (**D**) SEM image shows the partially fused CaP particles in one channel (scale 10 μm).

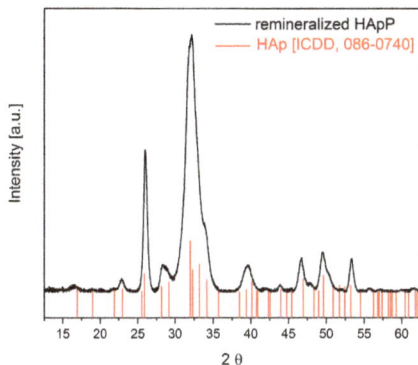

Figure 4. PXRD pattern of a crystallized HApP after incubation in the remineralization solution for seven days (pH 7.4).

4. Conclusions

We have developed a new micro-comb test system (MCTS) for the investigation of the infiltration and crystallization behavior of filling materials for any kind of porous systems in situ. The MCTSs are inexpensive and reusable and provide a very significant advantage over natural-based test systems, like artificial tooth lesions or aged stone samples, because they are transparent and the infiltration processes can, consequently, be directly observed by optical microscopy. Therefore, the MCTS is suitable to visually investigate important infiltration processes into micrometer-sized pores. By modifying the surface of the epoxy resin-based MCTS with a self-assembled monolayer (SAM) of MUO or a mixed SAM, composed of 70% MUO and 30% DDT (v/v), we were able to mimic the wettability of Pietra Serena (PS), as well as Carrara marble (CM) and human enamel. Beyond that, all contact angles between 26° and 104° are available by the appropriate mixture (see Figure 1D). Moreover, we demonstrated the potential of this powerful and straightforward test system by infiltrating a $CaCO_3$ crystallization solution and a hydroxyapatite precursor (HApP). It is possible to investigate the depth of infiltration and mineralization in situ and in real-time.

For focusing the complete filling of the MCTSs with the mineral precursors to realize a complete remineralization of human enamel lesions or cavities and pores in historical artifacts, we aim to increase the mineral content of the precursor solutions. Furthermore, the effect of a multiple-step infiltration will be investigated. A second part of further experiments will be the improvement of the MCTS by changing the production parameters to decrease the layer thickness and the channel sizes as well.

In the future, the possibility to vary the channel size of the MCTS and the surface properties by modification with different SAMs will enable the adaptation of the MCTS, combined with the infiltration setup, to several additional areas of applications, like micro-fissures in steel or high-performance plastics. Therefore, our developed MCTS provides a quantitative tool to observe and explain several important infiltration processes into different pores.

Acknowledgments: The authors thank the Deutsche Forschungsgemeinschaft for the framework of the Collaborative Research Center SFB-1214 "Anisotropic Particles as Building Blocks" and the included "Particle Analysis Center" (project Z1) for the PXRD measurements. We also thank Brigitte Bössenecker, Elana Harbalik, and Kay Hagedorn for their help in PXRD measurements and the calculation of the crystallite sizes. Furthermore, we want to thank Julian Opel for the graphical design of the infiltration setup.

Author Contributions: Helmut Cölfen, Stefan L. P. Wolf, and Dominik Gruber conceived and designed the experiments; Dominik Gruber, Stefan L. P. Wolf, Andra-Lisa M. Hoyt, and Julian P. Konsek performed the experiments and analyzed the data; and Dominik Gruber, Stefan L. P. Wolf, and Helmut Cölfen wrote the paper.

Conflicts of Interest: The authors declare no conflict of interest.

References

1. Cölfen, H.; Mann, S. Higher-order organization by mesoscale self-assembly and transformation of hybrid nanostructures. *Angew. Chem. Int. Ed.* **2003**, *42*, 2350–2365. [CrossRef] [PubMed]
2. Christenson, K.H. Confinement effects on freezing and melting. *J. Phys. Condens. Matter* **2001**, *13*, R95. [CrossRef]
3. Wang, Y.-W.; Christenson, H.K.; Meldrum, F.C. Confinement leads to control over calcium sulfate polymorph. *Adv. Funct. Mater.* **2013**, *23*, 5615–5623. [CrossRef]
4. Cantaert, B.; Beniash, E.; Meldrum, F.C. Nanoscale confinement controls the crystallization of calcium phosphate: Relevance to bone formation. *Chem. Eur. J.* **2013**, *19*, 14918–14924. [CrossRef] [PubMed]
5. Wolf, S.L.P.; Jahme, K.; Gebauer, D. Synergy of Mg^{2+} and poly(aspartic acid) in additive-controlled calcium carbonate precipitation. *CrystEngComm* **2015**, *17*, 6857–6862. [CrossRef]
6. Kellermeier, M.; Picker, A.; Kempter, A.; Cölfen, H.; Gebauer, D. A straightforward treatment of activity in aqueous $CaCO_3$ solutions and the consequences for nucleation theory. *Adv. Mater.* **2014**, *26*, 752–757. [CrossRef] [PubMed]
7. Tester, C.C.; Brock, R.E.; Wu, C.-H.; Krejci, M.R.; Weigand, S.; Joester, D. In vitro synthesis and stabilization of amorphous calcium carbonate (ACC) nanoparticles within liposomes. *CrystEngComm* **2011**, *13*, 3975–3978. [CrossRef]
8. Xu, A.-W.; Ma, Y.; Cölfen, H. Biomimetic mineralization. *J. Mater. Chem.* **2007**, *17*, 415–449. [CrossRef]
9. Picker, A.; Nicoleau, L.; Nonat, A.; Labbez, C.; Cölfen, H. Identification of binding peptides on calcium silicate hydrate: A novel view on cement additives. *Adv. Mater.* **2014**, *26*, 1135–1140. [CrossRef] [PubMed]
10. Kim, Y.-Y.; Hetherington, N.B.J.; Noel, E.H.; Kröger, R.; Charnock, J.M.; Christenson, H.K.; Meldrum, F.C. Capillarity creates single-crystal calcite nanowires from amorphous calcium carbonate. *Angew. Chem. Int. Ed.* **2011**, *50*, 12572–12577. [CrossRef] [PubMed]
11. Hamilton, B.D.; Ha, J.-M.; Hillmyer, M.A.; Ward, M.D. Manipulating crystal growth and polymorphism by confinement in nanoscale crystallization chambers. *Acc. Chem. Res.* **2012**, *45*, 414–423. [CrossRef] [PubMed]
12. Finnemore, A.S.; Scherer, M.R.J.; Langford, R.; Mahajan, S.; Ludwigs, S.; Meldrum, F.C.; Steiner, U. Nanostructured calcite single crystals with gyroid morphologies. *Adv. Mater.* **2009**, *21*, 3928–3932. [CrossRef]
13. Campbell, J.M.; Meldrum, F.C.; Christenson, H.K. Observing the formation of ice and organic crystals in active sites. *Proc. Natl. Acad. Sci. USA* **2017**, *114*, 810–815. [CrossRef] [PubMed]
14. Diao, Y.; Myerson, A.S.; Hatton, T.A.; Trout, B.L. Surface design for controlled crystallization: The role of surface chemistry and nanoscale pores in heterogeneous nucleation. *Langmuir* **2011**, *27*, 5324–5334. [CrossRef] [PubMed]
15. Ha, J.-M.; Wolf, J.H.; Hillmyer, M.A.; Ward, M.D. Polymorph selectivity under nanoscopic confinement. *J. Am. Chem. Soc.* **2004**, *126*, 3382–3383. [CrossRef] [PubMed]
16. Stephens, C.J.; Ladden, S.F.; Meldrum, F.C.; Christenson, H.K. Amorphous calcium carbonate is stabilized in confinement. *Adv. Funct. Mater.* **2010**, *20*, 2108–2115. [CrossRef]
17. Stephens, C.J.; Kim, Y.-Y.; Evans, S.D.; Meldrum, F.C.; Christenson, H.K. Early stages of crystallization of calcium carbonate revealed in picoliter droplets. *J. Am. Chem. Soc.* **2011**, *133*, 5210–5213. [CrossRef] [PubMed]
18. Wang, Y.-W.; Christenson, H.K.; Meldrum, F.C. Confinement increases the lifetimes of hydroxyapatite precursors. *Chem. Mater.* **2014**, *26*, 5830–5838. [CrossRef]
19. Gardeniers, H.J.G.E. Chemistry in nanochannel confinement. *Anal. Bioanal. Chem.* **2009**, *394*, 385–397. [CrossRef] [PubMed]
20. Markl, G. *Minerale und Gesteine*; Springer: Berlin/Heidelberg, Germany, 2015; Volume 3.
21. Fratini, F.; Pecchioni, E.; Cantisani, E.; Rescic, S.; Vettori, S. Pietra serena: The stone of the renaissance. *Geol. Soc. Lond. Spec. Publ.* **2014**, *407*. [CrossRef]
22. Baglioni, P.; Chelazzi, D. *Nanoscience for the Conservation of Works of Art*; RSC Publishing: Cambridge, UK, 2013.
23. Karagiannis, N.; Karoglou, M.; Bakolas, A.; Moropoulou, A. *New Approaches to Building Pathology and Durability*; Springer: Singapore, 2016.
24. Ambrosi, M.; Dei, L.; Giorgi, R.; Neto, C.; Baglioni, P. Colloidal particles of $Ca(OH)_2$: Properties and applications to restoration of frescoes. *Langmuir* **2001**, *17*, 4251–4255. [CrossRef]

25. Giorgi, R.; Dei, L.; Baglioni, P. A new method for consolidating wall paintings based on dispersions of lime in alcohol. *Stud. Conserv.* **2000**, *45*, 154–161.

26. Otero, J.; Charola, A.E.; Grissom, C.A.; Starinieri, V. An overview of nanolime as a consolidation method for calcareous substrates. *Ge-Conservación* **2016**, *11*, 71–78.

27. Kassebaum, N.J.; Bernabé, E.; Dahiya, M.; Bhandari, B.; Murray, C.J.L.; Marcenes, W. Global burden of untreated caries. *J. Dent. Res.* **2015**, *94*, 650–658. [CrossRef] [PubMed]

28. Bagramian, R.A.; Garcia-Godoy, F.; Volpe, A.R. The global increase in dental caries. A pending public health crisis. *Am. J. Dent.* **2009**, *22*, 3–8. [PubMed]

29. Selwitz, R.H.; Ismail, A.I.; Pitts, N.B. Dental caries. *Lancet* **2007**, *369*, 51–59. [CrossRef]

30. Palmer, L.C.; Newcomb, C.J.; Kaltz, S.R.; Spoerke, E.D.; Stupp, S.I. Biomimetic systems for hydroxyapatite mineralization inspired by bone and enamel. *Chem. Rev.* **2008**, *108*, 4754–4783. [CrossRef] [PubMed]

31. Dorozhkin, S.V.; Epple, M. Biological and medical significance of calcium phosphates. *Angew. Chem. Int. Ed.* **2002**, *41*, 3130–3146. [CrossRef]

32. Ten Cate, J.M. Remineralization of deep enamel dentine caries lesions. *Aust. Dent. J.* **2008**, *53*, 281–285. [CrossRef] [PubMed]

33. Ginebra, M.P.; Espanol, M.; Montufar, E.B.; Perez, R.A.; Mestres, G. New processing approaches in calcium phosphate cements and their applications in regenerative medicine. *Acta Biomater.* **2010**, *6*, 2863–2873. [CrossRef] [PubMed]

34. Suchanek, W.; Yoshimura, M. Processing and properties of hydroxyapatite-based biomaterials for use as hard tissue replacement implants. *J. Mater. Res.* **1998**, *13*, 94–117. [CrossRef]

35. Yamagishi, K.; Onuma, K.; Suzuki, T.; Okada, F.; Tagami, J.; Otsuki, M.; Senawangse, P. Materials chemistry: A synthetic enamel for rapid tooth repair. *Nature* **2005**, *433*, 819. [CrossRef] [PubMed]

36. Hannig, M.; Hannig, C. Nanomaterials in preventive dentistry. *Nat. Nanotech.* **2010**, *5*, 565–569. [CrossRef] [PubMed]

37. Denry, I.; Kelly, J.R. Emerging ceramic-based materials for dentistry. *J. Dent. Res.* **2014**, *93*, 1235–1242. [CrossRef] [PubMed]

38. Li, J.; Xie, X.; Wang, Y.; Yin, W.; Antoun, J.S.; Farella, M.; Mei, L. Long-term remineralizing effect of casein phosphopeptide-amorphous calcium phosphate (CPP-ACP) on early caries lesions in vivo: A systematic review. *J. Dent.* **2014**, *42*, 769–777. [CrossRef] [PubMed]

39. Reynolds, E.C. Remineralization of enamel subsurface lesions by casein phosphopeptide-stabilized calcium phosphate solutions. *J. Dent. Res.* **1997**, *76*, 1587–1595. [CrossRef] [PubMed]

40. Dei, L.; Mauro, M.; Baglioni, P.; Manganelli Del Fà, C.; Fratini, F. Growth of crystal phases in porous media. *Langmuir* **1999**, *15*, 8915–8922. [CrossRef]

41. Freire-Lista, D.M.; Gomez-Villalba, L.S.; Fort, R. Microcracking of granite feldspar during thermal artificial processes. *Period Min.* **2015**, *84*, 519–537.

42. Bortz, S.; Stecich, J.; Wonneberger, B.; Chin, I. Accelerated weathering in building stone. *Int. J. Rock Mech. Min. Sci.* **1993**, *30*, 1559–1562. [CrossRef]

43. Besinis, A.; De Peralta, T.; Tredwin, C.J.; Handy, R.D. Review of nanomaterials in dentistry: Interactions with the oral microenvironment, clinical applications, hazards, and benefits. *ACS Nano* **2015**, *9*, 2255–2289. [CrossRef] [PubMed]

44. Burwell, A.K.; Thula-Mata, T.; Gower, L.B.; Habeliz, S.; Kurylo, M.; Ho, S.P.; Chien, Y.C.; Cheng, J.; Cheng, N.F.; Gansky, S.A.; et al. Functional remineralization of dentin lesions using polymer-induced liquid-precursor process. *PLoS ONE* **2012**, *7*, e38852. [CrossRef] [PubMed]

45. Kirkham, J.; Firth, A.; Vernals, D.; Boden, N.; Robinson, C.; Shore, R.C.; Brookes, S.J.; Aggeli, A. Self-assembling peptide scaffolds promote enamel remineralization. *J. Dent. Res.* **2007**, *86*, 426–430. [CrossRef] [PubMed]

46. Wolf, S.L.P.; Caballero, L.; Melo, F.; Cölfen, H. Gel-like calcium carbonate precursors observed by in situ AFM. *Langmuir* **2017**, *33*, 158–163. [CrossRef] [PubMed]

47. Wang, P.; Yook, S.-W.; Jun, S.-H.; Li, Y.-L.; Kim, M.; Kim, H.-E.; Koh, Y.-H. Synthesis of nanoporous calcium phosphate spheres using poly(acrylic acid) as a structuring unit. *Mater. Lett.* **2009**, *63*, 1207–1209. [CrossRef]

48. Bain, C.D.; Troughton, E.B.; Tao, Y.T.; Evall, J.; Whitesides, G.M.; Nuzzo, R.G. Formation of monolayer films by the spontaneous assembly of organic thiols from solution onto gold. *J. Am. Chem. Soc.* **1989**, *111*, 321–335. [CrossRef]

49. Love, J.C.; Estroff, L.A.; Kriebel, J.K.; Nuzzo, R.G.; Whitesides, G.M. Self-assembled monolayers of thiolates on metals as a form of nanotechnology. *Chem. Rev.* **2005**, *105*, 1103–1170. [CrossRef] [PubMed]
50. Wu, W.; Nancollas, G.H. The relationship between surface free-energy and kinetics in the mineralization and demineralization of dental hard tissue. *Adv. Dent. Res.* **1997**, *11*, 566–575. [CrossRef] [PubMed]
51. De Jong, H.P.; van Pelt, A.W.J.; Arends, J. Contact angle measurements on human enamel—An in vitro study of influence of pellicle and storage period. *J. Dent. Res.* **1982**, *61*, 11–13. [CrossRef] [PubMed]
52. Huang, S.-C.; Naka, K.; Chujo, Y. A carbonate controlled-addition method for amorphous calcium carbonate spheres stabilized by poly(acrylic acid)s. *Langmuir* **2007**, *23*, 12086–12095. [CrossRef] [PubMed]
53. Drouet, C. Apatite formation: Why it may not work as planned, and how to conclusively identify apatite compounds. *BioMed Res. Int.* **2013**, *2013*, 12. [CrossRef] [PubMed]
54. Scherrer, P. Bestimmung der größe und der inneren struktur von kolloidteilchen mittels röntgenstrahlen. *Nachr. Ges. Wiss. Gött. Math.-Phys. Kl.* **1918**, *2*, 98.

minerals

MDPI

Article

Physicochemical and Additive Controls on the Multistep Precipitation Pathway of Gypsum

Mercedes Ossorio [1], Tomasz M. Stawski [2,*], Juan Diego Rodríguez-Blanco [3], Mike Sleutel [4], Juan Manuel García-Ruiz [1], Liane G. Benning [2,5,6] and Alexander E. S. Van Driessche [1,7,*]

[1] LEC, IACT, CSIC, Universidad de Granada, 18100 Granada, Spain; mercedes@lec.csic.es (M.O.); juanmanuel.garcia@csic.es (J.M.G.-R.)
[2] GFZ German Research Centre for Geosciences, Telegrafenberg, 14473 Potsdam, Germany; benning@gfz-potsdam.de
[3] Department of Geology, Museum Building, Trinity College Dublin, 2 Dublin, Ireland; j.d.rodriguez-blanco@tcd.ie
[4] Structural Biology Brussels, Flanders Interuniversity Institute for Biotechnology and Vrije Universiteit Brussel, 1050 Elsene, Belgium; msleutel@vub.ac.be
[5] Department of Earth Sciences, Free University of Berlin, 12249 Berlin, Germany
[6] School of Earth and Environment, University of Leeds, LS2 9JT Leeds, UK
[7] University Grenoble-Alpes, CNRS, ISTerre, F-38000 Grenoble, France
* Correspondence: stawski@gfz-potsdam.de (T.M.S.); alexander.van-driessche@univ-grenoble-alpes.fr (A.E.S.V.D.); Tel.: +33-4-7663-5194 (A.E.S.V.D.)

Received: 6 July 2017; Accepted: 3 August 2017; Published: 9 August 2017

Abstract: Synchrotron-based small- and wide-angle X-ray scattering (SAXS/WAXS) was used to examine in situ the precipitation of gypsum ($CaSO_4 \cdot 2H_2O$) from solution. We determined the role of (I) supersaturation, (II) temperature and (III) additives (Mg^{2+} and citric acid) on the precipitation mechanism and rate of gypsum. Detailed analysis of the SAXS data showed that for all tested supersaturations and temperatures the same nucleation pathway was maintained, i.e., formation of primary particles that aggregate and transform/re-organize into gypsum. In the presence of Mg^{2+} more primary particle are formed compared to the pure experiment, but the onset of their transformation/reorganization was slowed down. Citrate reduces the formation of primary particles resulting in a longer induction time of gypsum formation. Based on the WAXS data we determined that the precipitation rate of gypsum increased 5-fold from 4 to 40 °C, which results in an effective activation energy of ~30 kJ·mol^{-1}. Mg^{2+} reduces the precipitation rate of gypsum by more than half, most likely by blocking the attachment sites of the growth units, while citric acid only weakly hampers the growth of gypsum by lowering the effective supersaturation. In short, our results show that the nucleation mechanism is independent of the solution conditions and that Mg^{2+} and citric acid influence differently the nucleation pathway and growth kinetics of gypsum. These insights are key for further improving our ability to control the crystallization process of calcium sulphate.

Keywords: nucleation; multistep pathway; crystal growth; gypsum; additives

1. Introduction

Three crystalline phases are known in the $CaSO_4$–H_2O system, distinguishable by their degree of hydration: gypsum ($CaSO_4 \cdot 2H_2O$), bassanite ($CaSO_4 \cdot 0.5H_2O$), and anhydrite ($CaSO_4$). Natural calcium sulphate deposits, predominantly composed of gypsum and anhydrite, are found throughout the geological record [1], and these minerals are an important source material for the building industry (e.g., plaster of Paris) [2]. Furthermore, calcium sulphate minerals are among the most common scalants [3], causing serious reduction in the efficiency of for example desalination plants [4]. Due to their prominent role in both natural and industrial environments, the precipitation behaviour of $CaSO_4$ phases has been a much-studied topic (e.g., [5]). Until recently, the formation of gypsum from solution (the stable phase < 60 °C) was assumed to proceed via a single-step pathway (e.g., [6]). But, new experimental evidence has revealed that gypsum precipitation does not proceed via a direct nucleation pathway. Instead, a variety of multi-step pathways have been proposed including amorphous calcium sulphate and bassanite as possible precursor phases [7–10]. However, all these studies are mainly based on ex situ characterization of the precipitation reaction (e.g., electron microscopic imaging combined with quenching techniques to isolate the precipitates from solution for their subsequent characterization). But, the inherent instability of the precursor phases in solution and the rapid kinetics of crystallization make it difficult to define the early-stages of the precipitation of calcium sulphate. Thus, in order to fully characterize the precipitation products and kinetics, and to eliminate possible quenching artifacts, the application of in situ monitoring of the reaction is paramount.

Synchrotron-based time-resolved X-ray scattering is a powerful method for in situ characterization of the reaction products of solution based precipitation processes [11]. When the measured angular range is further extended to wide-angles (WAXS), the diffraction patterns of crystalline materials can be recorded allowing identification, and quantification of the formed phases and their evolution. In a previous study, using in situ SAXS/WAXS, we observed the initial stages of calcium sulphate nucleation and found that nanosized primary particles are formed immediately after a supersaturated solution is created. These particles aggregate and transform/re-organize into gypsum [12]. Building upon this initial study, we again used in situ SAXS/WAXS to follow the precipitation behavior of calcium sulphate phase(s) over a broad range of temperatures (4–40 °C) and initial calcium sulphate concentrations (50–150 mM). In addition, the influence of two inhibitors (Mg^{2+} and citric acid) on the precipitation behavior was studied. SAXS provided us with information on the precursor stage of the reaction, while reaction rates and the activation energy for calcium sulphate precipitation were derived from the WAXS data.

2. Materials and Methods

Three different calcium sulphate precipitation scenarios were designed to obtain a representative set of data: (I) The role of supersaturation was tested (at 21 °C) by carrying out experiments at four initial calcium sulphate concentrations (50, 75, 100 and 150 mmol·L^{-1} $CaSO_4$, solutions were obtained by mixing equimolar stock solutions of Na_2SO_4 and $CaCl_2$). (II) The influence of temperature on the precipitation behavior was analyzed by conducting experiments at five temperatures (4, 12, 21, 30, 40 °C) for a 50 mmol·L^{-1} $CaSO_4$ solution. (III) Finally, the effect of additives on the precipitation of calcium sulphate was studied for 50, 75, 100 and 150 mmol·L^{-1} $CaSO_4$ solutions at 21 °C. One additive, Mg, which enhances the solubility of $CaSO_4$, was introduced, replacing the Na^+ ions with Mg^{2+} ions in stoichiometric conditions ($MgSO_4 \cdot 7H_2O$ and $CaCl_2 \cdot 2H_2O$ were used as reactants, and a 1:1 Ca^{2+}/Mg^{2+} ratio was maintained for all cases). The second additive that was used, citric acid ($C_6H_8O_7$), is known to retard the nucleation/growth kinetics of calcium sulphate [13,14], and its effect was tested by adding 2.6 mmol·L^{-1} $C_6H_8O_7$ to the $CaSO_4$ solutions.

The initial solution saturation index, SI_{gyp}, for each experimental condition was calculated using the PHREEQC code [15] with the minteq.v4 database (Table 1). In the case of the experiments with inhibitors, their effect on the solubility of gypsum was taken into account.

Table 1. Initial solution saturation index of gypsum, SI_{gyp} calculated with PHREEQC using the minteq.v4 database.

[CaSO$_4$] (mmol·L^{-1})	[Mg^{2+}] (mmol·L^{-1})	[C$_6$H$_8$O$_7$] (mmol·L^{-1})	T (°C)	SI_{gyp}
50			4	0.59
50			12	0.56
50			21	0.53
50			30	0.49
50			40	0.46
75			21	0.75
75	75		21	0.67
75		2.6	21	0.74
100			21	0.90
100	100		21	0.83
100		2.6	21	0.9
150			21	1.13
150	150		21	1.05
150		2.6	21	1.13

2.1. Turbidity Experiments

Previous to small- and wide-angle X-ray scattering (SAXS/WAXS) experiments extensive testing of the experimental conditions was undertaken. Making use of a UV-vis spectrophotometer (Uvikon XL, SCHOTT Instruments, Mainz, Germany) the precipitation process was monitored measuring the change in the ratio of incident light/transmitted light ($\approx$turbidity) passing through the solution over time. Changes in turbidity of the solution were followed at $\lambda = 520$ nm for all experimental conditions and turbidity values were recorded every 0.017 s. Two experimental setups were employed: (I) equimolar solutions of the reactants, CaCl$_2$·2H$_2$O (Sigma, pure) and Na$_2$SO$_4$ (Sigma, >99%), were mixed in polystyrene cuvettes and stirred at a constant rate of 350 rpm, (II) 50 mL of each reactant (1:1 ratio) were added to a reaction beaker, positioned next to the spectrophotometer, and continuously stirred at 350 rpm. Immediately after the reactants were mixed the resulting solution was continuously circulated between the reaction beaker and a flow-through cuvette located inside of the spectrophotometer, using a peristaltic pump. The temperature of the beaker was maintained constant using a temperature controlled water bath and reactants were pre-heated before mixing. The cuvettes inside of the spectrophotometer where located inside of a temperature controlled holder.

2.2. Synchrotron In Situ Small- and Wide-Angle X-ray Scattering

In situ SAXS/WAXS measurements were conducted at the I22 beamline of the Diamond Light Source (UK). In Figure 1, a schematic representation of the experimental setup employed at the synchrotron facility is shown.

First, 40 mL of a CaCl$_2$·2H$_2$O solution stirred (350 rpm) in a reaction beaker was circulated using a peristaltic pump through a thin-walled capillary (GLAS, quartz glass capillaries, 1.5 mm diameter) aligned with the incoming X-ray beam. Then, 40 mL of a Na$_2$SO$_4$ solution (with or without additive) was injected by means of remotely controlled stopped-flow device (BioLogic SFM400, Bio-Logic Science Instruments, Seyssinet-Pariset, France). Reactants were preheated before mixing and the temperature during the precipitation reaction was controlled in the same way as in the off-line experiments.

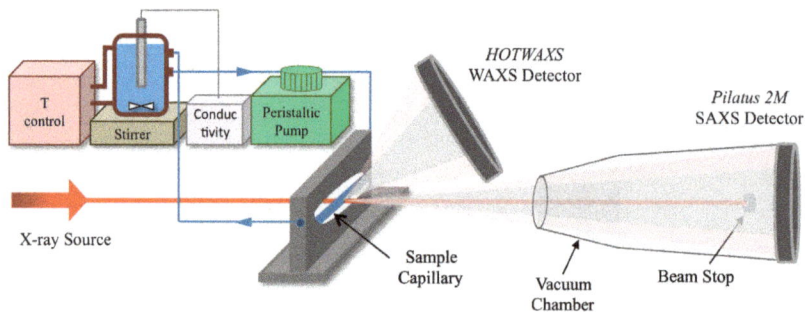

Figure 1. Experimental setup employed in synchrotron facility.

Samples flowing (pump rate ~10 mL/s) through the capillary were measured in transmission mode, using a monochromatic X-ray beam at 12.4 keV, and two-dimensional scattered intensities were collected at small angles with a Pilatus 2M (2D large area pixel-array detector, Dectris AG, Baden-Daettwil, Switzerland) and at wide angles with a HOTWAXS detector (a photo-counting 1D microstrip gas chamber detector). Transmission was measured with a photodiode installed in the beam-stopper of the SAXS detector. A sample-to-detector distance of 4.22 m was used, and the q-range at small angles was calibrated using silver behenate [16] and dry collagen standards [17], while for the WAXS detector a highly crystalline Si standard (NIST 640c) was used.

SAXS/WAXS measurements were started just before mixing of the reactants, and lasted for a maximum of 4 h/experiment, collecting 5–60 s/scan. The triggering of SAXS and WAXS frame acquisition was synchronized between the two detectors so that a given frame in SAXS corresponded to the one in WAXS. The scattered intensity was calibrated to absolute units using water as a reference. Diffraction patterns of commercial $CaSO_4 \cdot 2H_2O$ (Sigma Aldrich, $\geq$99%, St. Louis, MO, USA) were collected and used as standard for comparison with the reaction products. Prior to each new experimental run, a pattern of the capillary filled with water was taken in order to ensure that all $CaSO_4$ was removed from the previous experimental run.

SAXS and WAXS Data Processing

Typically the recorded 2D SAXS patterns were circular in shape, and thus independent of the in-plane azimuthal angle with respect to the detector, showing that the investigated systems could be considered isotropic. For those isotropic patterns, pixels corresponding to similar q regardless of their azimuthal angle where averaged together, reducing the 2D patterns to 1D curves. SAXS data processing and reduction included masking of undesired pixels, normalizations and correction for transmission, background subtraction and data integration to 1D. These steps were performed using the Data Analysis WorkbeNch (DAWN) software package (v. 1.3 and 1.4) according to I22 guidelines [18]. A more detailed account of the SAXS analysis strategy can be found in our previous works [12,19].

The initial WAXS data reduction consisted in the subtraction of the background (pattern of the cell filled with water or filled with a $CaCl_2 \cdot 2H_2O$ solution) from each of the datasets. Typically water and diluted salt solutions produced the same background characteristics. Subsequently, all the datasets were normalized with respect to the intensity of the incident beam (i.e., the incoming beam measured before the sample).

The diffraction patterns of commercial gypsum powder and those from the solid phase formed from a 50 mmol·L^{-1} $CaSO_4$ solution at 21 °C after 3.3 h of reaction (i.e., the final stage of the experiment) were compared and the reflections of both solid phases were identified as gypsum using the CIF PDF2: 33–311 (Figure S1). Diffraction patterns obtained during precipitation were added (maintaining the same proportion of added frames for the total course of the experiment) for each dataset in order to increase the signal-to-noise ratio. Whole-pattern-fitting was applied, using XFit Koalariet [20], to obtain time-resolved information about the precipitation of gypsum. Reaction rates of precipitation, k, were derived for each experimental condition by fitting the experimental data with a mathematical expression extracted from the kinetic model of Kolmogorov-Johnson-Mehl-Avrami (KJMA), commonly referred to as the Avrami equation:

$$\alpha = 1 - e^{-(k(t - t_{ind}))^n} \tag{1}$$

where α is the time-resolved peak area (i.e., the degree of crystallization of the precipitate at time t), k is the rate constant, t_{ind} is the induction time of reaction, and n is a parameter that varies according to the nucleation and growth mechanism defined within the KJMA kinetic model. This model applies to specific solid-state phase transformations. However, it has also been used to fit other nucleation-and-growth processes exhibiting sigmoidal conversion kinetics, such as crystal formation from solution and melts [21,22]. For solid-state phase transformations, the value of n is usually related to the mechanism of nucleation and the dimensionality of growth. But, since the application of the KJMA model to precipitation from solution is nonrigorous, no physical interpretation of the fitted values of this exponent is done. Similarly, k lacks a true physical definition for our system, but it is still useful for comparing growth rates among similar precipitation processes, and allows us to estimate the activation barrier for crystal growth.

The determination of the reaction rate k was subject to the adjustment of the induction time, t_{ind}, and of the time exponent n to fixed values in Equation (1) for fitting of the experimental data; t_{ind} for each experimental condition was taken as the time at which the WAXS signal started to develop. The time exponent n was obtained through the fitting of the experimental data with Equation (1), with n as an unfixed parameter. For all the experimental series, the obtained values of n were approximate 1, thus for the calculation of the reaction rate n was fixed at 1.

For the experiments (50 mmol·L^{-1} CaSO$_4$) performed at different temperatures (4–40 °C) the activation energy (E_A) for CaSO$_4$ formation was derived using the Arrhenius's equation:

$$k = A \cdot exp^{-E_A/RT} \tag{2}$$

where k is the reaction rate estimated from our experimental data, A is an pre-exponential factor, R is the ideal gas constant, and T is the temperature. WAXS patterns were also used to estimate the crystallite size of gypsum, but on average particles sizes were outside the application range of the Scherrer equation (For details see Supplementary Section S2, Tables S1 and S2, Figure S2 and [23]).

2.3. AFM Experiments

In situ atomic force microscopy (AFM) observations were carried out to characterize the surface topography of growing {010} faces of gypsum in the presence of Mg^{2+} and Citric Acid. Measurements were done with a commercial microscope equipped with a closed liquid cell with an inlet and outlet (Nanoscope IIIa, Bruker, Billerica, MA, USA). AFM images were obtained in Tapping mode using silicon cantilevers (Tap-300, BudgetSensors, Sofia, Bulgaria).

Freshly cleaved gypsum samples were immobilized using vacuum grease on a glass disc glued to a metal puck and mounted onto the AFM scanner. The gypsum surface was first observed in air in order to locate areas with low edge density. Thereafter the fluid cell was filled with a supersaturated CaSO$_4$ solution (with and without additives) using a syringe. Directly after injection of a solution the evolution of the surface topography was monitored over time. In order to avoid evaporation effects during the measurements, the solution of the liquid cell was changed every 5 to 10 min. During imaging, the temperature inside of the liquid-cell was ~30 ± 1 °C.

3. Results

3.1. The Initial Stages of Gypsum Formation

For all experimental conditions, diffraction peaks started to develop in the time-resolved WAXS patterns after a certain induction time. The length of this period is a function of supersaturation, temperature and the presence/concentration of additives (Figures 2 and 3). Regardless of the induction time derived from the WAXS signal, our time-resolved SAXS patterns indicated that as soon as a supersaturated solution was created, i.e., after mixing of both stock solutions, the presence of small scatterers beyond the size of ions/ion pairs (Figure 2) were immediately observed. This is in line with the precipitation pathway of gypsum discussed in our previous works [12,19], where the following stages leading to the crystallization of gypsum from a 50 mmol·L^{-1} CaSO$_4$ solution (at 12 °C) are distinguished: (I) formation of nanosized (<3 nm) primary species composed of Ca–SO$_4$ cores, (II) assembly of these primary species into loose domains (the interparticle distance is larger than 2 times the radius of gyration of the primary species), (III) densification of the loose domains into aggregates of primary species, and finally (IV) growth and reorganization of the primary units within the aggregates. Simultaneously with the onset of stage IV, diffraction peaks start to appear in the WAXS patterns, indicating that gypsum is being formed.

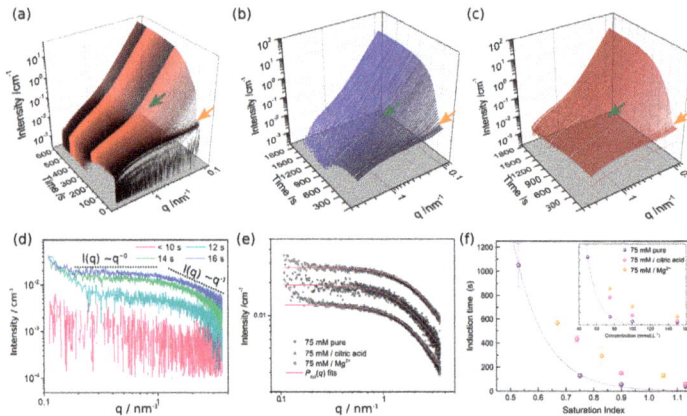

Figure 2. Time-resolved SAXS patterns during gypsum precipitation at 21 °C from: (**a**) a 75 mmol·L^{-1} CaSO$_4$ solution, (**b**) a 75 mmol·L^{-1} CaSO$_4$ solution with 75 mmol·L^{-1} of Mg^{2+} and, (**c**) a 75 mmol·L^{-1} CaSO$_4$ solution with 2.6 mmol·L^{-1} citric acid. Orange and green arrows indicate the onset of aggregation and transformation of primary species, respectively. (**d**) SAXS patterns during the mixing of 150 mmol·L^{-1} stock solutions of Na$_2$SO$_4$ and CaCl$_2$. (**e**) SAXS patterns after 30 s of mixing, fitted using a cylindrical form factor [12]. (**f**) Induction times for gypsum formation as a function of supersaturation, for pure and additive containing solutions. The inset shows the induction time as a function of initial CaSO$_4$ concentration.

In Figure 2a time-resolved SAXS patterns (1 s time-frames) of a 75 mmol·L^{-1} CaSO$_4$ solution at 21 °C are shown. SAXS/WAXS measurements were triggered when the Na$_2$SO$_4$ solution was injected (~15 s) into the CaCl$_2$ solution and thus the first patterns show scattering from the solution followed by a rapid increase in intensity and the formation of scattering features evidenced through $I(q) \propto q^{-1}$ for $q > \sim1.0$ nm^{-1} and $I(q) \propto q^0$ for $q < \sim1.0$ nm^{-1} (Figure 3d). Almost simultaneously with their formation, these primary species start to aggregate forming larger scattering features (denoted by an increase of the intensity at $q < \sim1.0$ nm^{-1} leading to $I(q) \propto q^{-4}$ (orange arrow Figure 2a). Eventually, after ~140 s the primary particles start to grow/rearrange (green arrow Figure 2a). This moment coincides with the appearance of the first diffraction peaks in the WAXS patterns, and thus the formation of gypsum.

For the experiments at 4 °C no proper SAXS analysis was possible, due to technical problems with the background subtraction and data normalization.

For the experiments with additives, i.e., 75 mmol·L^{-1} CaSO$_4$ + citric acid/Mg^{2+} at 21 °C, the general precipitation pathway described above remains valid (Figure 3b,c), but some noteworthy differences exist with respect to the additive free experiments. Figure 2e shows the scattering pattern for the three different types of 75 mmol·L^{-1} solutions at 30 s. At these early stages we already observe a small increase in intensity at $q > 0.3$ nm^{-1} indicating the onset of the aggregation processes. However, because the increase was rather minor we obtained the best fit using a cylindrical form factor [12], from which the following geometrical parameters were extracted for the primary species: pure 75 mmol·L^{-1} CaSO$_4$, $R = 0.22$ nm and $L = 2.55$ nm; 75 mmol·L^{-1} CaSO$_4$ + Mg^{2+}, $R = 0.24$ nm and $L = 2.03$ nm; 75 mmol·L^{-1} CaSO$_4$ + citric acid, $R = 0.26$ nm and $L = 2.45$ nm. Since a similar size of the primary species was obtained for all conditions, we can use direct extrapolation of the scaling parameter of the form factor (Figure 2e) towards $q = 0$ as an approximate indication of the volume fraction (i.e., concentration) of the primary species, assuming that the density of these particles did not change in the presence of additives. Thus, the higher the intensity of the plateau $I(q) \propto q^0$ for $0.3 < q < {\sim}0.4$ nm^{-1}, the more primary species are present in the solution. For precipitation in the presence of 2.6 mmol·L^{-1} of citric acid, the intensity of the plateau is significantly lower than in the pure case (0.0125 cm^{-1} vs. 0.0191 cm^{-1}, Figure 2e). This directly indicates that fewer primary species are formed when citric acid is present in the solution. In addition, the onset of stage IV is delayed with respect to the pure case, from ~140 to ~450 s (green arrows Figure 2a,b,f). On the other hand when Mg^{2+} is present, we find that the intensity of the plateau $I(q) \propto q^0$ for $q < {\sim}1$ nm^{-1}, is considerably higher than that in the pure case (0.0274 cm^{-1} vs. 0.0191 cm^{-1}, Figure 3e). Notwithstanding the higher concentration of primary species the onset of stage IV (i.e., transformation/rearrangement of the primary species) is delayed compared to the pure case (green arrows Figure 2a,c,f) from ~140 s to ~575 s.

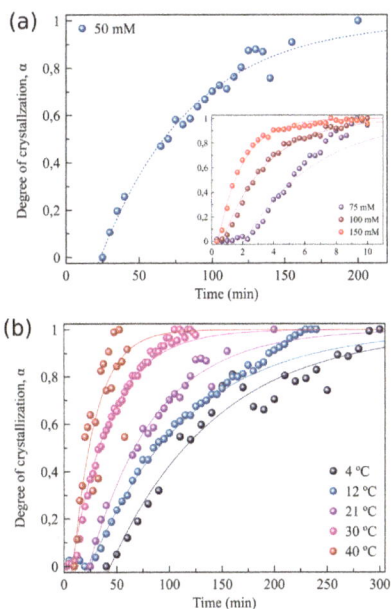

Figure 3. *Cont.*

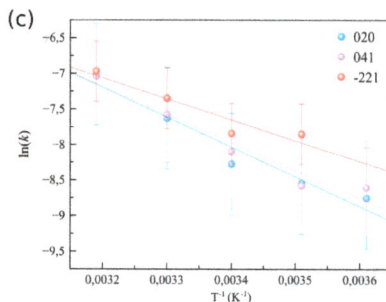

Figure 3. Precipitation kinetics of gypsum as function of initial $CaSO_4$ concentration and temperature. (a) Evolution of the normalized peak area α of the (020) reflection of gypsum for 50–150 mmol·L^{-1} $CaSO_4$ solutions at 21 °C, (b) α evolution for 50 mmol·L^{-1} $CaSO_4$ solutions from 4 to 40 °C. Solid lines correspond to fittings of the experimental data using Equation (1). (c) $\ln(k)$ versus $(1/T)$ for a 50 mmol·L^{-1} $CaSO_4$ solution corresponding to three reflections of gypsum: (020), (041) and (−221).

Thus, for all the tested conditions in this study, our SAXS data indicate the presence of nanosized particles, <3 nm, prior to the formation of crystalline gypsum, conforming our previously reported results [12]. When precipitation dynamics are slower, i.e., 50 mmol·L^{-1} $CaSO_4$ and 12 °C, the pathway leading to gypsum formation can be divided into 4 stages as was mentioned before. While, at faster reaction rates, i.e., >12 °C and $\geq$50 mmol·L^{-1} $CaSO_4$, the time-resolved SAXS patterns revealed that stages I to III are merged into one continuous stage, although the nanosized primary species can be recognized in the SAXS patterns until the onset of stage IV. Thus, for the range of experimental conditions explored in this work, neither temperature nor the initial $CaSO_4$ concentration alter the intrinsic characteristics of the pathway; only the temporal persistence of the different pre-gypsum stages is a function of these parameters. Likewise, the presence of additives in the solution does also not alter the pathway. But, citric acid reduces the number of primary particles (Figure 2e), which eventually leads to a delay in the onset of stage IV (Figure 2b). This seems to indicate that the volume fraction of primary particles plays a key role in determining the induction period of gypsum formation. Contrary to citric acid, the presence of Mg^{2+} increases the concentration of primary species and also delays the onset of stage IV. This suggests that Mg^{2+} ions/complexes hamper the transformation/re-organization of the primary particles needed for the crystallization of gypsum. Despite these insights gained from our in situ experiments, the molecular mechanisms by which citric acid and Mg^{2+} reduce/promote primary particle formation is not clear yet, and will be the subject of a forthcoming detailed study.

3.2. The Effect of Supersaturation and Temperature on the Crystallization Kinetics

Gypsum was the only crystalline phase that could be identified from the time-resolved WAXS patterns. Even though SAXS data indicated the presence of nanoparticles prior to the formation of gypsum as well as large aggregates of those particles (>100 nm), neither these nano-sized scatterers nor their aggregates led to any characteristic peaks in the WAXS patterns. Hence, WAXS data do not allow us to reveal the precise nature of this precursor phase. As such, the WAXS data were used to study the crystallization kinetics of gypsum as a function of (i) initial calcium sulphate concentration (from 50 to 150 mmol·L^{-1} $CaSO_4$) at 21 °C and (ii) as a function of temperature (4–40 °C) for a 50 mmol·L^{-1} $CaSO_4$ solution. Figure 3a shows the evolution of the normalized peak area for the (020) reflection of gypsum as a function of reaction time for different concentrations. As expected, the precipitation rate of gypsum increases with increasing $CaSO_4$ concentration. This same trend is found for the (041) and (−221) reflections. From all three reflections the precipitation rate was derived through fitting the experimental data using Equation (1) and a summary of the results is shown in Supplementary Section S3 (Tables S3–S6).

The second series of experiments probed the precipitation kinetics of a 50 mmol·L^{-1} CaSO$_4$ solution at five different temperatures (4, 12, 21, 30, and 40 °C). Figure 3b shows the evolution of the normalized peak area as a function of the reaction time for the (020) reflection. Despite the fact that the supersaturation with respect to gypsum slightly decreases with increasing temperature, from SI_{Gp} = 0.59 at 4 °C down to 0.46 at 40 °C, the precipitation rate significantly increased at higher temperatures (from 1.58 × 10^{-4} s^{-1} at 4 °C to 8.98 × 10^{-4} s^{-1} at 40 °C, Figures 3c and S3, Table S3). This is also reflected in the t_{ind} for gypsum formation, which decreased with increasing temperature, from ~2000 s at 4 °C to ~200 s at 40 °C (Supplementary Section S3, Table S3).

By applying the Arrhenius equation (Equation (2)) the effective activation energies, E_A, for gypsum crystallization were determined considering the reaction rates k obtained from the three main gypsum reflections, (020), (041) and (−221), observed in the WAXS patterns. The resulting effective E_A for gypsum crystallization ranged between ~25–35 kJ·mol^{-1}, depending on the reflection of gypsum chosen for its calculation (Figure 3c and Supplementary Section S4). These values are approximately half of the previously reported activation energies determined by different methods (Table 2). Previous studies mainly used seeded experiments, monitoring the growth of micrometer-sized crystal seeds, where growth was controlled by the advancement of single steps on the different faces due to ion-by-ion addition to kink sites. In our study growth kinetics were recorded from the very beginning of the formation of gypsum, which, based on our SAXS data, are formed by the transformation/re-organization of primary particles in the already existing aggregates. It is generally assumed that desolvation/rearrangement of water molecules is the dominant barrier upon incorporation of ions into a crystal (e.g., [24]). Hence, if we take into account that the reorganization/transformation of the precursor phase is dominant during the early stages of gypsum growth, and that the ions inside of this precursor phase are already partially desolvated, compared to dissolved ions, it follows that during the early stages the activation barrier for growth should be lower than during the latter stages of gypsum growth that will be controlled by the attachment rate of fully solvated Ca^{2+} and SO$_4$$^{2-}$ ions. This hypothesis is supported by the activation energy obtained from in situ measurements of the advancement rate of steps on the {010} face of gypsum. This study showed that steps grow by the incorporation of ions and resulted in an effective activation barrier of ~70 kJ·mol^{-1} [25]. A similar observation was made for the growth of the {120} face [26], leading to an effective barrier of ~55–63 kJ·mol^{-1}.

Table 2. An overview of activation energies E_A for gypsum reported in the literature.

Reference	E_A/kJ·mol^{-1}	Observations
[27]	62.8	Crystallization of gypsum on the addition of seed crystals to stable supersaturated solutions was studied from 15–45 °C.
[28]	62.8	Conductometric study of seeded crystallization at four temperatures and several ionic compositions
[29]	58.6	Growth of gypsum seed crystals from supersaturated solutions from 60–105 °C.
[30]	60.0	A suspension of gypsum crystals was grown from electrolyte solution using the constant composition technique
[31]	184.2	Potentiometric measurements in supersaturated solutions of calcium sulphate.
[32]	46.0–67.0	The spontaneous precipitation of calcium sulphate in supersaturated solutions over the temperature range between 25.0 and 80.0 °C was investigated by monitoring the solution specific conductivity during desupersaturation.
[26]	55.0–63.0	Experimental data of the growth rate of the (120) face from electrolytic solutions on a heated metal surface.
[25]	70.7	Step kinetics on the {010} face were measured as a function of supersaturation at different temperatures
This work	25.0–35.0	(020) E_A = 35.2 ± 0.6 kJ·mol^{-1}; (041) E_A = 33.1 ± 0.6 kJ·mol^{-1}; (−221) E_A = 24.9 ± 0.7 kJ·mol^{-1}

3.3. Crystallization Kinetics in the Presence of Additives

We extended our study also to explore the precipitation kinetics of gypsum in the presence of two additives at 21 °C: citric acid (2.6 mmol·L^{-1}, 75 to 150 mmol·L^{-1} CaSO$_4$) and Mg^{2+} (75 to 150 mmol·L^{-1} Mg^{2+} for 75 to 150 mmol·L^{-1} CaSO$_4$). In Figure 4a the evolution of the normalized peak area α of the (020) reflection of gypsum is plotted for the three experimental conditions. Both additives significantly slowed down the formation of gypsum, which is evidenced by the lower reaction rates (Figure 4b) and the longer induction times (Figure 3f) for all studied supersaturations (SI_{Gyp} was calculated taking into account the additives, Table 1).

The plot of k versus SI_{Gyp} (Figure 4b) indicates that the effect of Mg^{2+} on the reaction rate was more pronounced than that of citric acid. For all precipitation scenarios the dependence of the reaction rate on supersaturation is reasonably well fitted using a linear dependence (Figure 4b). From those fits, different trends for each additive can be inferred:

(I) In the case of citric acid, the slope of the linear fit remains constant with respect to that of the pure case, but the intersection at $k = 0$ is shifted to a significantly higher relative supersaturation value ~1.0. These very similar slopes indicate that citrate does not reduce the growth kinetics of gypsum through directly impeding crystal growth, but rather through the lowering of the effective driving force [33]. This occurs when for example additive molecules are incorporated into the growing crystals. Typically, these molecules will distort the crystal structure and thereby increase the internal energy of the solid through an enthalpic contribution [33]. The resulting increase in free energy is manifested as an increase in the effective solubility (K_{sp}) of the crystal, leading to a lower supersaturation. This will lead to a shift of the growth curves to higher equilibrium activities and the growth velocity will always remain below that of the pure system at the same supersaturation value. However, the unchanged slope of the reaction rate versus saturation index (SI) implies that there is no direct impact on the attachment kinetics at the attachment sites [33]. This is further corroborated by in situ AFM experiments performed on growing {010} faces of gypsum crystals in the presence of 2.6 mmol·L^{-1} citric acid. No significant change in the step morphology is observed compared to those of steps growing from a pure solution (Figure 5a,b), i.e., citrate molecules do not induce step pinning. Similar observations were made by Bosbach and Hochella [34] for Na-citrate, who did not find any pinning of growth steps on the {010} face of gypsum. They argued that the retarding effect of Na-citrate on the precipitation of gypsum is caused by the formation of complexes between the –COO$^-$ groups and the Ca^{2+} present in solution. But based on PHREEQC calculation this complexation is only minor for citric acid and cannot account for the observed reduction in the growth rate (in Figure 4b, supersaturation was calculated taking into account this complexation).

(II) For the Mg^{2+} experiments, the slope of the reaction rate dependence on supersaturation is strongly reduced compared to that of the pure case, but the intersection at $k = 0$ occurs at roughly the same value (relative supersaturation ~0.35, supersaturation was calculated based on the K_{sp} of gypsum in the presence of Mg^{2+}). This pronounced change in the slope indicates that Mg^{2+} ions/complexes lower the growth kinetics of gypsum, probably through the temporarily blocking of kink sites (e.g., [33]). From IC analyses (Supplementary Section S5) it can be inferred that no significant quantities of Mg^{2+} are removed from solution during gypsum precipitation, and thus but Mg^{2+} does not get incorporated into the crystal structure. This is further corroborated by a previous work, which studied the induction times of gypsum formation in the presence of different ions, and showed that Mg^{2+} only adsorbs to the newly forming surfaces of the growing gypsum crystals [35]. In situ AFM observations show that the presence of Mg^{2+} ions in the solution does not change the morphology of step on growing {010} faces of gypsum compared to the those formed in pure solutions (Figure 5a,c). This also concurs with the hypothesis that these ions mainly block kink sites (for kink blocking no obvious step pinning is expected (e.g., [33])).

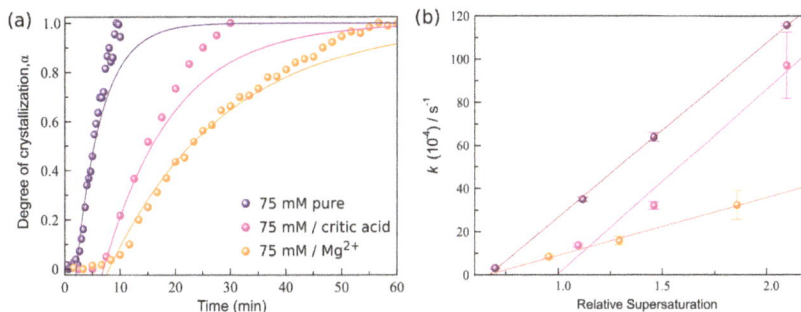

Figure 4. The effect of Mg^{2+} and citric acid on the precipitation of gypsum at 21 °C. (**a**) Evolution of the normalized peak area α of the (020) reflection of gypsum growth from a pure 75 mmol·L^{-1} CaSO$_4$ solution (purple dots), in the presence of 75 mmol·L^{-1} Mg^{2+} (orange dots) and 2.6 mmol·L^{-1} citric acid (pink dots). (**b**) Averaged reaction rates as a function of the relative supersaturation (IAP/K$_{sp}$).

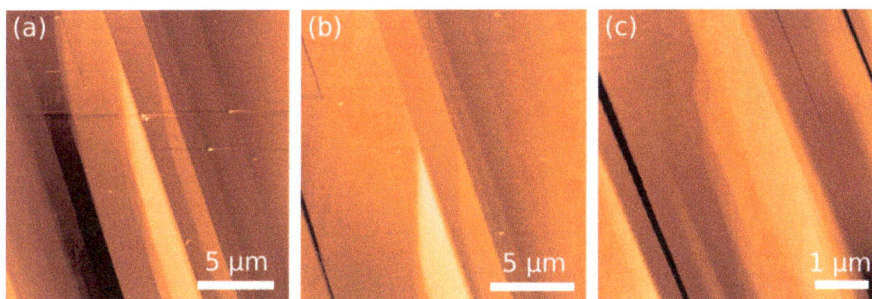

Figure 5. In situ AFM height deflection images of steps on the {020} face of gypsum crystals growing from (**a**) pure 50 mmol·L^{-1} CaSO$_4$ solutions, (**b**) 50 mmol·L^{-1} CaSO$_4$ solutions with 2.6 mmol·L^{-1} Citric Acid and (**c**) 50 mmol·L^{-1} CaSO$_4$ solutions with 50 mmol·L^{-1} Mg^{2+}.

4. Concluding Remarks

Our SAXS data confirm the presence of precursor CaSO$_4$ species in supersaturated gypsum solutions, which aggregate and eventually transform/re-organize leading to the formation of gypsum for a broad range of supersaturations and temperatures. These experimental observations demonstrate the generality of our previously reported nucleation pathway [12]. Moreover, simultaneously collected WAXS data indicate that the reaction rate for the precipitation of gypsum increases with increasing supersaturation and temperature. The estimated activation energy for gypsum precipitation from solution accounts for ~30 kJ·mol^{-1}. This value is lower than previously reported values and seems to indicate that gypsum growth during the early stages is partially controlled by a particle-based mechanism. During the latter stages ion-by-ion addition would then become the dominant mechanism.

Importantly, in this work we have also shown that two types of additives, Mg^{2+} and citric acid, increase and decreases, respectively, the concentration of precursor particles through which the formation of gypsum is delayed. Both additives also influenced the growth of gypsum: Mg^{2+} hampered crystal growth most likely by directly blocking kink sites, while citric acid seems to reduce the effective supersaturation during gypsum growth. But, despite the insights obtained on the role of additives during gypsum precipitation in this work, the precise molecular retardation mechanisms of these additives are still unclear and further experiments are warranted.

Supplementary Materials: The following are available online at www.mdpi.com/2075-163X/7/8/140/s1, Figure S1: Diffraction patterns of commercial gypsum (gray) used as standard, together with a diffraction pattern (blue) of the solid phase formed from a 50 mmol·L^{-1} CaSO$_4$ solution at 21 °C after 3.3 h of reaction. The analyzed reflections of gypsum (black; pattern PDF2: 33–311) are indicated, Figure S2: Relation between peak broadening and crystallite size for two of the main reflections of gypsum; (020) at $2\theta = 11.59°$, and (021) at $2\theta = 20.72°$. The variations in the slope (°/nm) of these curves with increasing crystallite size are also indicated (in red). Figure S3: Variation in the concentration of Ca^{2+} and Mg^{2+} ions analyzed by IC (dots), during the off-line precipitation from a 50 mmol·L^{-1} CaSO$_4$ solution at 21 °C, Table S1: Estimation of the approximate errors associated to crystallite sizes from 10 to 85 nm, Table S2: Estimation of the crystallite size for the (020) reflection of gypsum, Table S3: 50 mmol·L^{-1} CaSO$_4$·2H$_2$O. Precipitation from pure aqueous solution, Table S4: 75 mmol·L^{-1} CaSO$_4$·2H$_2$O, at 21 °C. Precipitation in pure aqueous solution and in presence of additives in solution, Table S5: 100 mmol·L^{-1} CaSO$_4$·2H$_2$O, at 21 °C. Precipitation in pure aqueous solution and in presence of additives in solution, Table S6: 150 mmol·L^{-1} CaSO$_4$·2H$_2$O, at 21 °C. Precipitation from a pure solution and in the presence of additives.

Acknowledgments: This work has been carried out within the framework of the project CGL2010-16882 of the Spanish MINECO. Mercedes Ossorio acknowledges the JAE-Predoc fellowship. This work was partially supported by a Marie Curie grant from the European Commission in the framework of the MINSC ITN (Initial Training Research network), Project No. 290040. We thank Diamond Light Source for access to beamline I22 through a grant to Liane G. Benning, Alexander E.S. Van Driessche, Mercedes Ossorio and Juan Diego Rodríguez-Blanco This research was partially made possible by Marie Curie grant from the European Commission in the framework of NanoSiAl Individual Fellowship, Project No. 703015 to Tomasz Stawski. Tomasz Stawski and Liane G. Benning also acknowledge the financial support of the Helmholtz Recruiting Initiative. Mike Sleutel acknowledges financial support by the FWO under project G0H5316N.

Author Contributions: A.E.S.V.D., L.G.B. and J.M.G.-R. conceived and designed the experiments; M.O., J.D.R.-B., M.S., A.E.S.V.D. and L.G.B. performed the experiments; M.O. and T.M.S. analyzed the data; M.O., A.E.S.V.D. and T.M.S. wrote the paper; all authors discussed the results and commented on the manuscript.

Conflicts of Interest: The authors declare no conflict of interest.

References

1. Warren, J.K. *Evaporites: Sediments, Resources and Hydrocarbons*; Springer: Berlin, Germany, 2006.
2. Sharpe, R.; Cork, G. Gypsum and Anhydrite. In *Industrial Minerals & Rocks*, 7th ed.; Kogel, J.E., Ed.; Society for Mining, Metallurgy and Exploration Inc.: Englewood, CO, USA, 2006; p. 519.
3. Stumm, W.; Morgan, J.J. *Aquatic Chemistry: Chemical Equilibria and Rates in Natural Waters*, 3rd ed.; Wiley: New York, NY, USA, 1995.
4. Mi, B.; Elimelech, M. Gypsum scaling and cleaning in forward osmosis: Measurements and mechanisms. *Environ. Sci. Technol.* **2010**, *44*, 2022–2028. [CrossRef] [PubMed]
5. Van Driessche, A.E.S.; Stawski, T.M.; Benning, L.G.; Kellermeier, M. *New Perspectives on Mineral Nucleation and Growth*; Van Driessche, A.E.S., Kellermeier, M., Benning, L.G., Gebauer, D., Eds.; Springer: Berlin, Germany, 2017; pp. 227–256.
6. Singh, N.B.; Middendorf, B. Calcium sulphate hemihydrate hydration leading to gypsum crystallization Prog. *Cryst. Growth Charact. Mater.* **2007**, *53*, 57–77. [CrossRef]
7. Wang, Y.W.; Kim, Y.Y.; Christenson, H.K.; Meldrum, F.C. A new precipitation pathway for calcium sulfate dihydrate (gypsum) via amorphous and hemihydrate intermediates. *Chem. Commun.* **2012**, *48*, 504–506. [CrossRef] [PubMed]
8. Van Driessche, A.E.S.; Benning, L.G.; Rodríguez-Blanco, J.D.; Ossorio, M.; Bots, P.; García-Ruiz, J.M. The Role and Implications of Bassanite as a Stable Precursor Phase to Gypsum Precipitation. *Science* **2012**, *336*, 69–72. [CrossRef] [PubMed]
9. Jones, F. Infrared investigation of barite and gypsum crystallization: Evidence for an amorphous to crystalline transition. *Cryst. Eng. Commun.* **2012**, *14*, 8374–8381. [CrossRef]
10. Saha, A.; Lee, J.; Pancera, S.M.; Bräeu, M.F.; Kempter, A.; Tripathi, A.; Bose, A. New Insights into the transformation of calcium sulfate hemihydrate to gypsum using time-resolved cryogenic transmission electron microscopy. *Langmuir* **2012**, *28*, 11182–11187. [CrossRef] [PubMed]
11. Stawski, T.M.; Benning, L.G. SAXS in Inorganic and Bioispired Research. *Methods Enzymol.* **2013**, *532*, 95–127. [PubMed]

12. Stawski, T.M.; Van Driessche, A.E.S.; Ossorio, M.; Rodriguez-Blanco, J.D.; Besselink, R.; Benning, L.G. Formation of calcium sulfate through the aggregation of sub-3 nanometre primary species. *Nat. Commun.* **2016**, *7*, 11177. [CrossRef] [PubMed]

13. Tadros, M.E.; Mayes, I.J. Linear growth rates of calcium sulfate dihydrate crystals in the presence of additives. *J. Colloid Inter. Sci.* **1979**, *72*, 245–254. [CrossRef]

14. Prisciandaro, M.; Lancia, A.; Musmarra, D. The Retarding Effect of Citric Acid on Calcium Sulfate Nucleation Kinetics. *Ind. Eng. Chem. Res.* **2003**, *42*, 6647–6652. [CrossRef]

15. Parkhurst, D.L.; Appelo, C.A.J. *User's Guide to PHREEQC: A Computer Program for Speciation, Reaction-Path, 1D-Transport, and Inverse Geochemical Calculations*; Water-Resources Investigations Report 99-4259; U.S. Geological Survey: Denver, CO, USA, 1999.

16. Huang, T.C.; Toraya, H.; Blanton, T.N.; Wu, Y. X-ray Powder Diffraction Analysis of Silver Behenate, a Possible Low-Angle Diffraction Standard. *J. Appl. Crystallogr.* **1993**, *26*, 180–184. [CrossRef]

17. Fratzl, P.; Misof, K.; Zizak, I.; Rapp, G.; Amenitsch, H.; Bernstorff, S. Fibrillar structure and mechanical properties of collagen. *J. Struct. Biol.* **1998**, *122*, 119–122. [CrossRef] [PubMed]

18. Smith, A. Software Manuals for SAXS. Available online: http://confluence.diamond.ac.uk/display/SCATTERWEB/Software+Manuals+for+SAXS (accessed on 3 November 2013).

19. Besselink, R.; Stawski, T.M.; Van Driessche, A.E.S.; Benning, L.G. Not just fractal surfaces, but surface fractal aggregates: Derivation of the expression for the structure factor and its applications. *J. Chem. Phys.* **2016**, *145*, 211908. [CrossRef]

20. Cheary, R.W.; Coelho, A.A. *Programs XFIT and FOURYA, Deposited in CCP14 Powder Diffraction Library*; Engineering and Physical Sciences Research Council: Swindon, UK; Daresbury Laboratory: Warrington, UK, 1996.

21. Yee, N.; Shaw, S.; Benning, L.G.; Nguyen, T.H. The rate of ferrihydrite transformation to goethite via the Fe (II) pathway. *Am. Mineral.* **2006**, *91*, 92–96. [CrossRef]

22. Wang, F.; Richards, V.N.; Shields, S.P.; Buhro, W.E. Kinetics and Mechanisms of Aggregative Nanocrystal Growth. *Chem. Mater.* **2014**, *26*, 5–21. [CrossRef]

23. Liebrecht, L.J.H. Sol-Gel Derived Barium Titanate Thin Films. Maste's Thesis, University of Twente, Enschede, The Netherlands, 2006.

24. Chernov, A.A. *Modern Crystallography III, Crystal Growth*; Springer: Berlin, Germany, 1984.

25. Van Driessche, A.E.S.; García-Ruiz, J.M.; Delgado-López, J.M.; Sazaki, G. In Situ Observation of Step Dynamics on Gypsum Crystals. *Cryst. Growth Des.* **2010**, *10*, 3909–3916. [CrossRef]

26. Linnikov, O.D. Investigation of the initial period of sulphate scale formation Part 2. Kinetics of calcium sulphate crystal growth at its crystallization on a heat-exchange surface. *Desalination* **2000**, *128*, 35–46. [CrossRef]

27. Liu, S.; Nancollas, G.H. The kinetics of crystal growth of calcium sulfate dihydrate. *J. Cryst. Growth* **1970**, *6*, 281–289. [CrossRef]

28. Smith, B.R.; Sweett, F. The crystallization of calcium sulfate dehydrate. *J. Colloid Inter. Sci.* **1971**, *37*, 612–618. [CrossRef]

29. Nancollas, G.H.; Reddy, M.M.; Tsai, F. Calcium sulfate dihydrate crystal growth in aqueous solution at elevated temperatures. *J. Cryst. Growth* **1973**, *20*, 125–134. [CrossRef]

30. Witkamp, G.J.; Van der Eerden, J.P.; Van Rosmalen, G.M. Growth of gypsum: I. Kinetics. *J. Cryst. Growth* **1990**, *120*, 281–289. [CrossRef]

31. Klepetsanis, P.G.; Koutsoukos, P.G. Spontaneous precipitation of calcium sulfate at conditions of sustained supersaturation. *J. Colloid Inter. Sci.* **1991**, *143*, 299–308. [CrossRef]

32. Klepetsanis, P.G.; Dalas, E.; Koutsoukos, P.G. Role of Temperature in the Spontaneous Precipitation of Calcium Sulfate Dihydrate. *Langmuir* **1999**, *15*, 1534–1540. [CrossRef]

33. De Yoreo, J.J.; Vekilov, P.G. Principles of Crystal Nucleation and Growth. *Mineral. Soc. Am.* **2003**, *54*, 57–93. [CrossRef]

34. Bosbach, D.; Hochella, M.F. Gypsum growth in the presence of growth inhibitors: A scanning force microscopy study. *Chem. Geol.* **1996**, *132*, 227–236. [CrossRef]
35. Rabizadeh, T.; Stawski, T.M.; Morgan, D.J.; Peacock, C.L.; Benning, L.G. The effects of inorganic additives on the nucleation and growth kinetics of calcium sulfate dihydrate crystals. *Cryst. Growth Des.* **2017**, *17*, 582–589. [CrossRef]

Article

Structural Transition of Inorganic Silica–Carbonate Composites Towards Curved Lifelike Morphologies

Julian Opel [1,2], Matthias Kellermeier [3], Annika Sickinger [1], Juan Morales [2,4], Helmut Cölfen [1,*] and Juan-Manuel García-Ruiz [2,*]

[1] Physical Chemistry, University of Konstanz, Universitätsstrasse 10, D-78457 Konstanz, Germany; julian.opel@uni-konstanz.de (J.O.); annika.sickinger@uni.kn (A.S.)
[2] Laboratorio de Estudios Cristalográficos, Instituto Andaluz de Ciencias de la Tierra (CSIC-UGR), Avenida de las Palmeras N° 4, E-18100 Armilla, Granada, Spain; jmoraless@udec.cl
[3] Material Physics, BASF SE, RAA/OS–B007, Carl-Bosch-Strasse 38, D-67056 Ludwigshafen, Germany; matthias.kellermeier@basf.com
[4] Instituto de Geología Económica Aplicada (GEA), Universidad de Concepción, 4030000 Concepción, Chile
* Correspondence: helmut.coelfen@uni-konstanz.de (H.C.); jmgruiz@ugr.es (J.-M.G.-R.); Tel.: +49-7531-884063 (H.C.); +34-958-230000 (J.-M.G.-R.)

Received: 2 February 2018; Accepted: 14 February 2018; Published: 18 February 2018

Abstract: The self-assembly of alkaline earth carbonates in the presence of silica at high pH leads to a unique class of composite materials displaying a broad variety of self-assembled superstructures with complex morphologies. A detailed understanding of the formation process of these purely inorganic architectures is crucial for their implications in the context of primitive life detection as well as for their use in the synthesis of advanced biomimetic materials. Recently, great efforts have been made to gain insight into the molecular mechanisms driving self-assembly in these systems, resulting in a consistent model for morphogenesis at ambient conditions. In the present work, we build on this knowledge and investigate the influence of temperature, supersaturation, and an added multivalent cation as parameters by which the shape of the forming superstructures can be controlled. In particular, we focus on trumpet- and coral-like structures which quantitatively replace the well-characterised sheets and worm-like braids at elevated temperature and in the presence of additional ions, respectively. The observed morphological changes are discussed in light of the recently proposed formation mechanism with the aim to ultimately understand and control the major physicochemical factors governing the self-assembly process.

Keywords: biomorphs; barium carbonate; silica; self-assembly; temperature; precipitation kinetics

1. Introduction

Silica biomorphs are an interesting type of self-assembling inorganic–inorganic composite material with remarkably complex architectures. They form in silica-rich solutions at high pH in the presence of alkaline earth metal cations like barium, strontium, and calcium under ambient conditions [1–6]. Upon diffusion of atmospheric carbon dioxide into the system, slow crystallisation of carbonate under the influence of dissolved silicate species results in the spontaneous formation of unusual ultrastructures that consist of uniform elongated carbonate nanocrystals (approximately 200–300 nm long), which maintain long-range co-orientation and are interspersed by certain amounts of amorphous silica; the whole structure is surrounded by a silica skin (typical Si/Ba ratios are 0.1–0.3) [4,7]. On the multimicron scale, these assemblies display intricate morphologies such as flat sheets, helicoids, or tightly wound worms, as shown in Figure 1. While most of the work on silica biomorphs has been carried out with witherite ($BaCO_3$) as the carbonate phase [4], there is increasing evidence that similar structures can also be obtained with other carbonates, including orthorhombic analogues like strontianite (SrCO$_3$) [8,9] or aragonite (CaCO$_3$) [10–12].

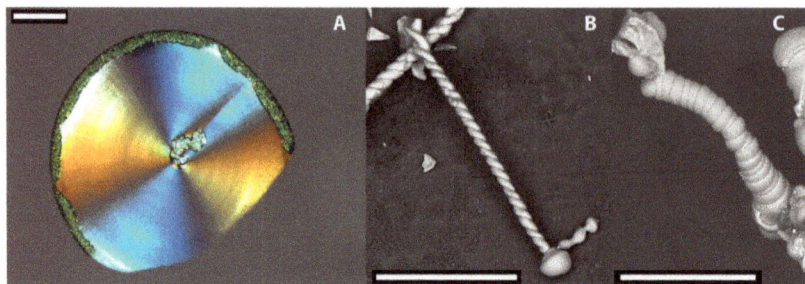

Figure 1. Polarised light microscopy (PLM) and SEM images of the most common architectures displayed by silica biomorph structures grown at room temperature from solutions containing 8.9 mM silica and 5 mM barium chloride: (**A**) flat sheet, (**B**) regular helicoid, (**C**) tightly wound worm. Scale bars are 100 μm.

The mechanisms enabling the structural and morphological complexity observed in this simple inorganic system have been investigated in detail over the past decade, focusing on both the molecular interactions driving self-assembly at the nanoscale and phenomenological aspects determining the final appearance at the micro- and nanoscales [6,13–16]. With respect to the latter, it was found that growth of biomorphs occurs in two general stages. At the beginning of the crystallisation process, small witherite crystals nucleate and grow in a more or less classical way. With time, oligomeric silica species induce fractal branching at the tips of the carbonate crystal, causing progressive bifurcation and ultimately resulting in closed cauliflowerlike spherulites [2,17–19]. At this point, the system passes into the second stage, which is characterised by polycrystalline growth, i.e. numerous nanocrystals are nucleated continuously and/or episodically and co-assemble into aggregates with shapes beyond any constraints of crystallographic symmetry. Typically, the first type of morphologies observed in this stage are flat sheets (Figure 1A), which grow along vessel walls or the solution–air interface. Randomly, these laminar structures start to curl at their rim and assume a new (orthogonal) growth direction along the perimeter of the sheet, which arrests radial advancement and gives birth to more complex three-dimensional shapes. The final morphology is then determined by the interplay of various local parameters, including the relative velocities of growth for different segments as well as their heights. The most common twisted forms obtained at ambient conditions are double helices (Figure 1B) and more tightly scrolled so-called worms (Figure 1C) [2]. In many cases, the carbonate-rich core of the aggregates becomes covered by a layer of amorphous silica, as a result of secondary silica precipitation in the later stages of growth when the bulk pH has been lowered (due to CO$_2$ uptake) to values where the solubility of silica is noticeably decreased [20].

At the molecular level, the unique behaviour of these inorganic precipitation systems has been ascribed to the pH-mediated coupling of the speciations of carbonate and silicate in solution [2]. As the two components have opposite trends in terms of solubility as a function of pH, it was proposed that they continuously stimulate each other's mineralisation by periodic changes in the local conditions at actively growing fronts: carbonate crystallisation reduces the pH in the microenvironment and thus increases the local supersaturation of silica, which in turn will precipitate and thereby re-increase the pH-shifting the local carbonate speciation to the side of CO$_3^{2-}$ and hence triggering a new event of carbonate nucleation. This coupled chemistry allows numerous silica-coated carbonate nanocrystals to be produced and integrated into the forming superstructures as building units. Indeed, the postulated local pH-cycling at the growing front could be verified experimentally in recent studies by using pH-sensitive fluorescent dyes [14,16].

Interestingly, most of the observations supporting the mechanisms described above were made for the formation of biomorphs in solution or gels under ambient conditions and within a narrow range of concentrations suitable for the growth of the most striking morphologies. However, it is well known already from early works [21] that the structural variety of silica biomorphs is not at all limited to the forms displayed in Figure 1, but can be significantly expanded by adjusting conditions such as pH, salinity and temperature, or by introducing certain additives that are able to interfere with the growth process [3,6,7,22–25]. In the present work, we have re-evaluated the role of temperature in biomorph formation, in order to discuss corresponding changes of structure and morphology in light of the current model of morphogenesis. We show that temperature variation is a straightforward means to obtain further interesting ultrastructures, if the conditions are carefully chosen. The resulting narrow distribution of morphologies highlights the possibility of shape control in these systems, which seems crucial for the targeted design of related materials with potential for actual applications [26,27]. Further experiments performed in the presence of added multivalent ions at room temperature expand the range of accessible morphologies and provide support for a morphogenetic scenario in which the relative rates of silica and carbonate mineralisation determine the evolution of the system on the micron scale.

2. Materials and Methods

2.1. Materials and Sample Preparation

Barium chloride dihydrate (>99%), sodium hydroxide (reagent grade, >98%), lanthanum chloride heptahydrate ($\geq$99.9%), and sodium silicate solution (commercial water glass, containing ~10.6% Na_2O and ~26.5% SiO_2, reagent grade, density: 1.39 g/mL) were purchased from Sigma-Aldrich and used without further purification. All solutions were prepared using MilliQ water with a conductivity of 18 μS/cm. Suitable silicate sols were obtained by diluting 1.39 g of water glass in 349 mL of water. The pH of the resulting sol was adjusted to 11.3 (at 25 °C) by adding aliquots of 0.1 M NaOH solution. Crystallisation experiments were carried out in 24-well plates (Linbro) by mixing 1 mL of the dilute silicate solution with 1 mL of $BaCl_2$ solution at different concentrations (5–500 mM). Optionally, 2 mM $LaCl_3$ was dissolved in the $BaCl_2$ solution prior to mixing with the silica sol. Subsequently, the well plates were covered loosely with a lid and stored, respectively, at room temperature (25 °C) or inside a temperature-controlled oven (30–50 °C) or a cooling chamber (5–20 °C) for 12 h in contact with air, to enable diffusion of CO_2 into the systems and slow crystallisation of $BaCO_3$ under the influence of silica species.

2.2. Analytical Methods

The pH of the mother solutions was measured by a pH meter (Eutech pH 510) before growth. The formed mineral structures were characterised routinely by means of polarised light microscopy (PLM), using a Nikon AZ100 microscope (Nikon Co., Tokyo, Japan) and an Imager M2m from Zeiss (Jena, Germany). In selected cases, scanning electron microscopy (SEM) was performed on a Zeiss EVO15 microscope (Carl Zeiss SMY Ltd., Cambridge, UK). To obtain suitable specimens for the SEM studies, biomorphs were grown on indium tin oxide (ITO) substrates (Osslia), which were placed in Petri dishes with either 35 or 60 mm diameter and covered with 4–8 mL of growth solution.

3. Results and Discussion

In order to investigate the effect of temperature on the formation of silica biomorphs, barium carbonate was crystallised using silica-containing solutions at a fixed composition ([SiO_2] = 17.8 mM and pH = 11.3 (at 25 °C) before mixing with $BaCl_2$ solution) and varying the temperature between 5 and 50 °C. Figure 2 provides an overview on characteristic structures formed at different temperatures and barium concentrations (i.e., a so-called morphodrome).

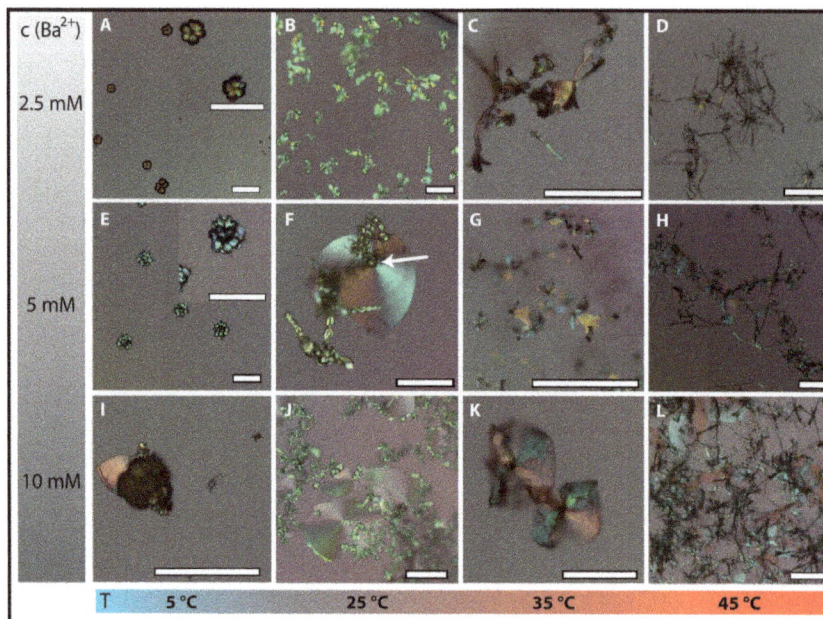

Figure 2. (**A–L**) PLM images of typical biomorphic morphologies obtained at varying barium concentrations and temperatures (as indicated) after a growth period of 12 h using a silica solution with $[SiO_2]$ = 8.9 mM and an initial pH of 11.3 (determined at 25 °C before mixing with barium chloride solution). Note the presence of peculiar 3D trumpetlike shapes in (**G**), as visualised by the optical depth of field. The arrow in (**F**) indicates a closed globular particle, from which a polycrystalline sheet-like aggregate has emerged. Scale bars are 200 μm.

Under "standard" conditions (25 °C and 5 mM BaCl₂, Figure 2F), extended flat sheets and helicoids (often intergrown) are observed, as expected and similar to the structures shown in Figure 1. These morphologies can thus be considered as a reference. We note that most of these complex structures emerge from small closed globular particles (or aggregates thereof, as indicated by the arrow in Figure 2F). They form in the fractal regime which has to be very short/fast for such small and discrete morphologies to be observed at the end of the chosen period of growth (12 h).

When the barium concentration is increased to 10 mM (Figure 2J), the total number of aggregates is higher and in particular the sheets grew slightly larger on average. In turn, a Ba^{2+} concentration of 2.5 mM results in fewer, smaller, and less defined structures (Figure 2B), but generally the morphologies are not strongly altered at both higher and lower amounts of cations in the given concentration range. By contrast, lowering the temperature to 5 °C changes the picture fundamentally, as most of the morphologies observed at 2.5 and 5 mM BaCl₂ were isolated globular or cauliflowerlike structures (Figure 2A,E), and only few poorly developed polycrystalline aggregates were formed at 10 mM even after 24 h (such as the small sheet in Figure 2I). This indicates that growth is terminated at the end of the fractal stage in most cases, i.e., the system fails to enter the second stage of dynamic nanocrystal formation and aggregation. In other words, the conditions prevailing in these experiments do not allow for chemically coupled co-precipitation to be initiated and/or maintained. Probably, this behaviour is related to temperature-dependent changes in the kinetics of carbonate and/or silica mineralisation, caused by differences in either diffusive transport of reactants or the relative rates of precipitation. Generally, the following trends and related consequences are expected with increasing temperature:

- Decrease in the solubility of $BaCO_3$ [28]: higher carbonate supersaturation, higher driving force for carbonate precipitation (thermodynamic factor);
- Decrease in the solubility of CO_2 in water [29]: reduced rate of CO_2 uptake from the atmosphere, slower carbonate precipitation due to diffusion limitation (kinetic factor);
- Acceleration of silica condensation kinetics [30]: reduced rate of silica polymerisation, faster SiO_2 precipitation (kinetic factor);
- Increased solubility of silica [31]: decrease in silica supersaturation, lower driving force for precipitation (thermodynamic factor);
- Increase in the ionic product of water [32]: higher effective pH at the same composition, leading to lower carbonate solubility and higher silica solubility (thermodynamic factor).

Based on these qualitative considerations, it seems reasonable to assume that the growth rates of both silica and carbonate are not sufficiently high to allow for complex ultrastructures to form at lower temperatures. This can be resolved to some extent by increasing the barium concentration (Figure 2I), but the influence of temperature appears to be dominant.

At higher temperatures (35 and 45 °C), there are also regions in the morphodrome where well-developed biomorphs are barely observed. This is particularly true for lower and higher barium concentration (Figure 2C,D,K,L). In the former case (2.5 mM $BaCl_2$), the slower rate of carbonate formation (lower supersaturation and reduced CO_2 uptake) can probably not supply enough building units to keep the pace of the accelerated silica condensation processes. In turn, at 10 mM $BaCl_2$ another effect comes into play, namely the electrostatic screening of negative charges on silicate species by dissolved divalent cations like Ba^{2+} [22]. This bridging interaction catalyses silica condensation and—along with the higher temperature—is expected to accelerate silica precipitation kinetics in a way that is less or not suitable for coupling with carbonate formation. As opposed to that, proper conditions for 3D self-assembly are still maintained at the "standard" barium concentration of 5 mM and temperatures up to 45 °C. Here, however, the resulting biomorphs show substantially different morphologies than at room temperature and 5 mM Ba^{2+} ion concentration, with large (>100 μm) trumpetlike forms representing the major population of ultrastructures. When the temperature is increased to 45 °C the biomorphic landscape changes again and highly branched crystal networks are observed, as shown in Figure 2D,H,L. These structures likely derive from a fractal route that shows substantially different branching behaviour than at room temperature. Here, the initial pseudo-hexagonal twinned crystal core transforms into an open star-like architecture that is poorly filled with silica–carbonate composite matter. In some cases, sheet-like domains are formed in between branches, but these structures are often heavily bended and limited in their growth by the mother crystal branches (*cf.* Figure 2H). This suggests that an increase in temperature changes the branching kinetics, presumably via modulated interactions between growing carbonate surfaces and dissolved silicate species that adsorb specifically on certain witherite faces. One potential scenario is that the accelerated carbonate precipitation gives the silica less time to interfere. Alternatively, the speciation of silica in solution may shift to larger oligomers that show different affinities to interact with specific carbonate faces, thus resulting in a different branching behaviour. In any case, high temperature in combination with high barium concentrations leads to a significant increase in the amount of precipitated matter (probably due to faster carbonate and silica formation). This is shown in Figure 2L, where large networks of tubular filaments and/or twisted ribbons are seen along with various other types of biomorphic morphologies. Interestingly, these forms are much thinner than biomorphs grown at ambient temperature, again indicating important changes in relative growth rates of the components. In summary, the most outstanding feature of the morphodrome established in our study was the formation of biomorphic ultrastructures with trumpetlike appearance. They can sporadically be obtained at temperatures above 30 °C and replace all sheet-like and most twisted morphologies at temperatures above 40 °C.

The formation process of the trumpetlike structures was analysed in more detail, in particular with respect to both differences and similarities compared to biomorph growth at room temperature, which leads to twisted (worms and helicoids) and flat sheet-like (as described in previous works [2–4] and shown in Figure 1). Figure 3 provides schematic representations of the occurring processes (panels A and B–E) along with selected SEM images showing the relevant stages of growth (panels F–H).

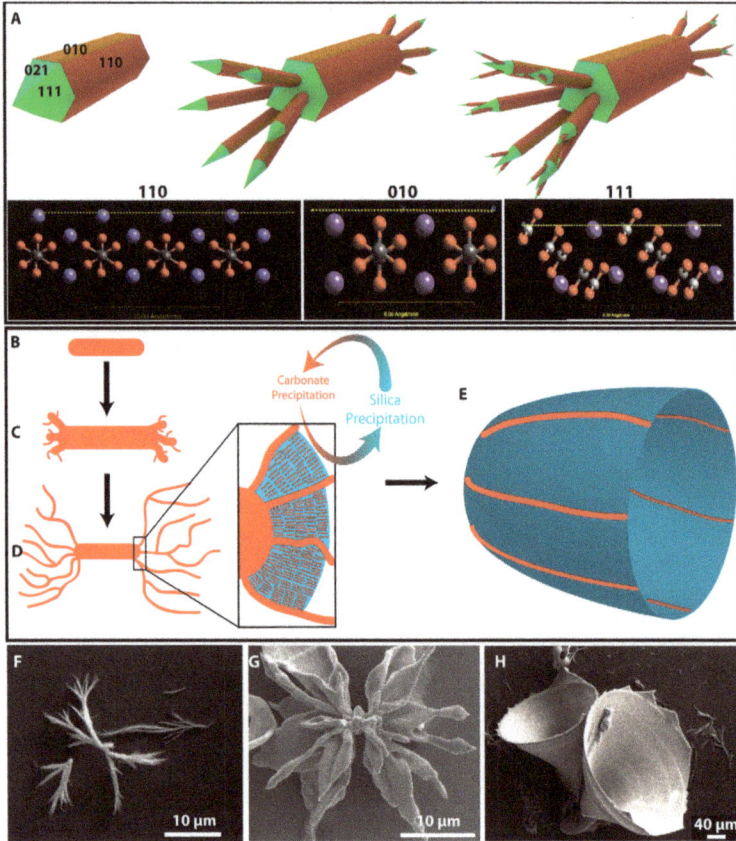

Figure 3. Formation of trumpet-like morphologies at 45 °C and 5 mM $BaCl_2$. (**A**) Scheme of a primary witherite crystal (top), which successively becomes branched due to specific interactions of silica with different faces of the carbonate crystal. Vacuum surface cuts of these faces (bottom) indicate that the (110) faces are positively charged and thus prone to be blocked by negatively charged silicate species. Also, (010) is polar and therefore also likely to interact with negative silica species. Further growth thus occurs through the neutral (021) and (111) faces, so that up to six branches can emerge from the pseudo-hexagonal base. Repetition of this process leads to multiple branching generations. (**B–E**) Schematic overview of the growth process of a trumpet-like structure: an initial elongated $BaCO_3$ twin crystal seed grows and undergoes fractal branching as described in (A). At high temperature, this leads to open architectures with large branches, between which polycrystalline growth (i.e., autocatalytic co-precipitation of silica/witherite nanoparticles) occurs and slowly fills up the empty space (inset). Depending on the orientation of the branching, trumpet-like aggregates are formed. (**F–H**) SEM micrographs of the initial stage of open fractal branching (**F**), followed by filling of the branches with biomorphic composite material (**G**), ultimately leading to trumpet-like forms (**H**).

Despite the absence of any of the three "standard" morphologies, the nature of the polycrystalline growth mechanism in the second stage of morphogenesis seems comparable to the corresponding processes at room temperature, based on the characteristic thickness oscillations observed for example in Figure 3H and detailed analysis of the texture of the aggregates at the nanoscale [3,15,16]. Consequently, the main difference leading to distinct final shapes must occur at the beginning of the structure formation. As at ambient conditions, morphogenesis starts with fractal branching of a twinned pseudo-hexagonal witherite crystal due to the poisoning influence of oligomeric silica species (Figure 3A). Closer analysis of the carbonate surfaces suggests that adsorption of silicate species (Bittarello et al. [33]) will mainly occur on the positively charged (110) faces but also on the polar (010) faces (cf. Figure 3A). At higher temperatures, the branching behaviour is different, though, and the first step is a splitting into up to six crystal arms, as silicate-covered (110) and (010) faces are blocked while the remaining (021) and (111) faces can serve for heterogeneous nucleation of the up to six crystal arms in a more or less unhindered fashion, as indicated in Figure 3A. This process repeats itself and produces second-generation branches. Increasing the temperature leads to more ordered and larger branched entities as drawn in Figure 3B–D, compared to the much smaller and heavily bifurcated dumbbell-shaped, spherical or raspberry-like forms observed at room temperature (cf. Figure 1). This suggests that branching at high temperature is more symmetric and of lower dimensionality than at ambient conditions, where a much higher branching density and thus much less symmetric architectures are observed. In other words, branching is more chaotic at low temperature and becomes well-defined at high temperature. This could be related to changes in silica condensation kinetics at higher temperatures, leading to different silica speciation and thus to different affinities of silicate species to adsorb on the different relevant carbonate faces, as already proposed above. That, together with changes in the growth rates of carbonate faces, could explain the different branching behaviour at elevated temperatures [28,30]. Once these open branched structures are formed (Figure 3F), the system enters into the second stage of morphogenesis, where the accelerated kinetics of the precipitation of both components leads to a fortification of the autocatalytic mechanism, which then seems to stay in the tolerant zone for coupled co-precipitation to work. However, the resulting nanocrystals do not assemble into the well-known shapes observed at room temperature, but seem to fill the space in between the large branches formed in the first stage of fractal branching (Figure 3G), leading to selective biomorphic growth from open branched fractals into trumpet-like structures (Figure 3H) according to a mechanism as indicated schematically in Figure 3D,E. Thereby, each structure is unique and shows its own interesting shape, which seems to depend on the orientation of the branches. The pathway to the trumpet structure skips the formation of globular particles and leads directly to 3D architectures of silica–barium carbonate composites with curved surfaces. The dominant effect causing this behaviour could be the increased silica solubility, which allows the system more time to develop in a defined manner throughout the fractal stage. Then, the second stage of morphogenesis is entered before a high degree of branching would lead to closed globular structures. Interestingly, the multiple limbs lead to only one final trumpet structure in most cases, suggesting that the different smaller leaf-like domains fuse to one surface or only the most pronounced leaf develops to the final shape.

During the secondary growth process, the orientation of the growing crystals is crucial. When the experiments are conducted under the same conditions but in petri dishes, where the surface/volume ratio is much higher than in the linbro plates, the reaction leads to trumpet-like structures with wider spreading exhausts. Some examples of these forms are shown in Figure 4A,C.

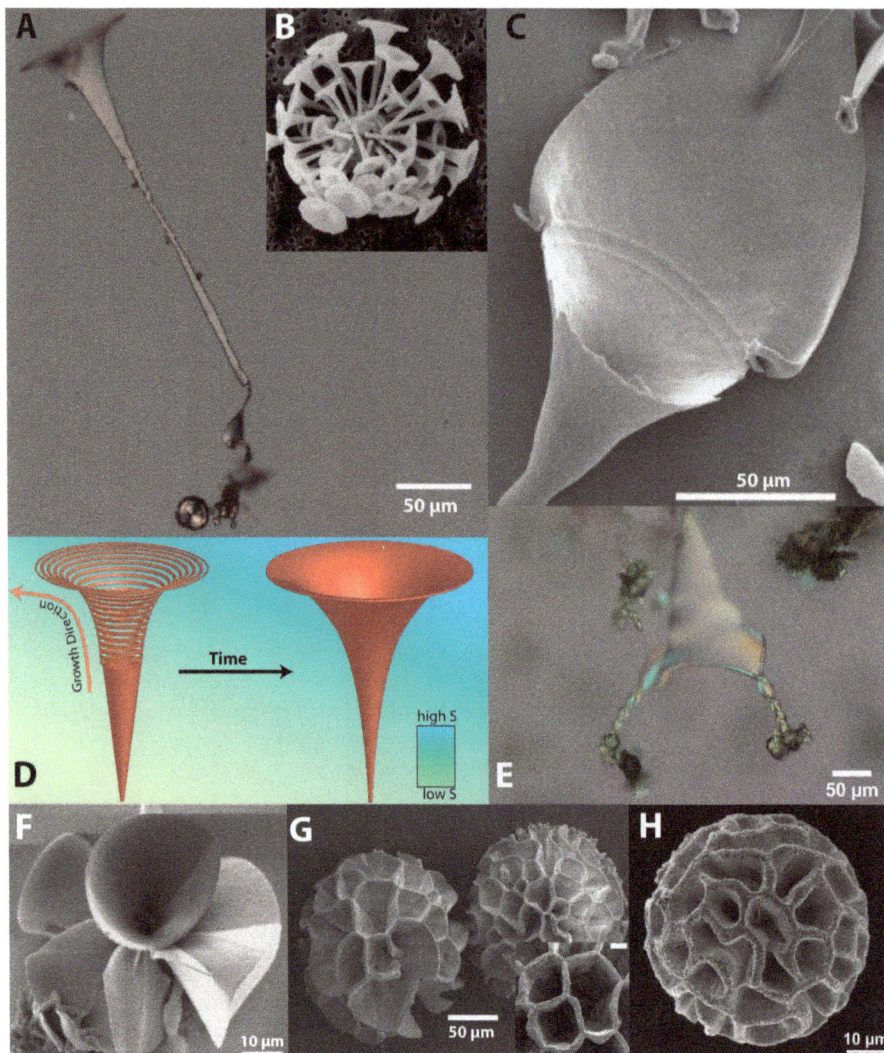

Figure 4. Biomorphs with lifelike morphologies. (**A,C**) PLM and SEM images of biomorphs in the shape of a marine coccolith. Formed from solution, containing 5 mM Ba^{2+} and 9 mM silica at 40 °C. (**B**) SEM image of a natural marine coccolith (Reproduced with permission from ref. [24], Copyright 2012 Wiley-VCH Verlag GmbH & Co KGaA). (**D**) Schematic illustration of the formation process of coccolith morphologies illustrating the supersaturation gradient into the solution. (**E**) PLM micrograph of a biomorph structure grown at higher temperature (40 °C) showing a trumpet structure exfolding into two helicoidal structures. (**F**) SEM micrograph of a flower-like biomorph grown from a 5 mM Ba^{2+} solution containing alkaline silica sol at 50 °C. (**G**) SEM micrograph of a coral-like biomorph structure grown at higher Ba^{2+} concentration (250 mM; scale bar of the inlet: 10 μm) (**H**) SEM micrograph grown from a 5 mM Ba^{2+} solution containing alkaline silica sol with 1 mM La^{3+} ions.

As it can be noticed, these structures look similar to a marine coccolith element (*cf.* Figure 4B) [34]. A model of the possible mechanism is presented in Figure 4D. When linbro plates are used, supersaturation is lower and biomorphic aggregates have a narrower shape. Since the petri dish setup favours the availability of CO_2 and therefore the carbonate in the growing front, a higher supersaturation with respect to barium carbonate and a consequently sharper drop of pH (and higher supersaturation with respect to silica) would be explained. This phenomenon is even more evident once the precipitation occurs nearer to the solution/atmosphere interface. It would also explain the spreading of the branches due to a faster (secondary) growth stage. A similar behaviour was already observed in higher concentrated solution and the spreading could be triggered with a CO_2 burst [35]. Therefore, it can be assumed that this sharp saturation gradient explains this behaviour when completely upright structures are forming. When tilted structures form under the influence of the CO_2, biomorphic aggregates develop irregular exhaust orientation as the one shown in Figure 4C. Additionally, a continuum of different morphologies can be detected in Figure 4A,C,F. In Figure 4A a transformation from a sheet over a small ribbon to the final trumpet can be found while C shows a reverse exfoliation from a rolled-up trumpet to a 2D sheet-like structure while Figure 4E demonstrates a continuum of shapes and a transition of a trumpet at elevated temperatures into two helicoidal structures.

Further lifelike morphologies can be produced by increasing the barium concentration (Figure 4G; $[Ba^{2+}]$ = 250–500 mM), which shift the formation mechanism as well. [36]. The obtained brain coral-like structures skip the formation of a globular centre, have a nearly perfect spherical shape, and consist of a network of interpenetrating sheet-like units which merge out of the leaf-like parts liked at the seed crystal (similar pathway presented in Figure 3). Their formation is already described by Hyde et al. [36] and reviewed here with consideration of newly obtained knowledge. The formation of these structures follows a similar pathway compared to the trumpets because their formation skips the production of a globular centre. Coral-like architectures can be obtained in a modified way with 5 mM Ba^{2+} ions in solution by using modifiers such as cationic surfactants like CTAB [7] or multivalent ions like La^{3+}. Using such modifiers, similar but more homogenous morphologies were obtained and are shown in the SEM micrograph in Figure 4H. The formation is again influenced in the first stage of the formation which is in good accordance with time-resolved experiments with CTAB additions. An addition of CTAB after a few minutes does not lead to coral structures and "standard" biomorphs will be obtained. At first sight, it is comprehensible that an increase of the pure number of multivalent ions like Ba^{2+} (tested before) or La^{3+} has a similar effect on the system despite the effect on silica precipitation, which is in good accordance with the Schulze–Hardy law [37]. A faster silica formation and a reduced colloidal stability by multivalent cations show similarity to the ultrastructures obtained at elevated temperature. Multivalent ions are known as efficient agents to bridge charged colloids or oligomers like silica [38]. Furthermore, these ions will not be free charged ions within this system due to their tendencies to form complexes with hydroxide, carbonate, and silicate ions to reduce their effective charge. In the end it should be noted that La^{3+} as an additive has less effect on the carbonate precipitation rate and that no significant lanthanum adsorption within the $BaCO_3$ lattice was detected. Additionally, a higher ionic strength depresses the CO_2 uptake rate and its solubility in water, which again leads to a deterioration of the carbonate formation [39]. Therefore, we can assume an extreme of a reduced fractal regime time in the case of coral-like structures. As shown in literature, upon the addition of sodium chloride (note the single charge of Na^+) to the standard procedure, the process will not enter the second stage at all and the crystal formation stops after the formation of the globular particles formed in the first stage (described by Eiblmeier et al. [22]). A reason for this might be the shifting of the equilibrium of free and sodium-coordinated silicate species to the sodium terminated side [40]. Therefore, less silica is available for the autocatalytic process, which is in line with observations made with decreased silica concentrations, and the resident time in the fractal regime is much longer and globules are formed.

4. Conclusions

The presented work describes the influence of several parameters on biomorph growth and highlights the formation process of reported and unreported lifelike morphologies. The individual impact of the temperature, supersaturation, and ions as additives on carbonate and silica formation is discussed in detail and meaningful, plausible explanations were given for the formation mechanism of the strongly bended morphologies. We showed a linkage between coral- and trumpet-like structures and that a continuum of shapes is possible. Influences of the various and further tuning parameters like the effects of the growth direction and surface/volume ratio were tested. The most exciting growth conditions to lifelike morphologies were highlighted and allow a superior reproducibility.

Acknowledgments: The research leading to these results has received funding from the European Research Council under the European Union's Seventh Framework Programme (FP7/2007-2013)/ERC grant agreement no. 340863.

Author Contributions: Julian Opel, Juan Morales and Matthias Kellermeier conceived and designed the experiments; the experiments and analytics were performed by Julian Opel, Annika Sickinger and Matthias Kellermeier; the results were discussed by Julian Opel, Juan Morales, Juan-Manuel Garcia-Ruiz, Helmut Cölfen and Matthias Kellermeier; Julian Opel wrote the paper under supervision of Juan-Manuel Garcia-Ruiz, Helmut Cölfen and Matthias Kellermeier.

Conflicts of Interest: The authors declare no conflict of interest.

References

1. García-Ruiz, J.M.; Hyde, S.T.; Carnerup, A.M.; Christy, A.G.; Van Kranendonk, M.J.; Welham, N.J. Self-assembled silica-carbonate structures and detection of ancient microfossils. *Science* **2003**, *302*, 1194–1197. [CrossRef] [PubMed]

2. García-Ruiz, J.M.; Melero-García, E.; Hyde, S.T. Morphogenesis of self-assembled nanocrystalline materials of barium carbonate and silica. *Science* **2009**, *323*, 362–365. [CrossRef] [PubMed]

3. Kellermeier, M.; Colfen, H.; Garcia-Ruiz, J.M. Silica biomorphs: Complex biomimetic hybrid materials from "sand and chalk". *Eur. J. Inorg. Chem.* **2012**, *32*, 5123–5144. [CrossRef]

4. Kellermeier, M.; Melero-Garcia, E.; Glaab, F.; Eiblmeier, J.; Kienle, L.; Rachel, R.; Kunz, W.; Garcia-Ruiz, J.M. Growth behavior and kinetics of self-assembled silica-carbonate biomorphs. *Chem.-A Eur. J.* **2012**, *18*, 2272–2282. [CrossRef] [PubMed]

5. García-Ruiz, J.M. On the formation of induced morphology crystal aggregates. *J. Cryst. Growth* **1985**, *73*, 251–262. [CrossRef]

6. Bittarello, E.; Aquilano, D. Self-assembled nanocrystals of barium carbonate in biomineral-like structures. *Eur. J. Mineral.* **2007**, *19*, 345–351. [CrossRef]

7. Kellermeier, M.; Glaab, F.; Carnerup, A.M.; Drechsler, M.; Gossler, B.; Hyde, S.T.; Kunz, W. Additive-induced morphological tuning of self-assembled silica-barium carbonate crystal aggregates. *J. Cryst. Growth* **2009**, *311*, 2530–2541. [CrossRef]

8. Voinescu, A.E.; Kellermeier, M.; Carnerup, A.M.; Larsson, A.-K.; Touraud, D.; Hyde, S.T.; Kunz, W. Co-precipitation of silica and alkaline-earth carbonates using teos as silica source. *J. Cryst. Growth* **2007**, *306*, 152–158. [CrossRef]

9. Terada, T.; Yamabi, S.; Imai, H. Formation process of sheets and helical forms consisting of strontium carbonate fibrous crystals with silicate. *J. Cryst. Growth* **2003**, *253*, 435–444. [CrossRef]

10. Imai, H.; Terada, T.; Miura, T.; Yamabi, S. Self-organized formation of porous aragonite with silicate. *J. Cryst. Growth* **2002**, *244*, 200–205. [CrossRef]

11. Bittarello, E.; Roberto Massaro, F.; Aquilano, D. The epitaxial role of silica groups in promoting the formation of silica/carbonate biomorphs: A first hypothesis. *J. Cryst. Growth* **2010**, *312*, 402–412. [CrossRef]

12. Voinescu, A.E.; Kellermeier, M.; Bartel, B.; Carnerup, A.M.; Larsson, A.-K.; Touraud, D.; Kunz, W.; Kienle, L.; Pfitzner, A.; Hyde, S.T. Inorganic self-organized silica aragonite biomorphic composites. *Cryst. Growth Des.* **2008**, *8*, 1515–1521. [CrossRef]

13. Kaplan, C.N.; Noorduin, W.L.; Li, L.; Sadza, R.; Folkertsma, L.; Aizenberg, J.; Mahadevan, L. Controlled growth and form of precipitating microsculptures. *Science* **2017**, *355*, 1395–1399. [CrossRef] [PubMed]

14. Montalti, M.; Zhang, G.; Genovese, D.; Morales, J.; Kellermeier, M.; García-Ruiz, J.M. Local ph oscillations witness autocatalytic self-organization of biomorphic nanostructures. *Nat. Commun.* **2017**, *8*, 14427. [CrossRef] [PubMed]

15. Nakouzi, E.; Ghoussoub, Y.E.; Knoll, P.; Steinbock, O. Biomorph oscillations self-organize micrometer-scale patterns and nanorod alignment waves. *J. Phys. Chem. C* **2015**, *119*, 15749–15754. [CrossRef]

16. Opel, J.; Hecht, M.; Rurack, K.; Eiblmeier, J.; Kunz, W.; Colfen, H.; Kellermeier, M. Probing local pH-based precipitation processes in self-assembled silica-carbonate hybrid materials. *Nanoscale* **2015**, *7*, 17434–17440. [CrossRef] [PubMed]

17. García-Ruiz, J.M.; Carnerup, A.; Christy, A.G.; Welham, N.J.; Hyde, S.T. Morphology: An ambiguous indicator of biogenicity. *Astrobiology* **2002**, *2*, 353–369. [CrossRef] [PubMed]

18. Kellermeier, M.; Eiblmeier, J.; Melero-García, E.; Pretzl, M.; Fery, A.; Kunz, W. Evolution and control of complex curved form in simple inorganic precipitation systems. *Cryst. Growth Des.* **2012**, *12*, 3647–3655. [CrossRef]

19. Kunz, W.; Kellermeier, M. Beyond biomineralization. *Science* **2009**, *323*, 344–345. [CrossRef] [PubMed]

20. Alexander, G.B.; Heston, W.; Iler, R.K. The solubility of amorphous silica in water. *J. Phys. Chem.* **1954**, *58*, 453–455. [CrossRef]

21. García-Ruiz, J.M. Inorganic self-organization in precambrian charts. *Origins Life Evol. Biosphere* **1994**, *24*, 451–467. [CrossRef]

22. Eiblmeier, J.; Dankesreiter, S.; Pfitzner, A.; Schmalz, G.; Kunz, W.; Kellermeier, M. Crystallization of mixed alkaline-earth carbonates in silica solutions at high pH. *Cryst. Growth Des.* **2014**, *14*, 6177–6188. [CrossRef]

23. Nakouzi, E.; Knoll, P.; Steinbock, O. Biomorph growth in single-phase systems: Expanding the structure spectrum and pH range. *Chem. Commun. (Camb)* **2016**, *52*, 2107–2110. [CrossRef] [PubMed]

24. Melero-García, E.; Santisteban-Bailón, R.; García-Ruiz, J.M. Role of bulk pH during witherite biomorph growth in silica gels. *Cryst. Growth Des.* **2009**, *9*, 4730–4734. [CrossRef]

25. Eiblmeier, J.; Kellermeier, M.; Rengstl, D.; García-Ruiz, J.M.; Kunz, W. Effect of bulk pH and supersaturation on the growth behavior of silica biomorphs in alkaline solutions. *CrystEngComm* **2013**, *15*, 43–53. [CrossRef]

26. Opel, J.; Wimmer, F.P.; Kellermeier, M.; Cölfen, H. Functionalisation of silica–carbonate biomorphs. *Nanoscale Horiz.* **2016**, *1*, 144–149. [CrossRef]

27. Wang, G.; Zhao, X.; Möller, M.; Moya, S.E. Interfacial reaction-driven formation of silica-carbonate biomorphs with subcellular topographic features and their biological activity. *ACS Appl. Mater. Interfaces* **2015**, *7*, 23412–23417. [CrossRef] [PubMed]

28. Busenberg, E.; Plummer, L.N. The solubility of $BaCO_3$ (cr)(witherite) in CO_2-H_2O solutions between 0 and 90 °C, evaluation of the association constants of $BaHCO_3^+$ (aq) and $BaCO_3$ (aq) between 5 and 80 °C, and a preliminary evaluation of the thermodynamic properties of Ba^{2+}(aq). *Geochim. Cosmochim. Acta* **1986**, *50*, 2225–2233. [CrossRef]

29. Weiss, R.F. Carbon dioxide in water and seawater: The solubility of a non-ideal gas. *Mar. Chem.* **1974**, *2*, 203–215. [CrossRef]

30. Hurd, C.B.; Pomatti, R.C.; Spittle, J.H.; Alois, F.J. Studies on silicic acid gels. XV. The effect of temperature upon the time of set of alkaline gel mixtures. *J. Am. Chem. Soc.* **1944**, *66*, 388–390. [CrossRef]

31. Krauskopf, K.B. Dissolution and precipitation of silica at low temperatures. *Geochim. Cosmochim. Acta* **1956**, *10*, 1–26. [CrossRef]

32. Marshall, W.L.; Franck, E.U. Ion product of water substance, 0–1000 °C, 1–10,000 bars new international formulation and its background. *J. Phys. Chem. Ref. Data* **1981**, *10*, 295–304. [CrossRef]

33. Bittarello, E.; Massaro, F.R.; Rubbo, M.; Costa, E.; Aquilano, D. Witherite ($BaCO_3$)/α-quartz epitaxial nucleation and growth: Experimental findings and theoretical implications on biomineralization. *Cryst. Growth Des.* **2009**, *9*, 971–977. [CrossRef]

34. Mann, S. The chemistry of form. *Angew. Chem. Int. Ed.* **2000**, *39*, 3392–3406. [CrossRef]

35. Noorduin, W.L.; Grinthal, A.; Mahadevan, L.; Aizenberg, J. Rationally designed complex, hierarchical microarchitectures. *Science* **2013**, *340*, 832–837. [CrossRef] [PubMed]

36. Hyde, S.T.; Carnerup, A.M.; Larsson, A.K.; Christy, A.G.; García-Ruiz, J.M. Self-assembly of carbonate-silica colloids: Between living and non-living form. *Physica A* **2004**, *339*, 24–33. [CrossRef]

37. Hardy, W.B. A preliminary investigation of the conditions which determine the stability of irreversible hydrosols. *J. Phys. Chem.* **1899**, *4*, 235–253. [CrossRef]

38. Bergna, H.E. *The Colloid Chemistry of Silica*; American Chemical Society: Washington, DC, USA, 1994; Volume 234, p. 724.

39. Zeebe, R.E.; Wolf-Gladrow, D.A. *CO_2 in Seawater: Equilibrium, Kinetics, Isotopes*; Elsevier Sience: Amsterdam, The Netherlands, 2001.
40. Iler, K.R. *The Chemistry of Silica: Solubility, Polymerization, Colloid and Surface Properties and Biochemistry of Silica*; John Wiley & Sons, Inc.: New York, NY, USA, 1979.

minerals

MDPI

Review

About the Genetic Mechanisms of Apatites: A Survey on the Methodological Approaches

Linda Pastero, Marco Bruno *and Dino Aquilano

Dipartimento di Scienze della Terra, Università degli Studi di Torino, Via Valperga Caluso 35,
I-10125 Torino, Italy; linda.pastero@unito.it (L.P.); dino.aquilano@unito.it (D.A.)
* Correspondence: marco.bruno@unito.it; Tel.: +39-011-670-5124

Received: 15 June 2017; Accepted: 1 August 2017; Published: 5 August 2017

Abstract: Apatites are properly considered as a strategic material owing to the broad range of their practical uses, primarily biomedical but chemical, pharmaceutical, environmental and geological as well. The apatite group of minerals has been the subject of a huge number of papers, mainly devoted to the mass crystallization of nanosized hydroxyapatite (or carboapatite) as a scaffold for osteoinduction purposes. Many wet and dry methods of synthesis have been proposed. The products have been characterized using various techniques, from the transmission electron microscopy to many spectroscopic methods like IR and Raman. The experimental approach usually found in literature allows getting tailor made micro- and nano- crystals ready to be used in a wide variety of fields. Despite the wide interest in synthesis and characterization, little attention has been paid to the relationships between bulk structure and corresponding surfaces and to the role plaid by surfaces on the mechanisms involved during the early stages of growth of apatites. In order to improve the understanding of their structure and chemical variability, close attention will be focused on the structural complexity of hydroxyapatite (HAp), on the richness of its surfaces and their role in the interaction with the precursor phases, and in growth kinetics and morphology.

Keywords: apatites; structure; equilibrium shape; growth shape; supersaturation; nucleation

1. Introduction

As described in two excellent review papers [1,2] the term "Apatite" represents a group of minerals of a very wide interest. As a matter of fact, extraction of phosphor from apatite ores provides, to this day, fertilizers to feed the Earth population; apatites are used as well in geological dating techniques and studies of rare earth element variation in rocks. In the applications related to the Materials Science, apatite is the basic raw material used in the fluorescent lighting industry, while its unique crystal-chemical properties make it fundamental in the production of laser devices with controllable properties. The industrial process named PIMS (phosphate-induced-metal-stabilization) intervenes in the environmental remediation and the crystal structure properties of the used apatites favour the storage of radiactive waste as substituents in the Ca-sites. Concerning the research in the health field, it is worth outlining that two important members of apatites (Hydroxyapatite, HAp, and Carbonated apatite, CAp) represent both the inorganic pillar building hard tissues such as bones, theet, antlers of mammals and their pathological calcified tissues (i.e., those generated by various diseases). Having considered that apatites are studied since more that two centuries and according to their wide field of interest, it is not surprising that the total amount of currently available publications on the "Apatite" subject exceeds 40,000 with the annual increase for, at least, 2000 papers.

In this review, we do not claim to open a wide-range discussion on this subject, but we will confine our attention to the genetic mechanisms of apatite crystals, with a peculiar care for the relations between the bulk structure of the crystal and its corresponding face-by-face surface structures, in order to improve the knowledge of the interfaces that form when the crystal interacts with its growth medium. This way of methodological thinking represents the unavoidable consequence of the scale-change which affects the crystals when going from geological (cm, mm) to industrial (mm-µm) to biomaterial (µm-nm) size. In fact, surface kinetics becomes more and more important as much as crystal size decreases, and the features of the crystal surfaces dominate the kinetics of both nucleation and crystal growth in its early stages, since interfacial tension, adsorption and/or absorption, and epitaxy (if any) play a fundamental role in determining the anisotropy of the growth kinetics and then the final growth morphology [3–5].

Our reasoning will develop according to the following steps:

(1) A sketch of the apatite bulk structure, drawn according the most realistic related symmetry space groups, focusing mainly on Hydroxyapatite (HAp) and, secondarily, on its structurally derivated, the Carbonated apatite (CAp).

(2) A comparison between the stable surface profiles of the main crystallographic {*hkl*} forms, i.e., the hexagonal prism and pinacoids, and the monoclinic pinacoids.

(3) The ambiguities arising in interpreting the growth morphology, when the hexagonal symmetry of apatite is assumed "a priori" and the twinning occurrence is not considered as a "habit modifier".

(4) The role of the epitaxy when another phosphate works as a precursor of the apatite (e.g., the case of the Monetite assisted growth of micrometric HAp crystals).

(5) The growth of nanosized vs. micrometric HAp crystals.

2. Crystal-Chemistry and Bulk Structure of Apatites: Hydroxy-Apatite as a "Case Study"

According to Dorozhkin [1], apatites belong to the wide family of Calcium ortho-phosphates, characterized by a structure based on the tetrahedral ion PO_4^{3-}. From the crystal-chemical point of view, all apatites are defined by a Ca/P molar ratio (Ca/P, mr) ranging between 1.5 and 1.67, as illustrated in Table 1. Ca/P molar ratio can slightly exceed 1.67, when CO_3^{2-} ions enter the crystal lattice replacing OH^- and/or PO_4^{3-} ions in CAp [6].

All other Ca orthophosphates have 0.5 < (Ca/P, mr) < 1.5, while the amorphous calcium phosphates, ACPs, [$Ca_xH_y(PO_4)_z \cdot nH_2O$, *n*= 3–4.5; 15%–20% H_2O] have 1.2 < (Ca/P, mr) < 2.2 and the Tetracalcium phosphate (TetCP), mineral hilgenstockite, $Ca_4(PO_4)_2O$, has (Ca/P, mr) = 2.0. Roughly speaking, from Table 2 it follows that the symmetry of apatites schematically belongs either to the hexagonal or to monoclinic system, HAp belonging to both, according to the crystallization temperature and chemical purity of the sample. This is the reason why we will focus our attention on HAp polymorphism, which is not yet fully understood but represents a fruitful case study for equilibrium and growth morphology of apatites.

Table 1. The main minerals of the Apatite group, their Ca/P molar ratio and chemical composition.

Ca/P Molar Ratio	Mineral Name	Chemical Composition
1.5–1.67	Calcium-deficient hydroxyapatite (Ca-def HAp)	$Ca_{10-x}(HPO_4)_x(PO_4)_{6-x}(OH)_{2-x} (0 < x < 1)$
1.67	Hydroxyapatite (HAp)	$Ca_{10}(PO_4)_6(OH)_2$
1.67	End-member, A-type, carbonated apatite	$Ca_{10}(PO_4)_6CO_3$
≥1.67	End-member, B-type, carbonated hydroxylapatite	$Ca_{10-x}[(PO_4)_{6-2x}(CO_3)_{2x}](OH)_2 (0 < x < 1)$
≥1.67	Mixed A-type and B-type carbonated apatite	$Ca_{10-x}[(PO_4)_{6-2x}(CO_3)_{2x}]CO_3 (0 < x < 1)$
1.67	Fluorapatite (FAp)	$Ca_{10}(PO_4)_6F_2$
1.67	Oxyapatite (OAp), mineral voelckerite	$Ca_{10}(PO_4)_6O$

Table 2. The main minerals of the Apatite group, along with their lattice parameters (in Å) and Space Groups, as it comes out from literature.

Structure	Space Group	a_0	b_0	c_0	α	β	γ	Reference
HAp (Hydroxy-apatite)								
hexagonal	$P6_3/m$	9.4302	9.4302	6.8911	90°	90°	120°	[7–19]
monoclinic	$P2_1/b$	9.4214	$2 \times a_0$	6.881	90°	90°	120°	[20–22]
FAp (fluor-apatite)								
hexagonal	$P6_3/m$	9.367	9.3973	6.8782	90°	90°	120°	[12,23–25]
OAp (oxy-apatite)								
hexagonal	$P\bar{6}$	9.432	9.432	6.88	90.3°	90°	119.9°	[1,23–25]

2.1. The Hexagonal Setting

(i) *The P6₃/m setting*. Itwas first assigned to a synthetic single crystal (by X-ray diffraction) without determining the position of H atoms [7]; later on, it was confirmed by neutron diffraction on a natural crystal containing ~0.3% of F replacing the OH groups [8]. Furthermore, this space group has been experimentally determined in crystal obtained by hydrothermal synthesis [9–11] and in natural samples [12] where the OH dipoles are partially replaced by simple cations such as F and Cl. As concerns the theoretical investigation on the surface properties, the $P6_3/m$ setting has been preferentially adopted to predict the theoretical growth morphology [13], to determine the surface relaxation [14], to simulate the molecular adsorption on the most important HAp surfaces [15–18] and to evaluate the formation energies of Na and K replacing the Ca ions of HAp [19].

(ii) *The P6₃ setting*. The symmetry mirror m is not consistent with an ordered alignment of the OH dipoles, since the alternating O and H occupying special positions should be equidistant. The symmetry reduction $P6_3/m \rightarrow P6_3$ brings *all OH dipoles with the same orientation in every structure channel* (ferroelectric ordering); moreover, it is energetically favored and then has been systematically chosen, in the ab initio quantum-mechanical calculations, to simulate either the specific surface energies or the interaction of biomolecules with highly occurring {*hkl*} crystal forms [26–28].

2.2. The Monoclinic Setting

In the monoclinic HAp structure, the OH dipoles lie on the screw 2_1 axes which run along the structure channels. Owing to the action of the glide plane, the *OH groups point upward and downward in alternate nearest neighbor channels (antiparallel orientation)*. The number of investigations about the monoclinic HAp polymorph is small compared with that on the hexagonal one. This is surely due to the difficulty of obtaining the pure monoclinic phase, the relative percentage between the two polymorphs varying from a synthesis run to another [29,30] and to the high nucleation frequency of monoclinic twins [31].

It has been theoretically shown that the monoclinic polymorph is energetically favored at low temperature [15,32–35], with respect to the hexagonal one, the reversible transition between them occurring at 480.5 K in heating and at 477.5 K in the cooling process.

In Figure 1 the structures assumed by HAp are drawn, projected along the direction of the channels. In Figure 1a the $P6_3/m$ setting shows the perfect hexagonality of the [001] zone. In Figure 1b the setting $P2_1/c$ outlines the alternating OH dipoles in contiguous channels. Even if the two structures look very similar each other (the transition enthalpy between them being quite low: 130 J/mol [36,37]), a deep difference arises when comparing the homologous faces in the two settings.

a
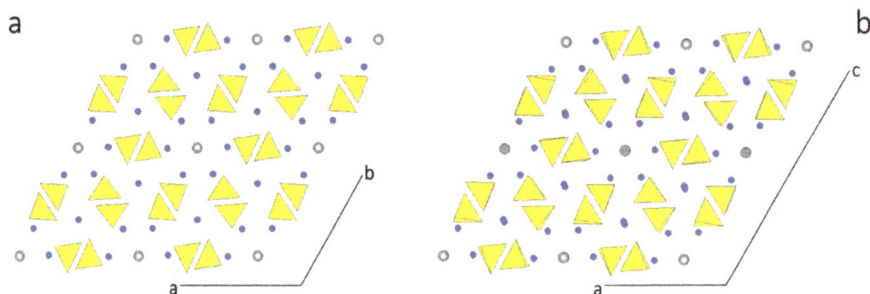
b

Figure 1. (**a**) The P6$_3$/m setting of a HAp crystal showing the perfect hexagonality of the [001] zone, being the axial parameters a$_0$ = b$_0$; (**b**) In the setting P2$_1$/c, the alternating up-down OH$^-$ dipoles in contiguous channels markedly outline the monoclinicity of the structure. The [010] direction coincides with the 2$_1$ axis and is perpendicular to the plane of the drawing, according to the convention adopted in [38].

2.3. The Complexity of Carbonated Apatites

Special care has to be deserved to the carbonated apatites, CAp, since much discussion and controversy still exist about how carbonate ions are incorporated into the apatite lattice and hence how its symmetry is affected. Carbonate in bones reaches approximately 7 wt %, while in tooth enamel is less than 3.5 wt %. When carbonate ions enter the structure channels, in the OH sites, the A-type substitution is obtained, while B-type is generated when carbonate ions enter in the PO_4^{3-} sites. As concerns the existence domains, it seems that A-type can be produced only through solid-state reactions (at 1000 °C), whereas the B-type can be found in biological apatite and obtained from solutions in the temperature range of 50–100 °C. Owing to the structural disorder intrinsic to the partial substitutions in both A- and B-types, and to the difficulty in preparing sufficient large crystals for an X-ray structural analysis, the precise configuration of the carbonate ions in the CAp structure has not been determined yet [6].

Thus, it should not be surprising if different space groups (belonging to the hexagonal or trigonal systems) have been hypothesized for CAp. As an example, three structural locations of type B were found for the carbonate ion in a complex, disordered type A-B carbonate apatite, with space group P6$_3$/m belonging to a crystal synthesized at 3 GPa and at 1400 °C [39].

Another CAp crystal, of type A, investigated by the same authors, was found domain-disordered having P$\bar{3}$ symmetry [40].

A third space group (P$\bar{6}$) was foundon a single crystal, grown by a CaCO$_3$ flux method, having chemical formula Ca$_{9.75}$[(PO$_4$)$_{5.5}$(CO$_3$)$_{0.5}$]CO$_3$. In this crystal all the A-sites, corresponding to OH-sitesin HAp were substituted by CO$_3$ ions [41].

In any case, suggestions are lacking to explain the growth mechanisms of CAp, i.e., to identify those crystal faces where the carbonated species present in the growth medium (as CO$_2$ or HCO$_3^-$, or CO$_3^{2-}$) preferentially react, in order to be absorbed in the growing crystal bulk. With the aim at giving a contribution to these aspects, in Section 5 we will describe how Ca-carbonates can affect both structure and morphology of growing CAp single crystals.

3. Solved and Unsolved Problems about the Relationship between Monoclinic and Hexagonal HAp Growth Behavior

The relationships between the "hypothesized" HAp hexagonal space groups (P6$_3$/m, P6$_3$, P$\bar{6}$) and the monoclinic one (P2$_1$/b) remains an open debate, to date. A non-exhaustive list of solved and unsolved problems are illustrated in the following:

(i) "HAp: hexagonal or monoclinic?" [42] and "Crystallographic structure of human tooth enamel by electron microscopy and X-ray diffraction: hexagonal or monoclinic?" [43] are representative titles of recent works concerning apatites growing in a bio-environment (bio-apatites).

(ii) Human single HAp nano-crystals (enamel and dentine) investigated by convergent beam electron- diffraction (CBED) and automated electron-diffraction tomography (ADT), recently showed [44] that $P6_3$ works instead of $P6_3/m$ S.G. and that this experimental evidence is of prime importance for understanding the influence of electric fields on the morphogenesis process of calcified tissues at the nano-scale [45].

(iii) An intriguing question has been recently asked about the "symmetry violation" or "symmetry breaking" by Tao et al. [46]: "*Why does apatite, which has a nominal hexagonal symmetry, form platelets with the c axis in the plane of the platelets, in violation of the underlying HAp crystallographic symmetry, and why do the platelets grow with their axes parallel to the fibril axis?*". A more precise problem [47] stands in line with this way of thinking: "*Focusing on nano-HAp morphology, needlelike [48–51] and platelike [52–54] nano-particles can be prepared, while in bone tissue only the second type seems to be present [55–57]. For both morphologies the prevailing surface terminations are of the {010} type, which in the case of needlelike nano-HAp are the lateral facets of the hexagonal nano-particles elongated along the crystallographic c-axis, while, for the platelike nano-particles they are the basal facets of nano-particles preferentially grown along both the c- and a- (or b-) axis [53,54,58] breaking the crystal symmetry through a mechanism still matter of investigation [52,58].*

(iv) Determining the outmost surface profiles of the most frequently observed {hkl} forms can be a risky attempt.

(a) From experiments: direct information from experiments are rare and doubtful. High resolution electron microscopy (HRTEM) was used to characterize the atomic structure of the "hexagonally shaped region" limited by the {01.0} faces [59]. The investigation on the crystalline-amorphous interfaces formed by electron-beam damage and on the grain boundaries parallel to the hexagonal prism, suggested that the HAp structure is terminated at the plane connecting the OH columns. However, the Authors "assumed" the crystal structure according to the hexagonal symmetry, on the ground that the differences between the monoclinic and the hexagonal phase are negligible in this kind of study. We will see, later on, that the deep differences between the surface profiles of hexagonal an monoclinic phases cannot be neglected, if one aims at understanding the complexity of growth kinetics ($\rightarrow$growth morphology) in pure growth medium and/or in the presence of habit modifiers. More recently, Ospina et al. [60] used the HRTEM technique to analyze the surface profiles of HAp nanoparticles (supposed to be structurally hexagonal . . .) precipitated from aqueous solution at 37 °C with preferential growth along the [001] direction and large (01.0) faces. The simulated nanocrystal was used to investigate the two different (01.0) faces observed in a single nanoparticle of 50 nm length and 8.49 nm width. This analysis permitted the characterization of the non-stoichiometric HAp surface terminations [59,61,62]. Starting from the HRTEM results, periodic density functional theory (DFT) was used to model and refine hydrated (01.0) HAp surfaces with the two different terminations in order to obtain information about the structural modification and chemical environment around the Ca-sites.

(b) From a theoretical point of view:

- The first paper in which the surface profiles of HAp were systematically examined and growth morphology was predicted and compared with the observed one, appeared on 1986. Terpstra et al. [13] applied the "connected net" method, based on the morphological theory by Hartman-Perdok [63] to the hexagonal ($P6_3/m$) HAp structure. Starting from a qualitative evaluation of the Ca–PO$_4$ and Ca–OH bonds, and assuming that the ordering of the OH dipoles has any influence on the Ca–OH bonds, the Authors argued that one can expect that " . . . monoclinic HAp should have the same morphological appearance of the hexagonal HAp"; accordingly, " . . . the lowering of the apatite symmetry ($P6_3/m \rightarrow P2_1/b$) due to the ordering of OH is not likely to be the cause of the occurrence of the plate-like apatitic crystals in calcified tissues".

- Surface relaxation on two opposite (100) free surfaces of a $P6_3/m$ HAp crystal, was invoked to explain the " . . . symmetry breaking effect on growth morphologies, producing equilibrium platelet morphologies even when these are inconsistent with the symmetry of the crystal" [14,64]. However, the Authors are conscious that platelet growth morphology does not occur when HAp is precipitated in vitro. Hence they speculated that " ... it may be possible that in vivo growth of apatite the body may manipulate the growth conditions (carbonate impurities, organic molecules found in bones) so as to allow the platelet growth mechanism to work".

- An interesting attempt was performed by De Leeuw and Rabone [17], who used molecular dynamics simulations to evaluate the interaction of citric acid with the hexagonal ($P6_3/m$) prismatic (01.0) and pinacoidal (00.1) surfaces in an aqueous environment. The prismatic surfaces were assumed as terminated by PO_4 groups (nothing being revealed about the termination of the pinacoid). Surface energies were calculated for dehydrated (*in vacuum*) surfaces, followed by those calculated in aqueous environment and, finally, with citric acid in an aqueous environment. The large difference in the adsorption energies for these two kind of surfaces indicates that " ... citric acid would be a much better growth inhibitor of the {01.0} prismatic form, slowing down its growth rate by binding to surface growth sites". As a consequence, the crystal shape of HAp grown in the presence of citric acid would become along [001], the direction of the OH channels, with expression of the prismatic form, as compared to the crystal morphology in the absence of citric acid. It is worth noting that this work, published on 2007, allowed outlining, for the first time, the influence of the different surface structure of two important HAp forms, even if the molecular adsorption is assumed within the "large constraints" of the Molecular Dynamic Simulation.

- Astala and Stott [61] carried out a first-principles study about water adsorption on the {01.0}, {00.1} and {10.1} forms of hexagonal ($P6_3/m$) HAp. In particular, they determined the surface energy *in vacuum* of the pinacoid {00.1} having considered the approach to modeling the surface profiles proposed by Rulis et al. [62]. In this model, the sole constraint imposed was that " ... surface must be constructed so that the corresponding slabs were charge neutral". Moreover, they considered that, in HAp, the orientation of the OH group along the [001] channels (*c* axis) reduces the symmetry such that the top and bottom of the 001 slab surfaces are no longer symmetric. Accordingly, these Authors introduced the term "symmetry breaking" limited to the form {00.1}. As concerns the other important form, the hexagonal {01.0} prism, the problem of the "symmetry breaking" does not longer exist since the OH dipoles lie parallel to the {01.0} surfaces. However, this form can terminate with three different profiles, according to the geometric 01.0 cut applied to the bulk of the crystal. One of them corresponds to a stoichiometric slab (Ca/P ratio being 1.67); the two out of three are non-stoichiometric: a Ca-rich slab (Ca/P ratio = 1.75) and a PO_4 rich slab (Ca/P ratio = 1.5). This is surely interesting, but something was wrong, not only from a crystallographic point of view but also because the growth mechanism of the flat faces (F-faces, in the sense of Hartman-Perdok [63]) was not respected. In fact, the space group ($P6_3/m$) implies that the slices of thickness d_{002} and d_{010} should be centrosymmetric and hence their total dipole moment should vanish (owing to the electrical stability of the slice). Hence, the symmetry breaking attributed to the steps growing on both sides of the {00.1} form is not self-consistent.

- A sensible step forward was taken by the Ugliengo et al. [19], who carried out a periodic B3LYP study of hexagonal $P6_3$ HAp (00.1) surface modelled by thin layer slabs [19]. The (00.1) surface coming out from $P6_3$ group is intrinsically polar, owing to the OH dipoles, all iso-oriented perpendicularly to the 00.1 planes. Nevertheless, the convergence of $\gamma_{00.1}^{exagonal}$, the specific surface energy value for the hexagonal (00.1) form, has been ascertained for a slab thickness varying from a minimum of one layer (~7Å) to a maximum of 9 layers (~60Å). For the homologous monoclinic (non-polar) $P2_1/b$ HAp (00.1) surface the convergence is, obviously, more rapidly obtained. For the sake of comparison, the two asymptotic values are: $\gamma_{00.1}^{exagonal}$ = 1107 and $\gamma_{001}^{monoclinic}$ = 1337 erg cm^{-2}. In a successive paper [26], it has been explained why the right choice of the $P6_3$ group has been made: as a matter of fact, " ... the quantum-mechanical simulation of the hexagonal HAp cannot be performed within $P6_3/m$ space group because of the unphysical duplication of each OH group by the *m* mirror plane". In the same paper, the surface profiles of hexagonal prismatic {01.0} and pyramidal {10.1} forms were investigated as well. On the ground of Astala and Stott [61] and according to the Authors way of thinking, the {01.0} non-polar surfaces can be imagined as a stacking of electro-neutral layers ... -ABA-ABA- ... where the A-type has the composition $Ca_3(PO_4)_2$ while the B-type corresponds to the composition $Ca_4(PO_4)_2(OH)_2$. The specific surface energy related to these "stoichiometric" surfaces is $\gamma_{(01.0)\ stoichiometric}^{exagonal}$ = 1709 erg cm^{-2}. On the ground of recent findings [59,61], two other "non-stoichiometric" {01.0} surfaces were also considered: [B-AA-B-AA-B], Ca-rich, being Ca/P = 1.71 and [AA-B-AA-B-AA], P-rich, being Ca/P = 1.62. Regrettably, the corresponding $\gamma_{(01.0)\ non-stoichiometric}^{exagonal}$ values were not calculated, since thesurface energy cannot be evaluated using the standard formula, adopted by the Authors, that is valid for stoichiometric slab only.

Notwithstanding, all these surfaces (both stoichiometric and non-stoichiometric) were usefully adopted to calculate the interaction energy of adsorbed single molecules such as: water, glycine, glutamic acid, lysine and carbon oxide [26,27].

- It is worth recollecting as well a remarkable paper recently publishedby Putnis et al. [65] about the growth kinetics measured by AFM on a (010) prismatic face of a synthetic HAp crystal growing from aqueous solution [65]. It was observed that HAp crystallization occurred by either classical spiral growth or non-classical particle-attachment from various supersaturated solutions at near-physiological conditions, suggesting these mechanisms do not need to be mutually exclusive. Moreover, this work represents " ... the first evidence of time-resolved morphology evolution during precursor-particle attachment processes, ranging from primary spheroidal particles of different sizes to triangular and hexagonal solids formed by kinetically accessible organized assembly and aggregation".

When summarizing, the critical points illustrated in the present Section show that the largest majority of the researches has been done on the hexagonal HAp polymorphs, especially those concerning the interfaces crystal/growth medium (both in vacuum and in the presence of a condensed mother phase, with and without specific impurities). Nevertheless, any satisfactory explanation has not been done, till now, about the "symmetry breaking". In other words, if the space groups is $P6_3/m$ or $P6_3$, how can a hexagonal prism (built by six equivalent faces) degenerate in a set of six faces where two out of them dominate the growth morphology, so determining the final appearance of platelets [001] elongated, instead of the expected hexagonal rods?

A few simple questions arise. Are we sure that HAp crystals are always hexagonal, especially when nucleated at low temperatures from aqueous solution? Moreover, what should be the action of the very frequent twinning on the final morphology of the monoclinic polymorph?

According to our experience about adsorption phenomena in crystal growth, selective adsorption/absorption occurs only when the adsorbing surfaces are crystallographically different [66,67]: hence, something should be wrong when neglecting that monoclinic HAp structure could be "necessarily" more rich than the hexagonal one, at least for the variety of the surface profiles exposed to the mother phase.

In the following Section we will deal with the method of investigation we propose to outline the marked difference between the surfaces generated by monoclinic and hexagonal HAp polymorphs, along with the related consequences on their equilibrium and growth morphologies. To do this, it would be more fruitful to begin with the more complex monoclinic phase.

4. Approaching the Morphology of the Monoclinic HAp through the Hartman-Perdok Method

Recently [38], we determined the ab initio theoretical equilibrium shape (ES) of the monoclinic HAp, in vacuum and at 0K; neither vibrational nor configurational entropy were taken into account. Stable surface profiles of the different {hkl} forms were obtained by applying the constraints intrinsic to the Hartman-Perdok method [63] (HP-method, hereinafter). To do this: (i) the character (flat-F, stepped-S, kinked-K) of the {*hkl*} forms was identified, through the search of the most important periodic bond chains (PBC) building up the crystal structure; (ii) the corresponding crystal slices of thickness d_{hkl} were considered, respecting the space group constraints; (iii) all the possible surface profiles related to every d_{hkl} slice were drawn, respecting the conditions of both electro-neutrality and vanishing of the dipole moment component perpendicular to the *hkl* plane; (iv) the specific surface energy (γ_{hkl}) was calculated, for each *hkl* surface profile, to draw the ES of the single crystal, by applying the selective rules required by the Gibbs-Wulff's theorem [3].

It is worth outlining as well that the steps (i) → (iv) were needed to investigate both crystallography and genetic aspects of the HAp twins, to find their original composition planes (OCP) and to evaluate the energy values associated to their formation [68]. Moreover, steps (i) → (iii) were also needed as the preliminary requirement to interpret the epitaxial interactions between HAp and monetite ($CaHPO_4$), used as a precursor phase favoring the 2D-heterogeneous nucleation that allows 3D HAp to appear beyond its usual growth conditions [69].

Finally, we really must spend a few words in favor of applying the HP method: it is somewhat laborious, since it cannot be successfully automated; nevertheless, its results obtained for a long time are reassuring, especially when dealing with low symmetry and complex crystal structures.

Computational details are described in reference [38]. The parameters (in Å) of the adopted optimized cell were: $a_0 = 9.3253$; $b_0 = 6.9503$; $c_0 = 18.6436$; $\beta = 119.972°$, referred to the space group $P2_1/c$ and to 4 unit formulas [i.e., $Ca_{20}(OH)_4(PO_4)_{12}$]. The setting $P2_1/c$ has been preferred to $P2_1/b$ because the diad traditionally coincides with the *y*-axis, in the monoclinic system.

4.1. The Surface Profiles of the Faces in the [010] Zone, i.e., Parallel to the HAp Channels

According to the HP analysis, three flat forms can be found in the [010] zone: {100}, {$\bar{1}$02} and {001}, related to the thicknesses d_{100}, $d_{\bar{1}02}$ and d_{002}, respectively. The thickness of these slices fulfills the systematic extinction rules coming out from the space group $P2_1/c$. In Figure 2 the monoclinic HAp structure is projected along the [010] direction and the surface profiles of the three forms are drawn, as comes out from choosing one of the possible [010] PBC, i.e., the PBC [010]$_A$. At first sight, one could say that the surface profiles of the three forms are practically indistinguishable: then, one could expect that both their surface and attachment energies should also have similar values. Consequently, it is not surprising that the ensemble built by the three monoclinic pinacoids {100}, {$\bar{1}$02} and {001} might be easily confused with the "really hexagonal" prism related to the hexagonal HAp polymorph. However, from a deeper examination of Figure 2 it follows that:

- Neither PO_4 tetrahedra nor Ca ions lie on the ideal separation surface between two adjacent slices and hence both ions are not shared between adjacent slices. Further, the d_{100}, $d_{\bar{1}02}$ and d_{002} slices do not show dipole moments perpendicular to their surfaces. Accordingly, the surface profiles of the three pinacoids {100}, {$\bar{1}$02} and {001} do not have to be reconstructed.
- The OH ions in the channels are screened from the mother phase by the outmost layers of each slice populated by PO_4 and Ca ions not directly bound to the OH in the channels.
- The d_{100} and $d_{\bar{1}02}$ slices are center-symmetric, while the d_{002} slice contains only the 2_1 screw axes parallel to the [010] channels: this implies that a growing HAp monoclinic crystal shows two types of d_{002} slices, according to whether the OH dipoles lying in the middle of the slice are oriented along the positive or negative sense of the *y* axis. On the contrary, d_{100} and $d_{\bar{1}02}$ slices do not show polarity parallel to the slice itself, which could unavoidably affect the difference of both surface and attachment energies of {100} and {$\bar{1}$02} with respect to the {001} form.

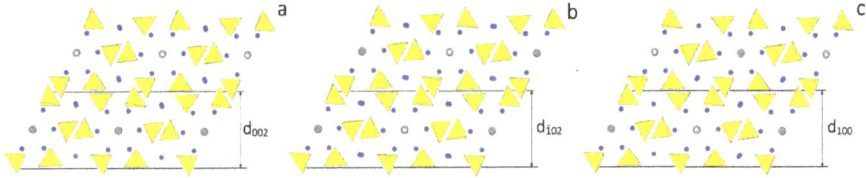

Figure 2. Projection along the structure channels of the HAp polymorph. The space group P2$_1$/c has been chosen, as in Figure 1b. Hence the channels run along the [010] direction parallel to the 2$_1$ diad axis. Slices are defined respecting the systematic extinction rules of the P2$_1$/c space group. (**a**) The structure of the slice of thickness d_{002}, limiting the pinacoid {001}, shows that in two adjacent slices the OH dipoles are oriented in opposite sense. On the contrary, in (**b**,**c**) the structures of the slices $d_{\bar{1}02}$ and d_{100}, limiting the pinacoids {$\bar{1}$02} and {100}, respectively, show a strong similarity (they are not identical) since, within each slice, the OH dipoles alternate up-down.

There are two other ways of building a PBC running along the channel axis [010]: they are both associated to a general PBC [010]$_B$, one being center-symmetric while the other one runs along the axis 2$_1$ // [010]. Both these PBCs do not contain the OH dipoles in their centers, at variance with the [010]$_A$ PBC. Here we will confine our attention to the effect of this new choice on the terminations of the monoclinic HAp surfaces; for deeper details the reader is kindly invited to refer to the original paper [38]. Concerning the new alternative terminations:

- the outmost layers of the new surfaces are populated not only by the PO$_4$ tetrahedra, but also by Ca ions and OH dipoles;
- the occupancies of the outmost layers must be reduced by 50%, as follows from the constraints (symmetry, charge, stoichiometry) imposed by the frontiers between adjacent d_{002}, d_{100} and $d_{\bar{1}02}$ slices. In fact, these frontiers pass through the centers of mass of PO$_4$, Ca and OH ions.

Summing up, this is the way we have got over the problem of the surfaces exposed towards the vacuum: a few simple physical rules, expressed by HP method, were able to generate a lot of different (and self-consistent) surface profiles for each {*hkl*} form.

4.2. The Specific Surface Energies of the Monoclinic HAp and Its Equilibrium Shape (ES) Calculated at 0 K

In Table 3 both structural and energetic features of the {001} pinacoid are described. One can see that HP method allowed evaluation of the dispersion of the surface energy (γ_{001}) values. Nine {001} surface profiles can be obtained: one out of them is terminated by the PO$_4$ tetrahedra (see Figure 2a) and does not need to be reconstructed, while other eight different configurations of the outmost layer can be obtained when surface reconstruction is needed, owing to the structure of the two [010]$_B$ PBCs. In the latter case, the γ_{001} values are distributed in two sets, one half of them being lower and the other one higher than their mean value, 1692 erg cm^{-2}, which is only 1.1% lower than the one corresponding to the unique profile obtained from the PBC [010]$_A$.

Table 3. Surface features and corresponding specific surface energy values related to the {001} form*.

Monoclinic HApForm	PBC of Reference	Surface Termination	Is the Outmost Layer Shared between Adjacent Slices?	Surface Reconstruction Needed	γ_{001}(erg cm^{-1})
{001}	[010]$_B$	PO$_4$, Ca, OH	yes	yes	1546, 1613, 1666,1691
	[010]$_A$	PO$_4$	no	no	1712
	[010]$_B$	PO$_4$, Ca, OH	yes	yes	1738,1741,1742,1793

*The unique γ_{001} value in the row referred to the PBC [010]$_A$ was obtained from the d$_{002}$ slice terminated by PO$_4$ ions, as drawn in Figure 2a. First and third row γ_{001} values refer to the eight profiles that can be obtained from the different disposition of PO$_4$, Ca and OH ions when reconstructing the outmost layers of the alternative d$_{002}$ slices, coming out from the two PBCs [010]$_B$.

The just mentioned investigation procedure was applied to the two other pinacoids $\{\bar{1}02\}$ and $\{100\}$, to the basal $\{010\}$ pinacoid and to the three monoclinic prisms $\{012\}$, $\{110\}$ and $\{\bar{1}12\}$ (see Figure 3b, for indexing). It followed that:

- The surfaces of the three pinacoids $\{001\}$, $\{\bar{1}02\}$ and $\{100\}$, parallel to the OH channels, can be terminated either by only PO_4 or by the complete set composed by PO_4, Ca and OH ions.
- The basal $\{010\}$ pinacoid, which is perpendicular to the OH channels, can be terminated either by only Ca or by the set made by PO_4, Ca and OH ions.
- The three monoclinic prisms $\{012\}$, $\{110\}$ and $\{\bar{1}12\}$ can terminate either by Ca ions (as for the $\{012\}$ prism) or by PO_4 ions (as for the $\{110\}$ and $\{\bar{1}12\}$ prisms). It follows that the terms like "Ca rich" and "P rich" along with the terms like "stoichiometric" and "non-stoichiometric" lose meaning. On the contrary, "reconstructed, or non-reconstructed, outmost layer", "polar, or non-polar, d_{hkl} slice", "relative disposition of ions in a given reconstruction", assume a precise crystallographic meaning, unambiguous in order to describe both equilibrium and growth crystal properties.
- The lowest values of the calculated specific surface energies, for every $\{hkl\}$ form, was used to draw the ES of HAp at 0K. As shown in Figure 7 of reference [38], the monoclinic ES can be described by a pseudo-hexagonal "prism" truncated by the $\{010\}$ pinacoid, which reaches the lowest γ value of the entire crystal when its surface is Ca-terminated ($\gamma_{(010)}^{Ca-term} = 1041$ erg cm^{-2}). The pseudo-hexagonality of the "prism" arises from the closeness of the lowest γ values of the three pinacoids: $\gamma_{(001)} = 1546$ erg cm^{-2}; $\gamma_{(100)} = 1525$ erg cm^{-2}; $\gamma_{(10\bar{2})} = 1515$ erg cm^{-2}, all these values being obtained when the outmost layers of these forms are reconstructed and exhibit an outmost population made by PO_4, Ca and OH ions. The ES is completed by the presence of the small–sized quasi-equivalent prisms $\{012\}$, $\{110\}$ and $\{\bar{1}12\}$.

Finally, it is worth stressing once again that the $\{001\}$ form is the most anisotropic in the [010] zone, since it is the only one (at variance with the $\{\bar{1}02\}$ and $\{100\}$ forms) in which the OH dipoles alternate up/down within the successive d_{002} slices, irrespective of the termination of its outmost layer. This marked difference does not play an appreciable role at equilibrium (in the vacuum), as we just demonstrated. A different situation has to be envisaged when dealing with both foreign adsorption and growth kinetics, as we will discuss later on.

Summing up, we can assess that distinguishing the athermal ES of the monoclinic HAp from that of the hexagonal one should be quite difficult.

4.3. About the Growth Morphology of the Monoclinic HAp: A Comparison with the Hexagonal One

In Figure 3a we consider the shape of a micrometric HAp crystal we grew from pure aqueous solution under mild hydrothermal conditions. The sketch in Figure 3b represents a reasonable attempt at indexing its crystallographic forms: the pseudo-hexagonality could be misleading, but the different development of the $\{001\}$ and $\{100\}$ forms largely plays in favor of the monoclinic 2/m symmetry. A comparison with the theoretical shape of a crystal showing a 6/m symmetry is made in Figure 3c: all the six faces belonging to the hexagonal $\{01.0\}$ prism must have the same extension, along with the hexagonal $\{10.2\}$ bi-pyramid. This is to say that growth kinetics does not allow confusing the two HAp symmetry systems.

Let's consider, as an example, a face (001) belonging to the form $\{001\}$ and a contiguous face (100), belonging to the form $\{100\}$, as drawn in Figure 3a,b. It is theoretically known that they are flat-F faces and then can advance either by 2D-nucleation and/or by spiral growth; moreover, this kind of face kinetics was experimentally observed in the Putnis research team [65]. As a consequence, their growth rate will depend (a part the temperature, the supersaturation and the composition of the growth medium), from the *structure of the steps* running on their flat surfaces, because the thermodynamic parameter ruling layer-by-layer growth (2D or spiral) is the specific edge energy of the steps limiting either the 2D nuclei, or the spiral pattern, or both.

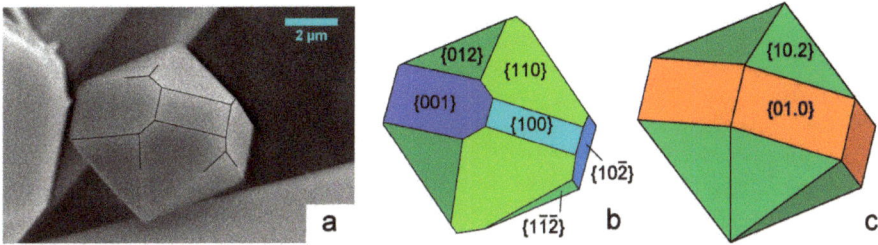

Figure 3. (a) SEM picture of our HAp crystal grown under mild hydrothermal conditions. It belongs to the monoclinic polymorph, and shows three consecutive pinacoids: {001}, {100} and {$\bar{1}$02}, all parallel to the [010] zone axis, along with the three corresponding prisms: {012}, {110}, and {$\bar{1}$12}, as drawn in (b); where the theoretical (2/m) morphology is represented with different sizes and colors associated to different crystallographic forms. The theoretical (6/m) morphology is drawn in (c). It is related to the space group P6$_3$/m, with the [001] direction parallel to the 6$_3$ axis. The comparison between the three frames makes evident that the growth shape of the monoclinic polymorph could be easily confused with the hexagonal one.

Concerning the form {001}, one can reasonably imagine that the shape of a 2D nucleus (and then of a spiral) developing on this face is a rectangle limited by the following kind of steps: two, different each other, parallel to the [010] direction and two, equivalent, parallel to the [100] axis. The surface structure of one out of the [010] steps is that of the {100} forms, while that of the other one is comes out from the {$\bar{1}$02} form; the surface structure of the two symmetry equivalent steps parallel to the [100] axis originates from the {012} form. Remembering that the surface structure of the {100} and {$\bar{1}$02} forms are practically the same, the 2D nuclei developing on the {001} form must have a quasi-*mm* symmetry.

This is no longer valid for nuclei (and/or spirals) which develop on the two {100} and {$\bar{1}$02} forms adjacent to {001}. In fact, the steps running along the [010] direction borrow the {001} and {$\bar{1}$02} structure, on the {100} form and the {001} and {100} structure, on the {$\bar{1}$02} form, respectively. Summing up, the two pinacoids {100} and {$\bar{1}$02} could have a very similar growth rate (due to the closeness of their surface patterns), while the pinacoid {001} could experience a different kinetic behavior.

As a consequence, the HAp growth shape (GS) viewed along the direction of the OH channels should be anisotropic with respect to the quasi-hexagonal ES: this means that, at constant temperature of crystallization and composition of the growth medium, the anisotropic aspect ratio will change with the supersaturation.

It is well known that F faces, crossed by screw dislocations outcrops, are characterized by a complex growth rate (R_F/kT) dependence on the supersaturation ($\Delta\mu$). In fact, the growth isotherms [(R_F/kT) vs.$\Delta\mu$] show: (i) a parabolic law, at low $\Delta\mu$ values; (ii) an exponential law when the 2D nucleation occurs in between the spiral steps, at medium-high supersaturation; (iii) and finally a linear law at higher $\Delta\mu$ values. Let's call $R_F^{(001)}$ and $R_F^{(100),(\bar{1}02)}$ the isotherms related to the form {001} and to the two other pinacoids {100} and {$\bar{1}$02}; owing to the reasons just exposed, these curves cannot have the same trend with respect to the supersaturation and then can cross at a given critical supersaturation $\Delta\mu_{cr}$. This implies that they sharply differ at high $\Delta\mu$ values, where $R_F^{(100),(\bar{1}02)} > R_F^{(001)}$, as we will show in a forthcoming paper: in this case the Growth Shape I can be obtained (Figure 4, center). In the supersaturation domains where $\Delta\mu$ approaches $\Delta\mu_{cr}$, $R_F^{(100),(\bar{1}02)} \cong R_F^{(001)}$, the aspect ratio tends to the Growth shape II (Figure 4, right side). Finally, when the growth system tends to the equilibrium ($\Delta\mu \rightarrow 0$), the crystal assumes a pseudo-hexagonal shape (Figure 4, left side).

Applying the same reasoning, we already demonstrated that both HAP rods and/or platelets should be strongly elongated along the direction of the OH channels [38].

A preliminary conclusion can be drawn: the formation of HAp platelets can be explained by the cooperation of two factors: (i) the asymmetrical behavior of the growth steps running on the form {001} with respect to those developing on the {100} and {$\bar{1}$02} couple; (ii) the role exerted by the supersaturation of the growth system. Hence, "symmetry breaking" cannot be invoked, because the platelets are originated as a monoclinic HAp phase which enhances its asymmetric properties during growth, contrarily to the hexagonal one which could not change its aspect ratio with supersaturation.

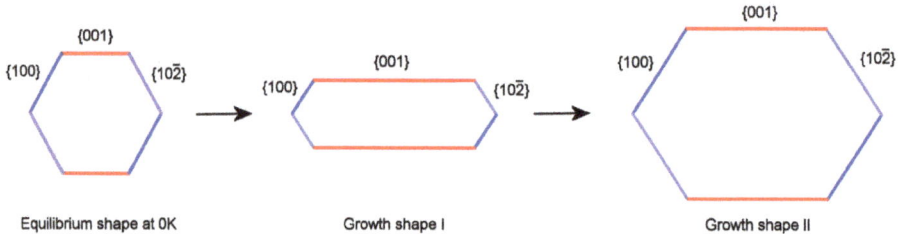

Figure 4. Sketch of a monoclinic $P2_1/c$ HAp crystal viewed along the OH channels direction. The equilibrium shape at 0 K is clearly pseudo-hexagonal, with the two {100} and {$\bar{1}$02} forms showing a quasi-identical surface profile (left side). The growth shape I does occur when the supersaturation is high (early stages of growth in a closed system), owing to the anisotropy of the growth kinetics between the {001} pinacoid and the two kinetically close {100} and {$\bar{1}$02} forms (center). The growth shape II, tending to the pseudo-hexagonality, is recovered when all the growth isotherms intersect, at a critical supersaturation value or when the system approaches the equilibrium (right side).

5. The Growth of Nanosized vs. Micrometric HAp Crystals, in the Frame of the Experimental Methods

As briefly mentioned in the Introduction, the research about the structure and properties of apatites was strongly sector-based since the early studies. In 1951, referring to the crowd of scientists interested in the field, Jaffe [70] wrote: "Although all these people have had the same problems, they have not always been sufficiently acquainted with each other's work". Then, she stated: "Each group of scientists concerned with the apatite minerals has its own problem which it has attempted to solve in its own way" denouncing the failing communication inside the scientific community. As in many other fields of sciences, these statements are still as relevant today.

We are firmly convinced that biomedical and geo-mineralogical research, even if their experimental method and goals are different, could master as a common language both crystal growth and structural crystallography. The intrinsic complementarity of these disciplines could represent the most efficient and rapid way to overcome the remaining unsolved problems between structure and growth of apatite, both in nature and laboratory.

As already mentioned, the discussion about the symmetry of HAp involved a large number of papers, the most extensively studied being the hexagonal polymorph. Only a few papers deal with the monoclinic form, relegating its appearance to the biologically precipitated HAp [42] or the cases of high-temperature crystallization of HAp [20,29–31].

The continuous research of a rigorous routine of synthesis of apatites (along with all the possible chemical and structural variations) is the topic of a huge number of multidisciplinary papers published since the early years of the XX century. The interest in the synthesis of apatites is mainly related to the importance of the mineral in biomineralization and its biocompatibility.

During the years, many solid-state reactions, melt growth, wet chemical methods and hydrothermal methods have been proposed in order to obtain reproducible results in apatite crystallization. In most cases, authors obtained the mass crystallization of nano-sized and often poorly crystalline HAp. Then, a wide range of techniques was applied to answer the queries arising from the natural variability of composition and structure of HAp crystals. A thorough collection of the scientific production prior to 1951 was reported by Jaffe [70].

Only a few efforts resulted in the hydrothermal growth of large HAp crystals for diffraction and morphology investigations. In 1956, Perloff and Posner [71] hydrolized monetite ($CaHPO_4$) working at low hydrothermal conditions (300 °C, water vapor pressure evaluated of 87 bar for 10 days). They modified a previous work by Schleede and coworkers [72] and obtained large HAp crystals about 300 μm long. In 1966, Jullman and Mosebach [73] obtained large hydroxyapatite crystals (2.35 mm long and 0.2 mm large) by direct precipitation from a solution of calcium nitrate, sodium hydroxyde, and potassium dihydrogen phosphate at 500 °C. In 1968, Kirn and Leidheiser [74] suggested a modification to the Perloff's method, hydrolyzing monetite at 350 °C and about 600 bar in a diluted phosphoric acid solution. The final pH ranged around 2.5, and it was found that, at this point, the growth reached its end-point.

In their paper, Schleede and coworkers [72] did not describe the size and crystal quality of HAp crystals obtained hydrolizing brushite ($CaHPO_4 \cdot 2H_2O$) under hydrothermal conditions but, recently, Ma and Liu [42] modified the original method by Schleede and demonstrated that monoclinic HAp could be obtained at low temperature without recourse to specific impurities.

5.1. The Role of the Epitaxy on the Growth of HAp from a Phosphate Precursor and the Control of Nucleation Frequency and Growth Rate by Supersaturation

Very recently [69], we described a procedure of growth of large HAp crystals with controlled size under mild hydrothermal conditions (210 °C and water vapor autogenic pressure evaluated of 20 bar). In this work, we followed the path suggested by Perloff and Kirn. As described in our paper, we obtained large, needle-like HAp crystals by the hydrolysis of monetite ($CaHPO_4$) in water and in diluted phosphoric acid.

Free calcium and phosphate ions, conductivity and pH were monitored during the reaction. We observed large and well-finished HAp crystals generated working at pH lower or near the lower limit of the stability field of the HAp (calculated at pH ≈ 4.5). Thus, the growth of HAp crystals cannot be explainedin terms of 3D homogeneous nucleation. We demonstrated both experimentally and theoretically that monetite acts as a heterogeneous substrate for the nucleation of HAp, the two phases showing excellent epitaxial agreement [69]. This finding allows us to stress that, in the presence of a suitable substrate, the heterogeneous nucleation of HAp on monetite can occur, even if the system is unsaturated with respect to the nucleating phase.

On the other hand, if brushite substitutes monetite as a precursor of HAp at the same T, p conditions, the mass crystallization of nanosized HAp occurs, as described by many authors, although in the case of brushite excellent epitaxial relationships with HAp are fulfilled. In a forthcoming paper [75], we will demonstrate that the interplay between excellent epitaxial relationships and supersaturation rules the quality of the crystals obtained. The rate of reaction of brushite in water results to be markedly higher with respect to that of monetite, leading to a sudden increase in the supersaturation of the solution with respect to HAp and therefore, to a rise in the nucleation frequency.

In fact, the sudden formation of a huge number of nuclei is, of course, related to the supersaturation reached during the procedure. In turn, the supersaturation value is related to the concentration of ionic species into the reactor and to the pH that could be varied in order to move the system nearly out of stability and, accordingly, to lower the related supersaturation. For that reason, at the end of our paper [69] we stressed the importance of the control over the pH value to improve both bulk and surface crystal quality. High supersaturation values lead to mass crystallization of small nuclei owing to the relationship between the supersaturation and the nucleation frequency of the phase. In light of that, the issue of the determination of trustworthy values of solubility becomes crucial to guarantee the effective control over the operational supersaturation. It is worth recollecting here that we adopted, as the thermodynamic supersaturation, the difference between the chemical potential of a growth unit in the mother phase and in the crystal: $\Delta\mu = (kT) \ln \beta$, where $\beta = (1 + \sigma_v) = [(a - a_{eq})/a_{eq}]$ represents the ratio between the activities in the volume of the mother phase and at equilibrium. Concerning J_{3D} (number of 3D nuclei per unit volume and time), i.e., thenucleation frequency of a 3D phase, we remember that $J_{3D} \approx \exp(-\Delta G_{3D}/(kT))$ where ΔG_{3D} indicates the activation energy for 3D nucleation: $\Delta G_{3D} = (f \times \Omega^2 \times \gamma^3)/(\Delta\mu)^2$. Here, f is a factor form depending on the shape of the nucleus, Ω is the volume of the growth unit in the crystal and γ is the averaged value of the specific surface energy of the crystallizing nucleus.

5.2. The Issue of the Solubility and Its Consequences on the Nucleation Frequency

The need for a reassessment of the solubility of calcium phosphates has been stressed by many authors [71] and we already pointed out [69] the attention to the lack of reliable data on the solubility of HAp from the literature. These data are of primary importance when growing crystals because of the relationship between the solubility of a solid phase, the supersaturation of the solution and, accordingly, the nucleation frequency. Many papers dealing with the solubility of HAp have been published during the years, being the HAp stability of crucial importance mainly for biomedical applications. The values of the solubility product vary over a range of 10^{11}, according to Van Wazer [76] and many other authors [77–80]. A unique value of solubility is controversial. In 1953 and successively in 1955, Neuman [81,82] and his collaborators declared the principle of solubility product non-enforceable in the case of HAp, due to the chemical variability of the solid calcium phosphate system. Other reports claim the system does not obey to the principle of the solubility product constant based on the stoichiometry of the salt. Some models entail the formation of surface complexes [83] or dependence from the pH of the surface stoichiometry [84,85] and surface reactions [86–88]. Following these models, the solid/solution equilibrium is achieved only by the external layers of the crystalline HAp, having a slightly different composition with respect to the bulk. This composition varies along with the chemical composition of the solution and all the possible chemical reactions at the surface. On the contrary, according to some authors, the effect of the surface on the HAp solubility has been disregarded [89].

Moreover, the preparation of pure HAp for solubility measurements is often not free from contamination. For example, small amounts of carbonate in solution modify a lot the value of the solubility product according to the papers by Greenwald [90–92]. Additionally, the precipitation of acid byproducts like OCP or $Ca_8H_2(PO_4)_5 \cdot 5H_2O$ is a recurring event, and their presence will modify the solubility product value.

Furthermore, the matter of the incongruent dissolution of calcium phosphates is widely debated by many authors [86–88,93]. In this case, the chemical composition of the solid and the solute may differ, leading to a misinterpretation of the solubility measurements. On the other hand, some reports claimed the existence of a single solubility product, as Clark did in his papers. In fact, in 1955, Clark and Peech [94] presented a solubility diagram for HAp in soils. At the same time, Clark [77] proposed a further paper, demonstrating that, within the variability due to analytical errors, HAp possesses a unique solubility product over a wide range of experimental conditions.

Besides the issue of the solubility, also the application ofthe right growth conditions knowing the stability field of HAp plays a major role in HAp synthesis. Many papers were dedicated to the subject. In 1974, Skinner [95] determined the equilibrium phase diagrams for the system $CaO\text{-}P_2O_5\text{-}H_2O$ growing HAp at temperature ranging from 300 to 600 °C in hydrothermal conditions. In 1997, Andrade and coworkers [96,97] come through an extensive thermodynamic analysis of the system $Ca\text{-}P\text{-}H_2O$. They developed the Eh-pH and Pa_{Ca}-pH diagrams varying the temperature and the activities of calcium and phosphorous in the system.

As one can infer from the charts, the HAp dominates at pH > 4 at any temperature, with a strong dependence on the activity of calcium in solution and on the Eh.

Summarizing, when referring to the issue of solubility, the traditional methods of solubility determination by the addition of large amounts of solid do not work properly in the calcium phosphate system. This may be related to many problems: (i) the incongruent dissolution of Ca-phosphates in water; (ii) the uncommonly high number of phases and equilibria involved in speciation of Ca-phosphates in solution, many of those discarded from calculations for simplification; (iii) the retrograde HAp solubility with respect to temperature, as reported by McDowell [98]; (iv) the Ca/P ratio. If non-stoichiometric HAp is used for solubility determination, the solubility value obtained will change with this ratio; (v) the chemical contamination by foreign ions; (vi) the surface effects on HAp solubility.

When obtaining HAp from direct precipitation by mixing of calcium and phosphate solutions, as reported by most of the papers about HAp precipitation, authors are compelled to work at high concentration of the former solutions. Moreover, in order to avoid the precipitation of the acid phases of calcium phosphate, authors usually force the system toward basic conditions (pH ranging between 8 and 10), moving the strongly concentrated solution into the field of HAp stability. A careful control of the solution supersaturation is hindered by the absence of reliable data about HAp solubility. The lack of monitoring of the supersaturation leads to the conventional mass crystallization of nano-sized HAp.

The growth of HAp by hydrolysis of a calcium phosphate (mainly monetite or brushite as in the papers by Perloff [71] and Schleede [72]) under hydrothermal conditions discourage, to some degree, the authors to work at very high supersaturation. In fact, the hydrolysis obtained at high temperature slows down the achievement of the supersaturation both due to the kinetics of dissolution of the precursor and to the retrograde behavior of HAp solubility with respect to temperature.

Summarizing, in our just quoted paper [69], we demonstrated the effectiveness of the low supersaturation growth obtained by the hydrolysis of monetite at low hydrothermal conditions. As stated in the previous paragraph, reducing the concentration of calcium and phosphate in solution and moving the system toward low pH values, near the low-pH boundary of the stability field of HAp lowers the supersaturation of the system with respect to HAp. This, in turn, lowers the HAp nucleation frequency allowing the crystallization of large individuals, which are needed for both structural and morphological investigations. Both in the case of monetite and brushite, we invoked the significance of excellent parametric coincidences between host and guest phases in order to promote the 2D heterogeneous nucleation and epitaxial growth (2D heterogeneous nucleation) of the latter, lowering the energy required for the nucleation. The amazing differences between the results are mainly due to the kinetics of dissolution and supersaturation.

Moreover, this approach let us demonstrate that, also during hydrolysis, at high dissolution rates, the first generation of crystals has a "tape-like" or "ribbon-like" [20] behavior typically ascribed to biological apatites. Therefore, the flat, undeniably monoclinic habit of small HAp crystals can be obtained when working at high supersaturation, which favors nucleation instead of crystal growth and without claiming the effect of either specific impurities or high crystallization temperature.

5.3. The Effect of the Presence of Carbonate Ions in Solution

The observation of the morphological effect of carbonated species present in the growth medium provides information about the crystal faces involved in the adsorption/absorption of those impurities.

It is experimentally demonstrated that the presence of carbonated species affects both the morphology and structure of apatites depending on their concentration and on the pH of the solution [75]. In fact, the pH strongly affects speciation of calcium carbonate in solution and, consequently, the surfaces that possibly interact with those surface specific impurities.

In the presence of a carbon dioxide saturated solution, the pH ranges in the stability field of the CO_2 and HCO_3^- carbon species. In such conditions, the growth rates of the {100} and {$\bar{1}02$} forms of monoclinic HAp differentiate with respect to the {001}. The interaction between apatite and impurities in solution results in a growth morphology strongly modified, from the typical pseudo–prismatic extensively described in a previous work [69] (Figure 5a) to a tabular one (Figure 5b). The tabular morphology is dominated by extended faces ascribable to the {001} pinacoid, the only one among the three kinetically strongly differentiated (see Figure 4).

Moving toward high pH values, in the stability field of the $CO_3^=$, the morphology of the apatite crystals (Figure 5c) is one more time strongly modified, with hexagonal prisms frequently twinned and terminated by flat pinacoids. In this case, carbonate enters the crystal modifying the crystal structure as well, as demonstrated by IR and Raman spectroscopic analisys [75].

Figure 5. HAp grown from CO_2—free solution (**a**) Apatite grown from CO_2 saturated solution (**b**) and CAp (**c**).

We suggest explaining the whole morphological and structural modification in the light of the cooperative effect [67,69,99] between Ca-carbonates in solution and Ca-apatite surfaces, considering both the modification of growth morphology and the stabilization of carbonate-containing apatites as the consequence of the epitaxy as the key mechanism during crystal growth.

6. Conclusions

Experimental and theoretical investigations developed over the last sixty years on the complex relationships between chemical composition, crystal structure and crystal morphology of apatites, showed that many unsolved problems arise owing to the structural anisotropy of the ions (OH^-, CO_3^{2-}) filling the channels of apatites. Hydroxyapatite (HAp) represents the case study in which the anisotropy, due to the presence of the OH^- dipoles in the structure channels, reduces the structure symmetry to the lowest one, both for hexagonal ($P6_3$) and monoclinic ($P2_1/c$) crystals. In this review, we focused our attention on the monoclinic HAp polymorph both because its structure does not show ambiguity [20,21] and because it should be more fruitful considering the lowest instead of the highest symmetry, when one has to interpret the wide morphological variations that HAp undergoes during growth.

The approach we adopted to predict the surface morphology of monoclinic HAp, both for equilibrium and growth, is that by Hartman-Perdok [63], that represents a self-consistent way not only to find the character of a growing face but also to obtain all the outmost surface profiles of a given {hkl} crystal form, either when the surface reconstruction is not necessary or when different reconstructed surface profiles are needed to render the surface compatible with the constraints of the electrical stability.

This allowed us to prove that the calculated equilibrium shape (ES) of a monoclinic HAp single crystal is markedly pseudo-hexagonal; as a consequence, this can be misleading when the growth shapes (GS) of both single and twinned grown crystals are considered [38,68]. As a matter of fact, careful examination of experimental growth shapes observed either in nature (both in geological and bio-mineralized samples) or in laboratory grown crystals from pure aqueous solutions, allows to say that the HAp hexagonality very often is nothing else but a manner of speaking. Moreover, the HAp platelets which grew elongated along the crystal channels, in the early stages of laboratory crystals and in mineralized biological tissues (tooth enamel, bones), are not due to the breaking of the hexagonal symmetry but to different growth rates of the different monoclinic pinacoids in zone with the crystal channels, according to the varying supersaturation values experienced by the growth solution in a closed system.

Furthermore, investigation about the surface profiles, carried out through the Hartman-Perdok method applied to both HAp and monetite ($CaHPO_4$), was also proved to be necessary as a preliminary treatment to determine which kind of interaction occurs between a pre-existing monetite and the 3D growth of HAp. In fact, it has been found (both experimentally and theoretically) that monetite can be reliably used as a precursor phase that favors the 2D epitaxial nucleation of HAp that allows the hydroxyapatite to form beyond its usual growth conditions [69]. This finding represents a useful pathway to interpret other relationships between a precursor and the HAp crystal, when the precursor is not an amorphous phase.

Acknowledgments: Special thanks are given to two anonymous reviewers for their useful suggestions and comments.

Author Contributions: All the authors conceived the work; Linda Pastero and Dino Aquilano wrote the paper; Marco Bruno made the figures.

Conflicts of Interest: The authors declare no conflict of interest.

References

1. Dorozhkin, S.V. Calcium orthophosphates (CaPO$_4$): Occurrence and properties. *Prog. Biomater.* **2016**, *5*, 9–70. [CrossRef] [PubMed]
2. Hughes, J.M. The many facets of apatite. *Am. Mineral.* **2015**, *100*, 1033–1039. [CrossRef]
3. Kern, R. The equilibrium form of a crystal. In *Morphology of Crystals: Part A: Fundamentals*; Sunagawa, I., Ed.; Terra Scientific Publishing Company/Tokyo; D. Reidel Publishing Company: Tokyo, Japan, 1987; pp. 77–206.
4. Kern, R.; Métois, J.J.; Le Lay, G. Basic mechanisms in the early stages of epitaxy. In *Current Topics in Materials Science*; Kaldis, E., Ed.; North-Holland Publishing Company: Amsterdam, The Netherland, 1979; Volume 3, pp. 131–419.
5. Kern, R. Adsorption, absorption, versus crystal growth. *Cryst. Res. Technol.* **2013**, *48*, 727–782. [CrossRef]
6. Wopenka, B.; Pasteris, J.D. A mineralogical perspective on the apatite in bone. *Mater. Sci. Eng. C* **2005**, *25*, 131–143. [CrossRef]
7. Posner, A.S.; Perloff, A.; Diorio, A.F. Refinement of the hydroxyapatite structure. *Acta Crystallogr.* **1958**, *11*, 308–309. [CrossRef]
8. Kay, M.I.; Young, R.A.; Posner, A.S. Crystal Structure of Hydroxyapatite. *Nature* **1964**, *204*, 1050–1052. [CrossRef] [PubMed]
9. Eysel, W.; Roy, D.M. Hydrothermal flux growth of hydroxyapatites by temperature oscillation. *J. Cryst. Growth* **1973**, *20*, 245–250. [CrossRef]
10. Arends, J.; Schuthof, J.; van der Lindwen, W.H.; Bennema, P.; van der Berg, P.J. Preparation of pure hydroxyapatite single crystals by hydrothermal recrystallization. *J. Cryst. Growth* **1979**, *46*, 213–220. [CrossRef]
11. Mengeot, M.; Harvill, M.L.; Gilliam, O.R. Hydrothermal growth of calcium hydroxyapatite single crystals. *J. Cryst. Growth* **1973**, *19*, 199–203. [CrossRef]
12. Hughes, J.M.; Cameron, M.; Crowley, K.D. Structural variations in natural F, OH, and Cl apatites. *Am. Mineral.* **1989**, *74*, 870–876.
13. Terpstra, R.A.; Bennema, P.; Hartman, P.; Woensdregt, C.F.; Perdok, W.G.; Senechal, M.L. F faces of apatite and its morphology: Theory and observation. *J. Cryst. Growth* **1986**, *78*, 468–478. [CrossRef]
14. Lee, W.T.; Dove, M.T.; Salje, E.K.H. Surface relaxations in hydroxyapatite. *J. Phys. Condens. Matter* **2000**, *12*, 9829–9841. [CrossRef]
15. De Leeuw, N.H. Molecular Dynamics Simulations of the Growth Inhibiting Effect of Fe^{2+}, Mg^{2+}, Cd^{2+}, and Sr^{2+} on Calcite Crystal Growth. *J. Phys. Chem. B* **2002**, *106*, 5241–5249. [CrossRef]
16. Filgueiras, M.R.T.; Mkhonto, D.; de Leeuw, N.H. Computer simulations of the adsorption of citric acid at hydroxyapatite surfaces. *J. Cryst. Growth* **2006**, *294*, 60–68. [CrossRef]
17. De Leeuw, N.H.; Rabone, J.A.L. Molecular dynamics simulations of the interaction of citric acid with the hydroxyapatite (0001) and (011¯0) surfaces in an aqueous environment. *CrystEngComm* **2007**, *9*, 1178. [CrossRef]
18. Matsunaga, K.; Murata, H. Formation Energies of Substitutional Sodium and Potassium in Hydroxyapatite. *Mater. Trans.* **2009**, *50*, 1041–1045. [CrossRef]
19. Corno, M.; Orlando, R.; Civalleri, B.; Ugliengo, P. Periodic B3LYP study of hydroxyapatite (001) surface modelled by thin layer slabs. *Eur. J. Mineral.* **2007**, *19*, 757–767. [CrossRef]
20. Elliott, J.C. Monoclinic Space Group of Hydroxyapatite. *Nature* **1971**, *230*, 72. [CrossRef]

21. Elliott, J.C.; Young, R.A. Conversion of Single Crystals of Chlorapatite into Single Crystals of Hydroxyapatite. *Nature* **1967**, *214*, 904–906. [CrossRef]
22. Dorozhkin, S.V. Calcium orthophosphates. *Biomatter* **2011**, *1*, 121–164. [CrossRef] [PubMed]
23. White, T.J.; Dong, Z.L. Structural derivation and crystal chemistry of apatites. *Acta Crystallogr. Sect. B Struct. Sci.* **2003**, *59*, 1–16. [CrossRef]
24. Mathew, M.; Takagi, S. Structures of biological minerals in dental research. *J. Res. Natl. Inst. Stand. Technol.* **2001**, *106*, 1035–1044. [CrossRef] [PubMed]
25. Elliott, J.C. *Structure and Chemistry of the Apatites and Other Calcium Ortophosphates*; Elsevier: Amsterdam, The Netherland, 1994; ISBN 9781483290317.
26. Corno, M.; Rimola, A.; Bolis, V.; Ugliengo, P. Hydroxyapatite as a key biomaterial: Quantum-mechanical simulation of its surfaces in interaction with biomolecules. *Phys. Chem. Chem. Phys.* **2010**, *12*, 6309. [CrossRef] [PubMed]
27. Bolis, V.; Busco, C.; Martra, G.; Bertinetti, L.; Sakhno, Y.; Ugliengo, P.; Chiatti, F.; Corno, M.; Roveri, N. Coordination chemistry of Ca sites at the surface of nanosized hydroxyapatite: Interaction with H_2O and CO. *Philos. Trans. Ser. A Math. Phys. Eng. Sci.* **2012**, *370*, 1313–1336. [CrossRef] [PubMed]
28. Sudarsanan, K.; Young, R.A. Significant precision in crystal structural details. Holly Springs hydroxyapatite. *Acta Crystallogr. Sect. B Struct. Crystallogr. Cryst. Chem.* **1969**, *25*, 1534–1543. [CrossRef]
29. Elliott, J.C.; Mackie, P.E.; Young, R.A. Monoclinic Hydroxyapatite. *Science* **1973**, *180*, 1055–1057. [CrossRef] [PubMed]
30. Ikoma, T.; Yamazaki, A.; Nakamura, S.; Akao, M. Preparation and Structure Refinement of Monoclinic Hydroxyapatite. *J. Solid State Chem.* **1999**, *144*, 272–276. [CrossRef]
31. Suetsugu, Y.; Tanaka, J. Crystal growth and structure analysis of twin-free monoclinic hydroxyapatite. *J. Mater. Sci. Mater. Med.* **2002**, *13*, 767–772. [CrossRef] [PubMed]
32. Treboux, G.; Layrolle, P.; Kanzaki, N.; Onuma, K.; Ito, A. Symmetry of posner's cluster. *J. Am. Chem. Soc.* **2000**, *122*, 8323–8324. [CrossRef]
33. Hochrein, O.; Kniep, R.; Zahn, D. Atomistic simulation study of the order/disorder (monoclinic to hexagonal) phase transition of hydroxyapatite. *Chem. Mater.* **2005**, *17*, 1978–1981. [CrossRef]
34. Haverty, D.; Tofail, S.A.M.; Stanton, K.T.; McMonagle, J.B. Structure and stability of hydroxyapatite: Density functional calculation and Rietveld analysis. *Phys. Rev. B* **2005**, *71*, 94103. [CrossRef]
35. Corno, M.; Busco, C.; Civalleri, B.; Ugliengo, P. Periodic ab initio study of structural and vibrational features of hexagonal hydroxyapatite $Ca_{10}(PO_4)_6(OH)_2$. *Phys. Chem. Chem. Phys.* **2006**, *8*, 2464. [CrossRef] [PubMed]
36. Suda, H.; Yashima, M.; Kakihana, M.; Yoshimura, M. Monoclinic ↔ Hexagonal Phase Transition in Hydroxyapatite Studied by X-ray Powder Diffraction and Differential Scanning Calorimeter Techniques. *J. Phys. Chem.* **1995**, *99*, 6752–6754. [CrossRef]
37. Ikoma, T.; Yamazaki, A.; Nakamura, S.; Masaru, A. Phase Transition of Monoclinic Hydroxyapatite. *Netsu Sokutei* **1998**, *25*, 141–149.
38. Aquilano, D.; Bruno, M.; Rubbo, M.; Massaro, F.R.; Pastero, L. Low Symmetry Polymorph of Hydroxyapatite. Theoretical Equilibrium Morphology of the Monoclinic $Ca_5(OH)(PO_4)_3$. *Cryst. Growth Des.* **2014**, *14*, 2846–2852. [CrossRef]
39. Fleet, M.E.; Liu, X. Location of type B carbonate ion in type A–B carbonate apatite synthesized at high pressure. *J. Solid State Chem.* **2004**, *177*, 3174–3182. [CrossRef]
40. Fleet, M.E.; Liu, X. Local structure of channel ions in carbonate apatite. *Biomaterials* **2005**, *26*, 7548–7554. [CrossRef] [PubMed]
41. Suetsugu, Y.; Takahashi, Y.; Okamura, F.P.; Tanaka, J. Structure Analysis of A-Type Carbonate Apatite by a Single-Crystal X-Ray Diffraction Method. *J. Solid State Chem.* **2000**, *155*, 292–297. [CrossRef]
42. Ma, G.; Liu, X.Y. Hydroxyapatite: Hexagonal or monoclinic? *Cryst. Growth Des.* **2009**, *9*, 2991–2994. [CrossRef]
43. Reyes-Gasga, J.; Martinéz-Piñeiro, E.L.; Brès, É.F. Crystallographic structure of human tooth enamel by electron microscopy and X-ray diffraction: Hexagonal or monoclinic? *J. Microsc.* **2012**, *248*, 102–109. [CrossRef] [PubMed]
44. Mugnaioli, E.; Reyes-Gasga, J.; Kolb, U.; Hemmerlé, J.; Brès, E.F. Evidence of noncentrosymmetry of human tooth hydroxyapatite crystals. *Chem. A Eur. J.* **2014**, *20*, 6849–6852. [CrossRef] [PubMed]

45. Busch, S.; Dolhaine, H.; Du Chesne, A.; Heinz, S.; Hochrein, O.; Laeri, F.; Podebrad, O.; Vietze, U.; Weiland, T.; Kniep, R. Biomimetic Morphogenesis of Fluorapatite-Gelatin Composites: Fractal Growth, the Question of Intrinsic Electric Fields, Core/Shell Assemblies, Hollow Spheres and Reorganization of Denatured Collagen. *Eur. J. Inorg. Chem.* **1999**, *1999*, 1643–1653. [CrossRef]

46. Tao, J.; Battle, K.C.; Pan, H.; Salter, E.A.; Chien, Y.-C.; Wierzbicki, A.; De Yoreo, J.J. Energetic basis for the molecular-scale organization of bone. *Proc. Natl. Acad. Sci. USA* **2015**, *112*, 326–331. [CrossRef] [PubMed]

47. Sakhno, Y.; Ivanchenko, P.; Iafisco, M.; Tampieri, A.; Martra, G. A step toward control of the surface structure of biomimetic hydroxyapatite nanoparticles: Effect of carboxylates on the {010} P-rich/Ca-rich facets ratio. *J. Phys. Chem. C* **2015**, *119*, 5928–5937. [CrossRef]

48. Yao, X.; Yao, H.; Li, G.; Li, Y. Biomimetic synthesis of needle-like nano-hydroxyapatite templated by double-hydrophilic block copolymer. *J. Mater. Sci.* **2010**, *45*, 1930–1936. [CrossRef]

49. Shuai, C.; Feng, P.; Nie, Y.; Hu, H.; Liu, J.; Peng, S. Nano-hydroxyapatite improves the properties of β-tricalcium phosphate bone scaffolds. *Int. J. Appl. Ceram. Technol.* **2013**, *10*, 1003–1013. [CrossRef]

50. Deng, Y.; Wang, H.; Zhang, L.; Li, Y.; Wei, S. In situ synthesis and in vitro biocompatibility of needle-like nano-hydroxyapatite in agar-gelatin co-hydrogel. *Mater. Lett.* **2013**, *104*, 8–12. [CrossRef]

51. Ito, H.; Oaki, Y.; Imai, H. Selective synthesis of various nanoscale morphologies of hydroxyapatite via an intermediate phase. *Cryst. Growth Des.* **2008**, *8*, 1055–1059. [CrossRef]

52. Kobayashi, T.; Ono, S.; Hirakura, S.; Oaki, Y.; Imai, H. Morphological variation of hydroxyapatite grown in aqueous solution based on simulated body fluid. *CrystEngComm* **2012**, *14*, 1143–1149. [CrossRef]

53. Bertinetti, L.; Tampieri, A.; Landi, E.; Ducati, C.; Midgley, P.A.; Coluccia, S.; Martra, G. Surface Structure, Hydration, and Cationic Sites of Nanohydroxyapatite: UHR-TEM, IR, and Microgravimetric Studies. *J. Phys. Chem. C* **2007**, *111*, 4027–4035. [CrossRef]

54. Sakhno, Y.; Bertinetti, L.; Iafisco, M.; Tampieri, A.; Roveri, N.; Martra, G. Surface Hydration and Cationic Sites of Nanohydroxyapatites with Amorphous or Crystalline Surfaces: A Comparative Study. *J. Phys. Chem. C* **2010**, *114*, 16640–16648. [CrossRef]

55. Eppell, S.J.; Tong, W.; Lawrence Katz, J.; Kuhn, L.; Glimcher, M.J. Shape and size of isolated bone mineralites measured using atomic force microscopy. *J. Orthop. Res.* **2001**, *19*, 1027–1034. [CrossRef]

56. Fratzl, P.; Gupta, H.S.; Paschalis, E.P.; Roschger, P. Structure and mechanical quality of the collagen–mineral nano-composite in bone. *J. Mater. Chem.* **2004**, *14*, 2115–2123. [CrossRef]

57. Olszta, M.J.; Cheng, X.; Jee, S.S.; Kumar, R.; Kim, Y.Y.; Kaufman, M.J.; Douglas, E.P.; Gower, L.B. Bone structure and formation: A new perspective. *Mater. Sci. Eng. R Rep.* **2007**, *58*, 77–116. [CrossRef]

58. Delgado-López, J.M.; Frison, R.; Cervellino, A.; Gómez-Morales, J.; Guagliardi, A.; Masciocchi, N. Crystal Size, Morphology, and Growth Mechanism in Bio-Inspired Apatite Nanocrystals. *Adv. Funct. Mater.* **2014**, *24*, 1090–1099. [CrossRef]

59. Sato, K.; Kogure, T.; Iwai, H.; Tanaka, J. Atomic-Scale {101⁻0} Interfacial Structure in Hydroxyapatite Determined by High-Resolution Transmission Electron Microscopy. *J. Am. Ceram. Soc.* **2002**, *85*, 3054–3058. [CrossRef]

60. Ospina, C.A.; Terra, J.; Ramirez, A.J.; Farina, M.; Ellis, D.E.; Rossi, A.M. Experimental evidence and structural modeling of nonstoichiometric (010) surfaces coexisting in hydroxyapatite nano-crystals. *Colloids Surf. B Biointerfaces* **2012**, *89*, 15–22. [CrossRef] [PubMed]

61. Astala, R.; Stott, M.J. First-principles study of hydroxyapatite surfaces and water adsorption. *Phys. Rev. B Condens. Matter Mater. Phys.* **2008**, *78*, 75427. [CrossRef]

62. Rulis, P.; Yao, H.; Ouyang, L.; Ching, W.Y. Electronic structure, bonding, charge distribution, and X-ray absorption spectra of the (001) surfaces of fluorapatite and hydroxyapatite from first principles. *Phys. Rev. B Condens. Matter Mater. Phys.* **2007**, *76*, 245410. [CrossRef]

63. Hartman, P. Structure and morphology. In *Crystal Growth: An Introduction*; Hartman, P., Ed.; North-Holland Publishing Company: Amsterdam, The Netherland, 1973; pp. 367–402.

64. Lee, W.T.; Salje, E.K.H.; Dove, M.T. Effect of surface relaxations on the equilibrium growth morphology of crystals: Platelet formation. *J. Phys. Condens. Matter* **1999**, *11*, 7385–7410. [CrossRef]

65. Li, M.; Wang, L.; Zhang, W.; Putnis, C.V.; Putnis, A. Direct Observation of Spiral Growth, Particle Attachment, and Morphology Evolution of Hydroxyapatite. *Cryst. Growth Des.* **2016**, *16*, 4509–4518. [CrossRef]

66. Bruno, M. The reconstruction of dipolar surfaces: A preliminary step for adsorption modeling. *Cryst. Res. Technol.* **2013**, *48*, 811–818. [CrossRef]

67. Aquilano, D.; Pastero, L. Anomalous mixed crystals: A peculiar case of adsorption/absorption. *Cryst. Res. Technol.* **2013**, *48*, 819–839. [CrossRef]
68. Aquilano, D.; Bruno, M.; Rubbo, M.; Pastero, L.; Massaro, F.R. Twin Laws and Energy in Monoclinic Hydroxyapatite, Ca$_5$(PO$_4$)$_3$(OH). *Cryst. Growth Des.* **2015**, *15*, 411–418. [CrossRef]
69. Pastero, L.; Aquilano, D. Monetite-Assisted Growth of Micrometric Ca-Hydroxyapatite Crystals from Mild Hydrothermal Conditions. *Cryst. Growth Des.* **2016**, *16*, 852–860. [CrossRef]
70. Jaffe, E.B. *Abstracts of the Literature on Synthesis of Apatites and Some Related Phosphates*; U.S. Geological Survey: Washington, DC, USA, 1951.
71. Perloff, A.; Posner, A.S. Preparation of Pure Hydroxyapatite Crystals. *Science* **1956**, *124*, 583–584. [CrossRef] [PubMed]
72. Schleede, H.A.; Schmidt, W.; Kindt, H. Zur Kenntnis der Calciumphosphate und Apatite. *Z. Elektrochem.* **1932**, *38*, 633–641.
73. Jullman, H.; Mosebach, R. Zur Synthese, Licht- und Doppelbrechung des Hydroxylapatits. *Z. Naturf. B* **1966**, *21*, 493–494. [CrossRef]
74. Kirn, J.F.; Leidheiser, H. Progress in efforts to grow large single crystals of hydroxyapatite. *J. Cryst. Growth* **1968**, *2*, 111–112. [CrossRef]
75. Pastero, L.; Bruno, M.; Rubbo, M.; Camara, F.; Aquilano, D. Growth of large Ca-Hydroxyapatite crystals from aqueous solution. In Proceedings of the IV Meeting of the Italian and Spanish Crystallographic Associations, Tenerife, Spain, 21–25 June 2016.
76. Van Wazer, J.R. Phosphorus and its Compounds, Bd. 1: Chemistry. *Angew. Chem.* **1961**, *73*, 552. [CrossRef]
77. Clark, J.S. Solubility criteria for the existence of hydroxyapatite. *Can. J. Chem.* **1955**, *33*, 1696–1700. [CrossRef]
78. Larsen, S. Solubility of Hydroxyapatite. *Nature* **1966**, *212*, 605. [CrossRef]
79. Chen, Z.F.; Darvell, B.W.; Leung, V.W.H. Hydroxyapatite solubility in simple inorganic solutions. *Arch. Oral Biol.* **2004**, *49*, 359–367. [CrossRef] [PubMed]
80. Pan, H.B.; Darvell, B.W. Calcium Phosphate Solubility: The Need for Re-Evaluation. *Cryst. Growth Des.* **2009**, *9*, 639–645. [CrossRef]
81. Neuman, W.F.F.; Neuman, M.W.W. The Nature of the Mineral Phase of Bone. *Chem. Rev.* **1953**, *53*, 1–45. [CrossRef]
82. Levinskas, G.; Neuman, W. The Solubility of Bone Mineral. I. Solubility Studies of Synthetic Hydroxylapatite. *J. Phys. Chem.* **1955**, *59*, 164–168. [CrossRef]
83. Rootare, H.M.; Deitz, V.R.; Carpenter, F.G. Solubility product phenomena in hydroxyapatite-water systems. *J. Colloid Sci.* **1962**, *17*, 179–206. [CrossRef]
84. Bell, L.C.; Mika, H. The pH dependence of the surface concentrations of calcium and phosphorous on hydroxyapatite in aqueous solutions. *J. Soil Sci.* **1979**, *30*, 247–258. [CrossRef]
85. Mika, H.; Bell, L.C.; Kruger, B.J. The role of surface reactions in the dissolution of stoichiometric hydroxyapatite. *Arch. Oral Biol.* **1976**, *21*, 697–701. [CrossRef]
86. Dorozhkin, S.V. Inorganic chemistry of the dissolution phenomenon: The dissolution mechanism of calcium apatites at the atomic (ionic) level. *Comments Inorg. Chem.* **1999**, *20*, 285–299. [CrossRef]
87. Dorozhkin, S. V Surface Reactions of Apatite Dissolution. *J. Colloid Interface Sci.* **1997**, *191*, 489–497. [CrossRef] [PubMed]
88. Dorozhkin, S.V. A review on the dissolution models of calcium apatites. *Prog. Cryst. Growth Charact. Mater.* **2002**, *44*, 45–61. [CrossRef]
89. Chuong, R. Experimental Study of Surface and Lattice Effects on the Solubility of Hydroxyapatite. *J. Dent. Res.* **1973**, *52*, 911–914. [CrossRef] [PubMed]
90. Greenwald, I. The solubility of calcium phosphate. I. The effect of pH and of amount of solid phase. *J. Biol. Chem.* **1942**, *143*, 703–710.
91. Greenwald, I. The solubility of calcium phosphate: II. The solubility product. *J. Biol. Chem.* **1942**, *143*, 711–714.
92. Greenwald, I. The effect of phosphate on the solubility of calcium carbonate and of bicarbonate on the solubility of calcium and magnesium phosphates. *J. Biol. Chem.* **1945**, *161*, 697–704. [PubMed]
93. Kaufman, H.W.; Kleinberg, I. Studies on the incongruent solubility of hydroxyapatite. *Calcif. Tissue Int.* **1979**, *27*, 143–151. [CrossRef] [PubMed]

94. Clark, J.S.; Peech, M. Solubility Criteria for the Existence of Calcium and Aluminum Phosphates in Soils1. *Soil Sci. Soc. Am. J.* **1955**, *19*, 171. [CrossRef]
95. Skinner, H.C.W. Studies in the basic mineralizing system, $CaO-P_2O_5-H_2O$. *Calcif. Tissue Res.* **1974**, *14*, 3–14. [CrossRef] [PubMed]
96. Andrade, M.C.; Ogasawara, T.; Silva, F.T. Hydrothermal Crystallization of HAp. *Proc 2nd Int. Symp. Apatite.* **1997**, *1*, 4147.
97. Byrappa, K.; Yoshimura, M. Hydrothermal growth of some selected crystals. In *Handbook of Hydrothermal Technology*; Noyes Publications/William Andrew Publishing, LLC: Norwich, NY, USA, 2001; pp. 287–299. ISBN 0-8155-1445-X.
98. McDowell, H.; Gregory, T.M.; Brown, W.E. Solubility of $Ca_5(PO_4)_3OH$ in the system $Ca(OH)_2-H_3PO_4-H_2O$ at 5,15,25 and 37 °C. *J. Res. Natl. Bur. Stand. Chem.* **1977**, *81A*, 273–281. [CrossRef]
99. Pastero, L.; Aquilano, D. $CaCO_3$ (Calcite)/Li_2CO_3 (zabuyelite) anomalous mixed crystals. Sector zoning and growth mechanisms. *Cryst. Growth Des.* **2008**, *8*, 3451–3460. [CrossRef]

minerals

MDPI

Article

Pyrophosphate-Inhibition of Apatite Formation Studied by In Situ X-Ray Diffraction

Casper Jon Steenberg Ibsen and Henrik Birkedal *

Department of Chemistry and iNANO, Aarhus University, Gustav Wieds Vej 14, 8000 Aarhus, Denmark;
cjsibsen@hotmail.com
* Correspondence: hbirkedal@chem.au.dk; Tel.: +45-2250-8475

Received: 28 August 2017; Accepted: 12 February 2018; Published: 13 February 2018

Abstract: The pathways to crystals are still under debate, especially for materials relevant to biomineralization, such as calcium phosphate apatite known from bone and teeth. Pyrophosphate is widely used in biology to control apatite formation since it is a potent inhibitor of apatite crystallization. The impacts of pyrophosphate on apatite formation and crystallization kinetics are, however, not fully understood. Therefore, we studied apatite crystallization in water by synchrotron in situ X-ray diffraction. Crystallization was conducted from calcium chloride (0.2 M) and sodium phosphate (0.12 M) at pH 12 where hydrogen phosphate is the dominant phosphate species and at 60 °C to allow the synchrotron measurements to be conducted in a timely fashion. Following the formation of an initial amorphous phase, needle shaped crystals formed that had an octacalcium phosphate-like composition, but were too small to display the full 3D periodic structure of octacalcium phosphate. At later growth stages the crystals became apatitic, as revealed by changes in the lattice constant and calcium content. Pyrophosphate strongly inhibited nucleation of apatite and increased the onset of crystallization from minute to hour time scales. Pyrophosphate also reduced the rate of growth. Furthermore, when the pyrophosphate concentration exceeded ~1% of the calcium concentration, the resultant crystals had reduced size anisotropy suggesting that pyrophosphate interacts in a site-specific manner with the formation of apatite crystals.

Keywords: apatite; in situ X-ray diffraction; crystallization; nucleation; pyrophosphate; biomimetic materials

1. Introduction

The formation of nanocrystalline apatitic biomineral in the skeleton remains far from understood and, indeed, many open questions remain about apatite formation in general [1]. Apatite biomineralization occurs under strict biological control enforced by the collective action of numerous additives, including small molecules, such as citrate [2–5], pyrophosphate, and polyphosphates [6,7], as well as larger macromolecules. Pyrophosphate is involved in controlling apatite formation during skeleton formation and is a potent inhibitor of apatite crystallization [8–21]. In spite of the important roles of pyrophosphate as a mineralization inhibitor in vivo, no quantitative studies of its impact on nanocrystal formation kinetics have been performed to the best of our knowledge. Herein, we use in situ synchrotron powder diffraction [22–25] to quantify the impact of pyrophosphate on apatite formation from amorphous calcium phosphate (ACP).

We have previously shown that the shape and size of the formed apatite nanocrystals depend strongly on whether sodium or potassium salts are used as precursors [26], the pH of the starting solution, e.g., whether it is phosphate or hydrogenphosphate dominated [22,26], and the action of various additives [5,22,24,25,27,28]. Herein, we formed ACP from a sodium phosphate solution (pH ~12.3) and the resultant ACP phase is, therefore, hydrogen phosphate-rich. Apatite was formed from hydrogen phosphate-rich ACP formed in situ by mixing a calcium chloride solution with a sodium phosphate solution to which various quantities of pyrophosphate were added according to:

$$10CaCl_2 \text{ (aq)} + 6 \, Na_3PO_4 \text{ (aq)} + 10x \, Na_4P_2O_7 \text{ (aq)} \rightarrow ACP \text{ (s)} \rightarrow \text{apatite (s)}$$

The counter ion determines the degree to which carbonate is incorporated into the lattice through co-substitution. In the case of potassium counter ion, carbonate incorporation into the lattice is reduced and we have previously shown that the apatite crystals obtained from such an ACP phase are long needles [22]. This was in contrast to the situation at higher pH (~12.5), at which the ACP phase is phosphate-dominated [26], where the crystals initially displayed almost spherical geometry and remained considerably less anisotropic, even at late stages of growth than the before mentioned long needles obtained at a slightly lower pH of 12.3. Therefore, the present work explores both the formation and morphology of apatite nanocrystals from hydrogen phosphate rich ACP with sodium as a counter ion and the impact of pyrophosphate additives thereon. In the case of potassium as a counter ion, we found, by a combination of diffraction and infrared spectroscopy, that the initial needle-shaped nanocrystals formed under the present pH conditions had octacalcium phosphate (OCP) stoichiometry, but without the 3D periodicity of OCP, because they were too thin [22].

2. Materials and Methods

We employed in situ X-ray diffraction [22–25,29,30] to follow the crystallization of apatite as in our previous work on other types of apatite formation [22,24,25]. We performed in situ synchrotron diffraction in a home built stopped flow apparatus with high temperature stability [24] by mixing equal volumes of a calcium chloride solution (0.4 M) with a sodium phosphate solution (0.24 M) loaded with varying quantities of sodium pyrophosphate (0–0.02 M). This gave a final calcium concentration in the mixture of 0.2 M, a total phosphate concentration of 0.12 M and a varying pyrophosphate concentration that we express as a percentage of the calcium concentration that thus varied from 0–5 mol% (i.e., concentrations of 0–0.01 M with the lowest pyrophosphate concentration being 1 mM). The pyrophosphate concentrations corresponded to 0.83–8.3 mol% of the phosphate concentration. This is higher than the values found in bone but approaches the values reported for resting cartilage ~0.7% [21,31]. The reactants were mixed by passing them through a liquid mixer and then directly into the reaction chamber (a single-crystal sapphire tube) where the reaction volume was investigated by synchrotron X-rays with a wavelength $\lambda = 0.6829$ Å at the Swiss Norwegian Beamlines at ESRF, Grenoble, France. Diffraction data were detected with a Dectris (CH) Pilatus 2 M detector placed at a sample-detector distance of 233.4 mm using 4 s exposure per data point. The present experiments were conducted at 60 °C. The 2D diffractograms were azimuthally integrated and corrected for beam-intensity fluctuations by normalizing the scattering intensity in the 1.12–1.44 Å^{-1} q-range as in our previous work [22].

The results reported herein result from Rietveld refinement of diffraction data [27] that was conducted in GSAS [32]. The Rietveld model assumed a needle morphology of the crystallites (extensive testing of other morphologies indicated that this was an appropriate model) and incorporated refinement of lattice constants, calcium occupancy [22], scale factor, and a Chebyshev polynomial background.

3. Results

According to our previous studies, crystallization progressed via the initial formation of an amorphous phase [5,22,24,25,27], which was also confirmed by inspection of the raw diffraction data (not shown). The Rietveld refined scale factor is a direct measure of the amount of crystalline material and, thus, monitors the progress of crystallization. Figure 1 shows results from the pyrophosphate-free system and thus addresses the question of crystallization process for sodium as a counter ion. Qualitatively, the trends mirror those seen with potassium as counter ion [22], but the numerical values are different. Crystals formed after 6 min, as revealed by a rapid increase of the scale factor.

Before ~6 min, the scale factor fluctuated strongly (Figure 1A) due to a lack of, or extremely low, crystalline signals. After formation of the first crystals, the scale factor continued to increase, albeit at a slower rate, until ~25 min, after which it remained constant, indicating that the total amount of crystalline material did not change thereafter. The initial crystals were highly calcium-deficient as reflected by the refined calcium occupancy in Figure 1B. The calcium occupancy increased until ~20 min, after which it stabilized at 0.917(8) (the average value of the last 100 s of the experiment with the number in parentheses representing the root mean square deviation (RMSD) around the mean). At early stages, the lattice constants differed from the values at maturity, Figure 1E,F with the *a*- and *c*-axes being, respectively, smaller and larger than the late-stage values. The crystallite sizes also evolve with time, Figure 1C,D. When first formed, the crystals had an aspect ratio of about 7, but becomes less anisotropic during growth ending up with a final average crystal size of 6.2(2) × 30.5(6) nm (aspect ratio 4.9). The average crystallite size parallel to the *c*-axis drops somewhat over the 8 min following the initial crystal formation after which it is essentially constant. During this time, the scale factor increases, meaning that the amount of crystalline material, and by implication the number density of crystals, increases. Thus, the first detected crystals are very long, while later-forming crystals are shorter, resulting in a decrease in the average crystallite length. The rapid formation of high aspect ratio high-nonstoichiometric needles with deformed lattice constants is in agreement with the observations on the hydrogenphosphate-rich potassium system [22], but in contrast to the behavior at higher starting pH (~12.5) where the ACP is phosphate-dominated and the initial crystal have a low aspect ratio [26].

We previously [22] deduced a growth mechanism for the system with potassium as counter ion in which carbonate substitutions in the lattice do not play a role. The initial crystals were determined by a combination of diffraction and Fourier transform infrared (FTIR) spectroscopy data including time resolved measurements to be OCP-like, but without the 3D stacking order needed for real OCP crystals because the initially-formed crystals were so thin. During growth the crystals in average became increasingly apatitic to result in the constant calcium occupancy and lattice constants accompanied by the reduced crystallite size aspect ratio. A similar evolution of Ca-occupancy and crystal shape/size is seen in the present case (Figure 1D) showing that the same growth mechanism acts in the present sodium counter-ion case.

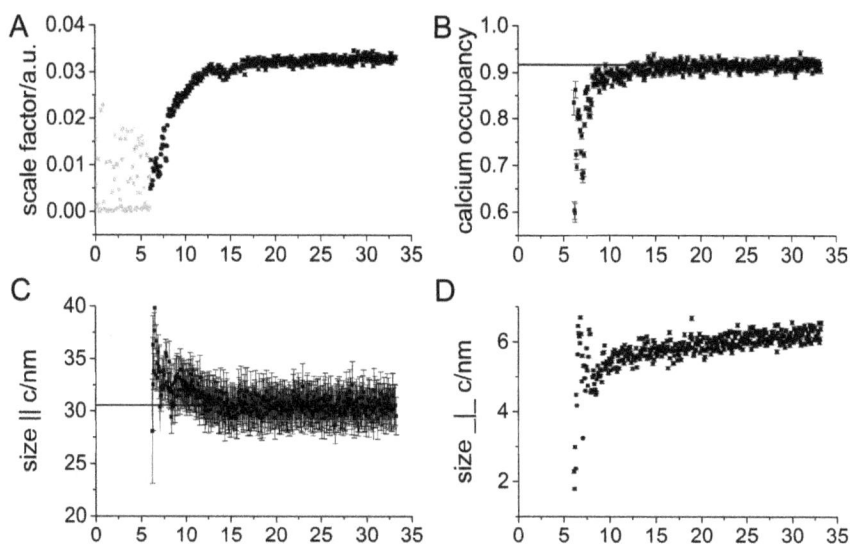

Figure 1. *Cont.*

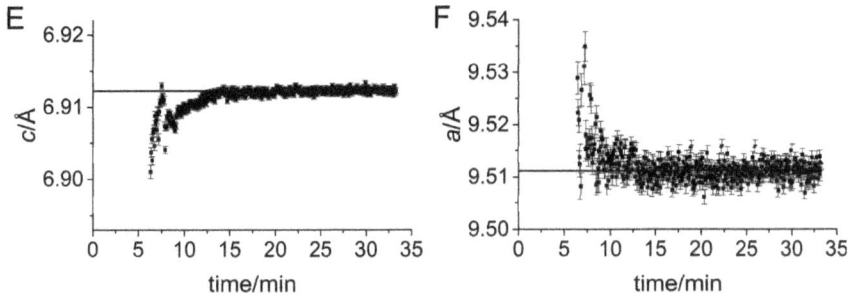

Figure 1. Results of Rietveld refinements of in situ diffraction data on the pyrophosphate free system. Horizontal lines represent the average of the last 100 s of the experiment. (**A**) Scale factor. Data points in gray are from refinements where the confidence in the results are lower due to convergence and/or local minima issues (see text); (**B**) refined calcium occupancy; (**C,D**) apparent crystallite sizes parallel and perpendicular to the *c*-axis, respectively; and (**E,F**) the refined lattice constants of the hexagonal apatite unit cell shown on the same relative scale. In panels (**B–F**), only results where the parameters were well-defined are shown; at earlier times the parameters were not determinable due to the extremely low or zero diffraction signal from apatite.

Impact of Pyrophosphate

Addition of pyrophosphate strongly impacted the crystallization kinetics. The results are summarized in Figures 2 and 3. Figure 2A compares the refined scale factors for different pyrophosphate concentrations with sigmoidal fits overlaid to further illustrate the trends. As evident from the data, pyrophosphate impacts both the time until first crystals are detected (nucleation time, Figure 2C), and the rate of transformation from amorphous to crystalline material, i.e., the slope of the curve (Figure 2B). The nucleation time was found to depend exponentially on pyrophosphate concentration, Figure 2C, featuring very significant stabilization of the ACP phase to the point of a retardation from minute to hour time scale. Results from repeated experiments are also shown to illustrate the excellent reproducibility of the data. This stabilization of ACP is in accordance with previous work by others at pH 8.5 measured ex situ [14]. The growth rate was likewise reduced as seen from the derivative of the scale factor time dependence, shown as the derivative of an analytical sigmoidal parameterization of the data in Figure 2B. In the presence of the additive, the crystallite size evolved qualitatively in the same manner as without additive, Figure 2D,E, i.e., with an aspect ratio that decreased with time. However, above a certain threshold concentration, in between 1.0% and 2.5% pyrophosphate, there was an abrupt change in the shape of the crystallites. The final length of the needles was reduced to 24.7(1) nm for all measured concentrations above 1.0%.

Concurrently, the size perpendicular to the long axis increased, to 9.0(1) nm, Figure 3A. This suggests that pyrophosphate binds specifically to specific crystal faces, probably the (002) surface inhibiting growth in this direction. Material was consequently redirected into thickening the crystallites, resulting in lowered aspect ratios (see Figure 3B). The fact that this behavior sets in above a threshold concentration suggests that the process is corporative and, thus, necessitates a critical pyrophosphate concentration to occur.

As in the pyrophosphate-free case, the calcium occupancy was also reduced during the initial growth stage when pyrophosphate was present indicating that the fundamental growth mechanism did not change, Figure 2I. With potassium as counter-ion, we previously deduced by in situ and ex situ diffraction, as well as time-resolved FTIR measurements that the crystals were hydrogen phosphate substituted with stoichiometries of the type $Ca_{10-x/2}(PO_4)_{6-x}(HPO_4)_x(OH)_2$ [22]. In the present case, the final calcium occupancy decreased with increasing pyrophosphate concentration, Figure 3D. Thus, the crystals become less stoichiometric when grown under the influence of pyrophosphate.

This is consistent with a scenario where a pyrophosphate substitute for two phosphates in the lattice structure (i.e., leaving also an anion vacancy). The resulting deficiency of negative charge could then conceivably be compensated by calcium vacancies resulting in stoichiometries of the type $Ca_{10-x/2-y}(PO_4)_{6-x-2y}(P_2O_7)_y(HPO_4)_x(OH)_2$ where additional carbonate substitutions have been left out. In this scenario, the lowest observed final calcium occupancy (for 5% pyrophosphate) of 0.86 in comparison the pyrophosphate-free value of 0.92 suggests that $x \sim 1.6$ and $y \sim 0.7$ so that the unit cell contains ~3.0 phosphates. This reflects the heavily disordered nature of these phases.

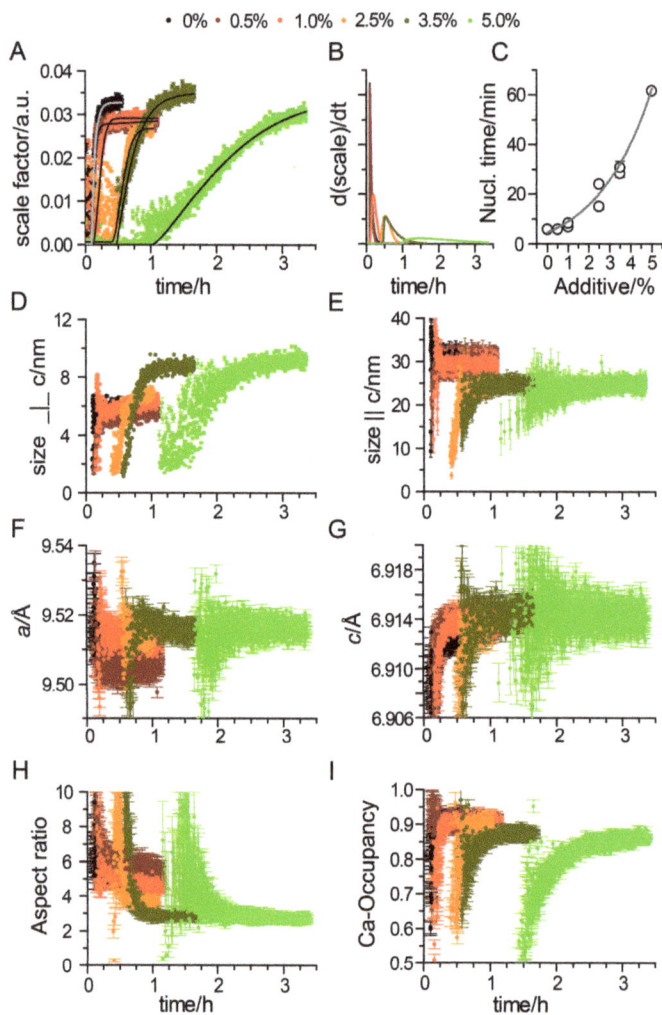

Figure 2. The impact of pyrophosphate on apatite formation kinetics derived from Rietveld refinements. The pyrophosphate concentration is given as molar percent of the calcium concentration. (**A**) The scale factor reflects the amount of crystalline material. The lines are sigmoidal fits to the data; (**B**) rate of growth in the amount of crystalline material as a function of time; (**C**) dependence of nucleation time on the amount of added additive; (**D–I**) time dependence of selected parameters; (**D,E**) crystallite sizes perpendicular and parallel to the *c*-axis, respectively; (**F,G**) apatite lattice constants; (**H**) crystallite size aspect ratio; and (**I**) calcium occupancy.

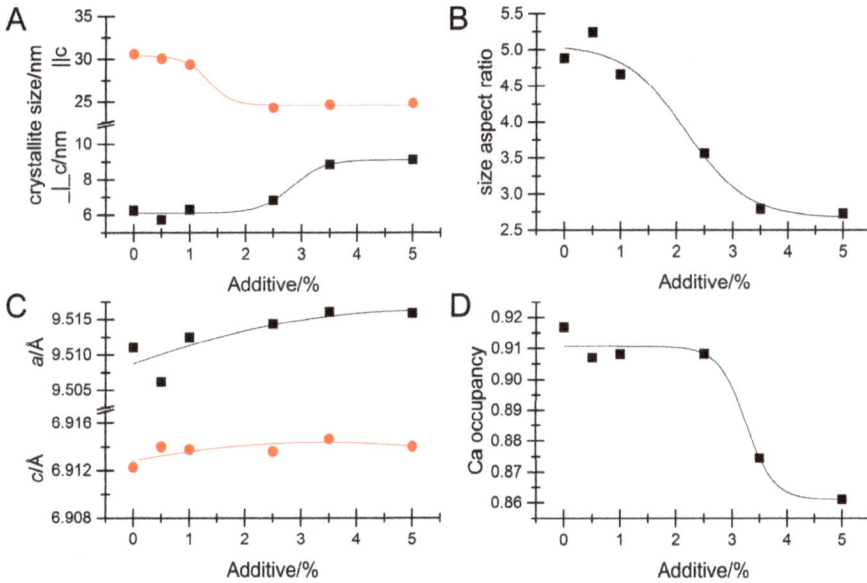

Figure 3. Values at long times of selected crystallographic parameters as a function of amount of added pyrophosphate. Lines are, in all cases, guides to the eye. The pyrophosphate concentration is given as a molar percent of the calcium concentration. (**A**) shows crystallite size along (red) and perpendicular (black to the *c*-axis); (**B**) shows the crystallite size aspect ratio (size ‖ *c*/size⊥*c*); (**C**) shows the lattice constants as a function of additive concentration while (**D**) gives the occupancy of the calcium sites in the apatite unit cell.

The lattice constants evolved in a manner similar to the additive free situation, but slowed down (see Figure 2F,G and Figure 3C). The final lattice constants do not display any significant dependency on pyrophosphate concentration indicating that only slight macrostrain is inflicted upon the crystals even though very specific interactions take place. This is in contrast to the impact of organic additives on $CaCO_3$ biominerals [33–37], ZnO [38,39], and apatite [25].

4. Discussion

The current use of in situ diffraction provided detailed information on the crystallization kinetics and behavior. We found, first of all, that the crystallization mechanism is akin to the one determined with potassium as a counter ion instead of sodium at the same pH [22]. Importantly, the initial crystallites did not attain the full 3D OCP structure because their thickness was of the same order as the stacking distance in OCP, which precludes the building of the 3D order and the telltale low-angle diffraction peak often used to detect OCP. The formation of an OCP-like intermediate was also found at a lower pH of 7.4 by Habraken et al. [40] suggesting that this pathway may be a general phenomenon in apatite formation when the solution pH is hydrogenphosphate dominated.

We further obtained insights into the inhibition of apatite formation due to pyrophosphate and found that pyrophosphate, as expected, strongly inhibits apatite crystallization and thereby increases the temporal stability of ACP. The fundamental mechanism of crystal formation via an initial OCP-like crystal was, however, unchanged, suggesting that this pathway may also occur in biogenic environments where additives abound.

Pyrophosphate additives resulted in nontrivial changes of crystal morphology above a threshold concentration suggesting a specific corporative interaction between pyrophosphate and the growing apatite nanocrystal.

Acknowledgments: We thank the Danish Agency for Science, Technology, and Innovation for funding (DANSCATT). The in situ diffraction experiments were performed on beamline BM01 (Swiss Norwegian Beam Lines) at the European Synchrotron Radiation Facility (ESRF), Grenoble, France. We are grateful to Dmitry Chernyshov at the ESRF for providing assistance in using beamline BM01. We further thank Vicki Nue for her assistance during the synchrotron experiments. Affiliation with the Center for Integrated Materials Research (iMAT) at Aarhus University is gratefully acknowledged.

Author Contributions: H.B. conceived the experiments. H.B. and C.J.S.I. designed the experiments, performed the experiments, and wrote the paper.

Conflicts of Interest: The authors declare no conflict of interest.

References

1. Birkedal, H. Phase transformations in calcium phosphate crystallization. In *New Perspectives on Mineral Nucleation and Growth: From Solution Precursors to Solid Materials*; Van Driessche, A.E.S., Kellermeier, M., Benning, L.G., Gebauer, D., Eds.; Springer International Publishing: Cham, Switzerland, 2017; pp. 199–210.

2. Davies, E.; Müller, K.H.; Wong, W.C.; Pickard, C.J.; Reid, D.G.; Skepper, J.N.; Duer, M.J. Citrate bridges between mineral platelets in bone. *Proc. Natl. Acad. Sci. USA* **2014**, *111*, E1354–E1363. [CrossRef] [PubMed]

3. Hu, Y.-Y.; Rawal, A.; Schmidt-Rohr, K. Strongly bound citrate stabilizes the apatite nanocrystals in bone. *Proc. Natl. Acad. Sci. USA* **2010**, *107*, 22425–22429. [CrossRef] [PubMed]

4. Reid, D.G.; Duer, M.J.; Jackson, G.E.; Murray, R.C.; Rodgers, A.L.; Shanahan, C.M. Citrate occurs widely in healthy and pathological apatitic biomineral: Mineralized articular cartilage, and intimal atherosclerotic plaque and apatitic kidney stones. *Calcif. Tissue Int.* **2013**, *93*, 253–260. [CrossRef] [PubMed]

5. Jensen, A.C.S.; Ibsen, C.J.S.; Birkedal, H. Transparent aggregates of nanocrystalline hydroxyapatite. *Cryst. Growth Des.* **2014**, *14*, 6343–6349. [CrossRef]

6. Omelon, S.; Georgiou, J.; Henneman, Z.J.; Wise, L.M.; Sukhu, B.; Hunt, T.; Wynnyckyj, C.; Holmyard, D.; Bielecki, R.; Grynpas, M.D. Control of vertebrate skeletal mineralization by polyphosphates. *PLoS ONE* **2009**, *4*, e5634. [CrossRef] [PubMed]

7. Omelon, S.; Grynpas, M.D. Relationships between polyphosphate chemistry, biochemistry and apatite biomineralization. *Chem. Rev.* **2008**, *108*, 4694–4715. [CrossRef] [PubMed]

8. Anderson, H.C.; Garimella, R.; Tague, S.E. The role of matrix vesicles in growth plate development and biomineralization. *Front. Biosci.* **2005**, *10*, 822–837. [CrossRef] [PubMed]

9. Anderson, H.C.; Shapiro, I.M. The epiphyseal growth plate. In *Bone and Development*; Bronner, F., Farach-Carson, M.C., Roach, H.I., Eds.; Springer: London, UK, 2010; pp. 39–64.

10. Anderson, H.C.; Sipe, J.B.; Hessle, L.; Dhamyamraju, R.; Atti, E.; Camacho, N.P.; Millán, J.L. Impaired calcification around matrix vesicles of growth plate and bone in alkaline phosphatase-deficient mice. *Am. J. Pathol.* **2004**, *164*, 841–847. [CrossRef]

11. Orriss, I.R.; Arnett, T.R.; Russell, R.G.G. Pyrophosphate: A key inhibitor of mineralisation. *Curr. Opin. Pharmacol.* **2016**, *28*, 57–68. [CrossRef] [PubMed]

12. Hessle, L.; Johnson, K.A.; Anderson, H.C.; Narisawa, S.; Sali, A.; Goding, J.W.; Terkeltaub, R.; Millán, J.L. Tissue-nonspecific alkaline phosphatase and plasma cell membrane glycoprotein-1 are central antagonistic regulators of bone mineralization. *Proc. Natl. Acad. Sci. USA* **2002**, *99*, 9445–9449. [CrossRef] [PubMed]

13. Fleisch, H.; Russell, R.G.G.; Straumann, F. Effect of pyrophosphate on hydroxyapatite and its implications in calcium homeostasis. *Nature* **1966**, *212*, 901–903. [CrossRef] [PubMed]

14. Fleisch, H.; Russell, R.G.G.; Bisaz, S.; Termine, J.D.; Posner, A.S. Influence of pyrophosphate on the transformation of amorphous to crystalline calcium phosphate. *Calcif. Tissue Res.* **1968**, *2*, 49–59. [CrossRef]

15. Fleisch, H.; Bisaz, S. Isolation from urine of pyrophosphate, a calcification inhibitor. *Am. J. Physiol. Leg. Content* **1962**, *203*, 671–675. [CrossRef] [PubMed]

16. Fleisch, H.; Straumann, F.; Schenk, R.; Bisaz, S.; Allgöwer, M. Effect of condensed phosphates on calcification of chick embryo femurs in tissue culture. *Am. J. Physiol.* **1966**, *211*, 821–825. [CrossRef] [PubMed]

17. Francis, M.D. The inhibition of calcium hydroxyapatite crystal growth by polyphosphonates and polyphosphates. *Calcif. Tissue Res.* **1969**, *3*, 151–162. [CrossRef] [PubMed]

18. Fleisch, H.; Neuman, W.F. Mechanisms of calcification: Role of collagen, polyphosphates, and phosphatase. *Am. J. Physiol.* **1961**, *200*, 1296–1300. [CrossRef] [PubMed]

19. Meyer, J.L.; McCall, J.T.; Smith, L.H. Inhibition of calcium phosphate crystallization by nucleoside phosphates. *Calcif. Tissue Res.* **1974**, *15*, 287–293. [CrossRef] [PubMed]

20. Meyer, J.L. Can biological calcification occur in the presence of pyrophosphate? *Arch. Biochem. Biophys.* **1984**, *231*, 1–8. [CrossRef]

21. Wuthier, R.E.; Bisaz, S.; Russell, R.G.G.; Fleisch, H. Relationship between pyrophosphate, amorphous calcium phosphate and other factors in the sequence of calcificationin vivo. *Calcif. Tissue Res.* **1972**, *10*, 198–206. [CrossRef] [PubMed]

22. Ibsen, C.J.S.; Chernyshov, D.; Birkedal, H. Apatite formation from amorphous calcium phosphate and mixed amorphous calcium phosphate/amorphous calcium carbonate. *Chem. Eur. J.* **2016**, *22*, 12347–12357. [CrossRef] [PubMed]

23. Jensen, A.C.S.; Hinge, M.; Birkedal, H. Calcite nucleation on the surface of PNIPAM-PAAc micelles studied by time resolved *in situ* PXRD. *Cryst. Eng. Commun.* **2015**, *17*, 6940–6946. [CrossRef]

24. Ibsen, C.J.S.; Birkedal, H. Influence of poly(acrylic acid) on apatite formation studied by in situ X-ray diffraction using an X-ray scattering reaction cell with high-precision temperature control. *J. Appl. Crystallogr.* **2012**, *45*, 976–981. [CrossRef]

25. Ibsen, C.J.S.; Birkedal, H. Modification of bone-like apatite nanoparticle size and growth kinetics by alizarin red S. *Nanoscale* **2010**, *2*, 2478–2486. [CrossRef] [PubMed]

26. Ibsen, C.J.S.; Leemreize, H.; Mikladal, B.F.; Skovgaard, J.; Eltzholtz, J.R.; Bremholm, M.; Iversen, B.B.; Birkedal, H. Crystallization kinetics of bone-like apatite nanocrystals formed from amorphous calcium phosphate in water by in situ synchrotron powder diffraction: Counter ions matter. 2018, in preparation.

27. Frølich, S.; Birkedal, H. Multiref: Software platform for automated and intelligent rietveld refinement of multiple powder diffractograms from in situ, scanning or diffraction tomography experiments. *J. Appl. Cryst.* **2015**, *48*, 2019–2025. [CrossRef]

28. Ibsen, C.J.S.; Gebauer, D.; Birkedal, H. Osteopontin strongly stabilizes metastable states prior to nucleation during apatite formation. *Chem. Mater.* **2016**, *28*, 8550–8555. [CrossRef]

29. Olliges-Stadler, I.; Rossell, M.D.; Suess, M.J.; Ludi, B.; Bunk, O.; Pedersen, J.S.; Birkedal, H.; Niederberger, M. A comprehensive study of the crystallization mechanism involved in the nonaqueous formation of tungstite. *Nanoscale* **2013**, *5*, 8517–8525. [CrossRef] [PubMed]

30. Jensen, G.V.; Bremholm, M.; Lock, N.; Deen, G.R.; Jensen, T.R.; Iversen, B.B.; Niederberger, M.; Pedersen, J.S.; Birkedal, H. Anisotropic crystal growth kinetics of anatase TiO_2 nanoparticles synthesized in a nonaqueous medium. *Chem. Mater.* **2010**, *22*, 6044–6055. [CrossRef]

31. Bisaz, S.; Russell, R.G.G.; Fleisch, H. Isolation of inorganic pyrophosphate from bovine and human teeth. *Arch. Oral Biol.* **1968**, *13*, 683–696. [CrossRef]

32. Larson, A.C.; Von Dreele, R.B. *Los Alamos National Laboratory Report Laur 86-748*; Los Alamos National Laboratory: Los Alamos, NM, USA, 2000.

33. Leemreize, H.; Eltzholtz, J.R.; Birkedal, H. Lattice macro and microstrain fluctuations in the calcified byssus of Anomia simplex. *Eur. J. Miner.* **2014**, *26*, 517–522.

34. Frølich, S.; Sørensen, H.O.; Hakim, S.S.; Marin, F.; Stipp, S.L.S.; Birkedal, H. Smaller calcite lattice deformation caused by occluded organic material in coccoliths than in mollusk shell. *Cryst. Growth Des.* **2015**, *15*, 2761–2767. [CrossRef]

35. Pokroy, B.; Fitch, A.N.; Lee, P.L.; Quintana, J.P.; Caspi, E.N.; Zolotoyabko, E. Anisotropic lattice distortions in mollusk-made aragonite: A widespread phenomenon. *J. Struct. Biol.* **2006**, *153*, 145–150. [CrossRef] [PubMed]

36. Pokroy, B.; Fitch, A.N.; Marin, F.; Kapon, M.; Adir, N.; Zolotoyabko, E. Anisotropic lattice distortions in biogenic calcite induced by intra-crystalline organic moleucles. *J. Struct. Biol.* **2006**, *155*, 96–103. [CrossRef] [PubMed]

37. Pokroy, B.; Quintana, J.P.; Caspi, E.N.; Berner, A.; Zolotoyabko, E. Anisotropic lattice distortions in biogenic aragonite. *Nat. Mater.* **2004**, *3*, 900–902. [CrossRef] [PubMed]

38. Brif, A.; Ankonina, G.; Drathen, C.; Pokroy, B. Bio-inspired band gap engineering of zinc oxide by intracrystalline incorporation of amino acids. *Adv. Mater.* **2014**, *26*, 477–481. [CrossRef] [PubMed]

39. Brif, A.; Bloch, L.; Pokroy, B. Bio-inspired engineering of a zinc oxide/amino acid composite: Synchrotron microstructure study. *Cryst. Eng. Comm.* **2014**, *16*, 3268–3273. [CrossRef]

40. Habraken, W.J.E.M.; Tao, J.; Brylka, L.J.; Friedrich, H.; Bertinetti, L.; Schenk, A.S.; Verch, A.; Dmitrovic, V.; Bomans, P.H.H.; Frederik, P.M.; et al. Ion-association complexes unite classical and non-classical theories for the biomimetic nucleation of calcium phosphate. *Nat. Comm.* **2013**, *4*, 1507. [CrossRef] [PubMed]

minerals

MDPI

Article

Carbonate Apatite Precipitation from Synthetic Municipal Wastewater

Jessica Ross, Lu Gao, Orysia Meouch, Essie Anthony, Divya Sutarwala, Helina Mamo and Sidney Omelon *

Department of Chemical and Biological Engineering, University of Ottawa, Ottawa, ON K1N 6N5, Canada; jross040@uottawa.ca (J.R.); lgao072@uottawa.ca (L.G.); omeou066@uottawa.ca (O.M.); eanth029@uottawa.ca (E.A.); dsuta007@uottawa.ca (D.S.); hmamo018@uottawa.ca (H.M.)
* Correspondence: somelon@uottawa.ca; Tel.: +1-613-562-5800

Received: 31 May 2017; Accepted: 18 July 2017; Published: 25 July 2017

Abstract: An important component of phosphorite (phosphate rock) is carbonate apatite, as it is required for phosphorous fertilizer production due to its increased phosphate solubility caused by carbonate substitution in the apatite mineral lattice. High phosphate concentrations in municipal wastewater treatment plants are commonly reduced by precipitating iron phosphate by addition of iron chloride. We investigated the possibility of precipitating carbonate apatite from a potential range of phosphate concentrations that could be available from municipal wastewater treatment plants with anaerobic digestion reactors (5 mM–30 mM). Synthetic phosphate solutions at neutral pH were mixed in batch experiments with a calcium carbonate solution produced by dissolving calcite in contact with carbon dioxide gas, with and without carbonate apatite seed. Batch experiments were used to identify the carbonate apatite supersaturation ranges for homogeneous and heterogeneous nucleation, and the precipitates analyzed with Raman spectroscopy, powder X-ray diffraction, inorganic carbon coulometry, and scanning electron microscopy. Some precipitates contained carbonate weight fractions within the range reported for geological phosphate rock (1.4–6.3 wt %). The precipitates were spherical, poorly crystalline carbonate apatite, suggesting an amorphous precursor transformed to a poorly crystalline carbonate apatite without changing morphology.

Keywords: amorphous precursor; nucleation; carbonate apatite; amorphous calcium carbonate phosphate

1. Introduction

The genesis of phosphorus-rich minerals such as carbonate apatite, which is a component of the valuable P-rich ore known as phosphorite or phosphate rock (PR), was an unanswered geological question for many years. Environmental conditions do not commonly generate inorganic orthophosphate (Pi) concentrations high enough for spontaneous phosphate mineral nucleation events. Theorized mechanisms for PR precipitation included inorganic precipitation and biologically-mediated precipitation [1,2]. One mechanism for Pi concentration in the marine environment is the accumulation and storage of polyphosphates (polyP: $(PO_3^-)_n$) within sulfide-oxidizing bacteria (genera *Thiomargarita* [3] and *Beggiatoa* [4]) during oxic environmental conditions. Periodic flow of anoxic waters into the bacterial mats causes the bacteria to switch their metabolism to a process that breaks down their polyP stores into Pi [5]. The dissolved Pi concentration in these extracellular matrices increases, and was measured to be 300 µM [6]—order of magnitudes above the average ocean Pi concentration of <1 µM [7]. When the Pi concentration increases upon polyP breakdown, the calcium and carbonate ions co-precipitate as apatite within the bacterial mat. The resultant high P content of ancient and modern apatite-rich ores qualifies them as PR. PR is a valuable resource because it is the only available phosphate mineral that can be economically used to produce phosphorus fertilizer, due to its increased solubility caused by its carbonate content [8]. PR is dissolved in sulfuric acid; this wet process produces phosphoric acid for phosphorous fertilizers [9]. The carbonate content of PR ranges from approximately 1.4–6.3 wt % CO_3^{2-} [10]. PR is a valuable and non-renewable resource [11,12].

A similar biological process of Pi concentration as polyP, followed by Pi release, Pi concentration increase, and precipitation is observed in municipal wastewater treatment plants. Biological phosphorus removal processes have been developed to enhance the uptake of Pi by polyP accumulating organisms [13]. In oxic conditions, the polyP-accumulating bacteria uptake Pi and generate intracellular polyP. Upon anoxic conditions within anaerobic digester reactors, these organisms hydrolyze their intracellular polyP and release Pi into solution. The resulting increased Pi concentration is decreased before treated water is released to the environment to avoid ecological upsets [14]. Soluble di- or trivalent metal chloride salts are added to reduce the Pi concentration by precipitating metal phosphate [15], which is a component of biosolids that are produced by many municipal wastewater treatment processes. Although sometimes applied to soils, the low solubility of iron phosphate in biosolids limits Pi availability to plants [16]. This processing results in a loss of Pi to the food chain.

We wished to investigate if a carbonate apatite could be precipitated from Pi at concentrations available within municipal wastewater treatment plants with anaerobic digestion reactors (ranging from 2.5 to 30 mM Pi), by precipitation with a dissolved calcium carbonate solution. Calcite, a calcium carbonate mineral, in equilibrium with an apatite-precipitation solution does not spontaneously precipitate a carbonate apatite product [17]. Therefore, a more concentrated calcium carbonate solution is required to generate a supersaturated solution with respect to carbonate appetite for our target Pi-concentration ([Pi]) solutions. Calcite solubility was increased by preparing a saturated solution of calcite in potable water, then contacting this slurry with gasses of different CO_2 fractions that simulated flue gas CO_2 concentrations that could be available at municipal wastewater treatment plants. Some municipal wastewater treatment plants with anaerobic digestion reactors produce a methane-rich biogas, and burn this biogas in an on-site electric power plant [18]. We investigated the possibility of mixing this calcium-bicarbonate-rich solution, which will be referred to as a saturated Ca-CO_3 solution, with representative [Pi] to precipitate a carbonate apatite mineral with a carbonate fraction similar to PR, with and without seed.

Seeding a crystallization process is expected to reduce the supersaturation required for precipitation [19]. Both protein-free biological carbonate apatite, and solids produced by homogeneous nucleation experiments were tested. This system is based on synthetic calcium, carbonate, and phosphate sources. The phosphate sources from a wastewater treatment plant contain a myriad of other chemical components which have unknown impacts on carbonate apatite precipitation with this strategy. This work was pursed to determine if the simplest case scenarios could generate carbonate apatite. Precipitation of carbonate apatite from Pi in municipal wastewater treatment streams could generate a potential useful resource for the wet-process phosphorous fertilizer industry.

2. Materials and Methods

2.1. Calcium Carbonate Solution Preparation

ACS grade calcium carbonate (identified as calcite by powder X-ray diffraction and Raman spectroscopy, MP Biomedicals, LLC, Solon, OH, USA) was mixed with potable water (pH 9, 0.2 mM Ca^{2+}, 0.2 mM inorganic carbon, City of Ottawa) at a concentration of 1 g CaCO$_3$/L overnight. N_2–CO_2 mixed gases with 7% CO_2 (representative of natural gas fired power plant flue gas composition [20]), 15% CO_2 (representative of the power plant flue gas composition powered by biofuel from a wastewater treatment plant [21]) and 100% CO_2 were mixed, then bubbled at a flow rate of ~0.1–0.4 L/min through 1 L of the calcium carbonate slurry in a closed, stirred tank reactor with a gas outlet ducted to a fume hood. The pH and calcium concentrations were monitored as function of time until the values stabilized. The pH was measured with a pH probe (AR10, Orion) and the calcium concentrations measured with colourimetry [22]. Samples were read from a 96 well plate by a spectrophotometer (Epoch Microplate Spectrophotometer, BioTek, Winooski, VT, USA).

2.2. Unseeded Carbonate Apatite Batch Precipitation

Inorganic phosphate solutions used throughout all experiments were prepared by dissolving Na_2HPO_4 (ACS grade, Sigma-Aldrich, Oakville, ON, Canada) in distilled and deionized water. The inorganic phosphate (Pi) concentrations used ranged from 2.5 mM Pi to 30 mM Pi, based on the Pi concentrations that are available from different municipal wastewater treatment process streams with aerobic and anaerobic sludge treatment. Pi concentrations were measured by colourimetry with the vanadomolybdate assay [23]. The batch precipitation experiments for 2.5, 5, 10, and 20 mM Pi solutions were repeated four times, and the 30 mM Pi solution batch experiment was repeated three times. The experiments are referred to with the initial Pi concentration ([Pi]) before mixing.

The different [Pi] solutions were mixed with an equal volume of a saturated Ca-CO_3 solution decanted from calcite in potable water that was mixed with 100% CO_2 for a minimum of 20 min contact time. 1:1, 250 mL volumes of the saturated Ca-CO_3 and Pi-solutions were mixed in a beaker at room temperature, sealed, and gently shaken. An aliquot was taken of each of the initial solutions prior to mixing to calculate the theoretical initial mixed concentration, and 5 days after mixing. A 5 day contact time was selected to ensure that precursor amorphous phases would recrystallize to apatite, as this transition has been reported to occur within 48 h [24,25]. The initial and final pH values were recorded. The pH was not controlled during these series of experiments. The difference in initial and final [Ca^{2+}] and [Pi] were used to calculate percentage changes, and to calculate the theoretical precipitate molar Ca/P ratio obtained by the reduction in [Ca^{2+}] and [Pi]. Observations regarding whether or not precipitate was immediately visible and if scale formed on the glassware surfaces were noted, as this would indicate whether the saturation state with respect to carbonate apatite was sufficient for heterogeneous nucleation on glass surfaces and other contaminants, or above the critical supersaturation for homogeneous nucleation. After five days, the slurries were vacuum-filtered through a glass filter, rinsed with deionized and distilled water, and vacuum-dried at 40 °C and 5 mm Hg.

The solubility product and the ion activity product are used to calculate the supersaturation of a given salt [26]. Since the exact composition and solubility product of carbonate apatite are unknown, a simplification of the supersaturation index will be used to compare the saturation states. The measured calcium concentration ([Ca^{2+}]) multiplied by the inorganic phosphate concentration ([Ca^{2+}] × [Pi]) will be used as a proxy saturation index, as the dissolved carbonate concentration is challenging to measure accurately. This is an oversimplification of a complex problem. As the carbonate substitution for phosphate or hydroxyl molecules is not known, stoichiometric coefficients will not be used in this saturation index, as used by Fleisch [27].

The goal of this work is to determine if mixing a saturated Ca-CO_3 solution produced by calcite in potable water that was in contact with CO_2 gas, with different [Pi] solutions generates a supersaturation large enough to spontaneously produce a synthetic PR analogue material through homogeneous nucleation. If possible, then heterogeneous precipitation in the presence of seed would also be possible, because this condition requires a lower supersaturation.

2.3. Seeded Carbonate Apatite Batch Precipitation

Given that the supersaturation required for heterogeneous nucleation is lower than that required for homogeneous nucleation, seeded, batch precipitation tests were undertaken to determine if crystal growth could occur with the lowest and highest supersaturation conditions that generated spontaneous precipitation (5 mM and 30 mM Pi). As the desired final precipitate is a carbonate apatite, it was assumed that biological apatite, which has a higher substituted carbonate fraction in carbonate apatite than PR (5–9 wt % $CO_3{}^{2-}$ [28,29]), could be a suitable seed material. The unseeded batch precipitation procedure was followed with the addition of 0.5 or 2 g/L bone mineral in the batch experiment. These seeded experiments were performed in duplicate.

To prepare the bone mineral, bovine bone was crushed, Tris-washed, lyophilized, and defatted in a 2:1 chloroform:methanol solution. The bone powder was deproteinized with bleach, rinsed with distilled and deionized water, and dried in a vacuum oven at less than 100 °C. The final product was ground with a mortar and pestle, and the fraction that passed below a 63 μm sieve was used as seed. A second set of batch precipitation experiments was undertaken, seeded with precipitates produced by an unseeded batch precipitation (referred to as "ppt"). This precipitation was induced by mixing 1 L of 27 mM Pi (similar to a known [Pi] in a local municipal wastewater treatment plant stream) with 1 L of 6 mM Ca^{2+} solution from $CaCO_{3(s)}$ dissolution in potable water (initial concentration 8 mg/L (0.2 mM) Ca^{2+}, pH 9) in equilibrium with 100% CO_2. The precipitate was vacuum filtered, rinsed with distilled and deionized water, and dried in a vacuum oven at 40 °C and 5 mm Hg. This seed was used in the 5 mM initial [Pi] test, as it was expected to provide surface chemistry with a closer match to solid phase favored to grow in these conditions. Precipitate produced from this seeded, batch crystallization experiment was used as seed for a 7 mM initial [Pi] batch crystallization experiment, which is characteristic of the lowest [Pi] in a local municipal wastewater treatment plant stream.

2.4. Precipitate Characterization

As bone mineral is composed of a higher carbonate weight fraction than PR, only unseeded precipitates and precipitates seeded with homogeneously nucleated seed were characterized in order to avoid confounding the new precipitate chemistry with the bone mineral seed chemistry.

2.4.1. Scanning Electron Microscopy

Samples were dried, mounted with carbon tape, and secondary electron images taken (Phenom PRO, Pheonom-World, Eindhoven, The Netherlands) at magnifications of 165×, 500×, and 1000×. The images were scanned at 10 kV and a resolution of 2048 dpi, with an exposure time of 26 s.

2.4.2. Raman Spectroscopy

Raman spectroscopy was undertaken with a WITec Alpha 300 confocal microscope with a 20× lens (WITec GmbH, Ulm, Germany). The laser source used was either 785 nm (XTRA High Power Single Frequency Diode Laser, Toptica Photonics, Victor, NY, USA), 488, or 532 nm (WITec GmbH, Ulm, Germany). Calibration was performed on the 520 nm Raman shift of Si.

Dry powdered samples were placed on a quartz slide, and data was collected with an integration time of 1 s for a minimum of 45 s. The Raman shifts of greatest interest were 960 cm^{-1} for the Pi identification in apatite, 1070 cm^{-1} for carbonate in apatite, and 1090 cm^{-1} for carbonate in calcite. Silicon, synthetic hydroxyapatite (HAP, Sigma-Aldrich, Oakville, ON, Canada), calcite ($CaCO_3$, MP Biomedicals, LLC, Solon, OH, USA), and bone powder were used as standard materials. The resulting spectra were filtered to remove cosmic ray interference. OPUS Spectroscopy software (v 6.5, Bruker Optiks, Ettlingen, Germany) was used to remove the baseline, smooth (9 smoothing points) slightly, and normalize the spectra that were then plotted with OriginPro 9.1 (OriginLab Corporation, Northampton, MA, USA).

2.4.3. Powder X-ray Diffraction

Precipitates were analyzed with a Multiflex D3 (Rigaku Americas Corporation, The Woodlands, TX, USA). The products were prepared on a low-background Si sample-holder, and scanned between 20–45° 2θ at a scan rate of 1°/min. The raw data were compared with ICDD PDF file for hydroxyapatite with PDXL 2 (Version 2.5.1.2, Rigaku Americas Corporation, The Woodlands, TX, USA).

2.4.4. Inorganic Carbon Coulometry

A carbon coulometer (Model #5012) (UIC Inc., Joliet, IL, USA) was used to measure the weight percent carbonate of the precipitates. A measured sample mass was dissolved in a CO_2-free chamber with 2 M perchloric acid ($HClO_4$, Sigma-Aldrich, Oakville, ON, Canada), and the CO_2 gas evolved from carbonate ions⁻ passed to another chamber to be quantified. Quantification of CO_2 (g) is accomplished by a titration reaction that is neutralized by redox reaction. The coulombs required for titration are proportional to the number of CO_2 (g) molecules that initiated the titration reaction. The coulometer calculates the mass of carbonate from titrated CO_2, allowing for the calculation of the weight percent carbonate.

2.4.5. Solubility

A few milligrams of precipitates were dissolved in 5 mL of distilled and deionized H_2O for one week under gentle mechanical shaking so they could reach equilibrium with the solution. The colorimetric methods were used to measure the resulting $[Ca^{2+}]$ and $[Pi]$ that were assumed to be in equilibrium with the precipitates. These values were used to calculate $[Ca^{2+}] \times [Pi]$ values, as well as the molar ratio of Ca/P in solution.

3. Results

3.1. Calcium Carbonate Solution Preparation

Figure 1 presents an example of initial conditions and the changes in pH and $[Ca^{2+}]$ with time during contact with 100% CO_2. Table 1 presents the equilibrium $[Ca^{2+}]$ and pH for the different % CO_2 values.

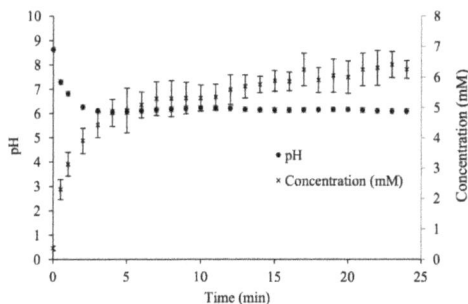

Figure 1. Changes in pH and calcium concentration with time in potable water mixed with $CaCO_{3(s)}$ while in contact with 100% CO_2 gas at room temperature.

Table 1. Effect of percent CO_2 on the equilibrium $[Ca^{2+}]$ and pH of saturated Ca-CO_3-potable water solutions.

% CO_2	$[Ca^{2+}]$ (mM)	pH
0	0.37 ± 0.1	8.0
7 ± 2	3.6 ± 0.7	6.7 ± 0.1
15 ± 2	4.8 ± 0.4	6.3 ± 0.1
100	5.7 ± 0.4	6.2 ± 0.1

Solutions from the 100% CO_2 experiments were used as calcium and carbonate sources in the subsequent carbonate apatite batch precipitation experiments. Ca-CO_3 saturated solutions were used as soon as possible after their production, as calcium carbonate precipitates from the decanted solutions as they shift to equilibrium with the atmosphere.

3.2. Unseeded Carbonate Apatite Batch Precipitation

The initial and final concentrations of calcium and Pi in the batch precipitation experiments, as well as the final pH, the theoretical percent calcium and Pi removed from solution, and the theoretical Ca/P in the precipitate are listed in Tables 2–4. Before mixing, the average $[Ca^{2+}]$ from $CaCO_3$ solutions used was 6 ± 1 mM, while the target [Pi]s were 2.5, 5, 10, 20, or 30 mM. Note that the sample names reflect the initial [Pi] before mixing with the saturated Ca-CO$_3$ solution.

Table 2. Summary of [Ca^{2+}] results for the batch, unseeded precipitation tests.

[Pi] Solution (mM)	Initial [Ca^{2+}] (mM)	Final [Ca^{2+}] (mM)	Percent [Ca^{2+}] Change
2.5	2.89 ± 0.43	2.07 ± 0.24	-17%
5	2.59 ± 0.50	1.14 ± 0.71	-56%
10	2.68 ± 0.40	0.36 ± 0.20	-87%
20	2.67 ± 0.38	0.15 ± 0.06	-94%
30	2.40 ± 0.18	0.05 ± 0.01	-98%

Table 3. Summary of [Pi] results for the batch, unseeded precipitation tests.

[Pi] Solution (mM)	Initial [Pi] (mM)	Final [Pi] (mM)	Percent [Pi] Change
2.5	1.24 ± 0.02	0.94 ± 0.28	-24%
5	2.45 ± 0.07	1.44 ± 0.32	-41%
10	5.07 ± 0.44	3.38 ± 0.19	-33%
20	11.10 ± 0.93	8.67 ± 0.29	-22%
30	15.06 ± 0.01	13.46 ± 0.26	-11%

Table 4. Summary of pH and theoretical Ca/P results for the batch, unseeded precipitation tests.

[Pi] Solution (mM)	Final pH	Theoretical Precipitate Ca/P
2.5	7.07 ± 0.19	1.84 ± 0.86
5	6.95 ± 0.21	1.47 ± 0.23
10	7.21 ± 0.15	1.46 ± 0.43
20	7.34 ± 0.18	1.08 ± 0.19
30	7.5 ± 0.08	1.48 ± 0.18

The concentration changes are shown graphically in Figure 2, where it becomes apparent with the different final [Ca^{2+}] (arrow head) that the limiting reagent is different above and below the case [Pi] = 5 mM. For [Pi] of 5 mM or less, only 56–17% of the initial mixed [Ca^{2+}] precipitated from the solution, whereas 98–94% of the [Ca^{2+}] is reduced when mixed with [Pi] > 10 mM. When the mixed [Pi] is higher than the [Ca^{2+}], the reaction is limited by the [Ca^{2+}]. The pH is likely buffered by the residual [Pi] and carbonate. The different theoretical Ca/P ratios calculated by the loss of these ions from solution also suggest that there is a range in this theoretically calculated chemical composition for the homogeneously nucleated precipitate phases, as the theoretical precipitate Ca/P values are not statistically significantly different (one-way ANOVA, with Bonferroni means comparison).

If the initial [Pi] was 10 mM or higher, the solution became cloudy immediately upon mixing with the Ca-CO$_3$ solution. This indicated that a supersaturation state above the critical supersaturation for homogeneous nucleation was achieved (S$_{critical}$, Figure 2). With initial [Pi] = 5 mM, there was no evidence of homogeneous nucleation, but precipitate formed on the beaker walls. This would suggest that the experimental conditions for the metastable zone, below S$_{critical}$ at an initial [Pi] = 5 mM, and above carbonate apatite saturation were identified. The 2.5 mM initial [Pi] experiment did not produce enough solids to be collected and quantified, but the measurable decrease in both [Ca^{2+}] and [Pi] suggests that heterogeneous precipitation occurred. It was assumed that S$_{critical}$ in this system lies between an initial [Pi] between 5 and 10 mM before mixing (2.5 and 5 mM after mixing), and [Ca^{2+}] of 3 mM after mixing (dashed vertical line in Figure 2). For this system, S$_{critical}$ with the simplified index [Ca^{2+}] $\times$ [Pi] lies between 7.5 and 15 mM^2.

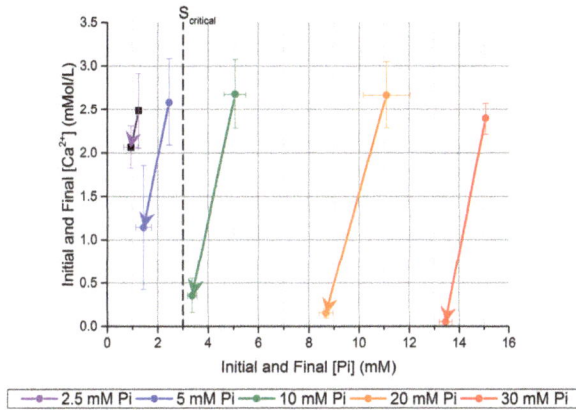

Figure 2. Change in [Ca^{2+}] and [Pi] after five days in unseeded batch precipitation. The samples are named after the [Pi] before mixing with the Ca-CO$_3$ solution.

3.3. Seeded Carbonate Apatite Batch Precipitation

The initial and final concentrations of calcium and Pi in the seeded batch precipitation experiments, initial and final pH, percent calcium and Pi removed from solution, and the calculated Ca/P in the precipitate are listed in Tables 5–7. The maximum [Ca^{2+}] from Ca-CO$_3$ solutions used was 6 ± 1 mM, while the [Pi] solution concentrations were 5, 7, or 30 mM before mixing with the Ca-CO$_3$ solution.

Table 5. Summary of [Ca^{2+}] results for the batch, seeded precipitation tests.

[Pi] Solution (mM)	[Seed] (g/L) & Type	Initial [Ca^{2+}] (mM)	Final [Ca^{2+}] (mM)	Percent [Ca^{2+}] Change
5 [1]	0.50 (bone)	3.06	1.11	−64%
5 [1]	2.0 (bone)	3.03	1.09	−64%
5 [1]	1.0 (ppt)	2.96	0.26	−91%
7 [2]	1.0 (ppt)	2.24 ± 0.14	0.01 ± 0.01	−100%
30 [1]	0.50 (bone)	3.10	0.12	−96%
30 [1]	2.0 (bone)	3.37	0.09	−97%

[1] $n = 2$, [2] $n = 4$.

Table 6. Summary of [Pi] results for the batch, seeded precipitation tests.

[Pi] Solution (mM)	[Seed] (g/L) & Type	Initial [Pi] (mM)	Final [Pi] (mM)	Percent [Pi] Change
5 [1]	0.50 (bone)	2.53	1.16	−54%
5 [1]	2.0 (bone)	2.53	1.11	−56%
5 [1]	1.0 (ppt)	2.29	0.73	−68%
7 [2]	1.0 (ppt)	3.56	2.9 ± 0.03	−41%
30 [1]	0.50 (bone)	15.30	11.59	−24%
30 [1]	2.0 (bone)	15.13	11.09	−27%

[1] $n = 2$, [2] $n = 4$.

Table 7. Summary of pH and theoretical Ca/P results for the batch, seeded precipitation tests.

[Pi] Solution (mM)	[Seed] (g/L) & Type	Final pH	Theoretical Precipitate Ca/P
5 [1]	0.5 (bone)	7.06	1.44
5 [1]	2 (bone)	7.06	0.95
5 [1]	1 (ppt)	7.56	1.74
7 [2]	1 (ppt)	8.23 ± 0.05	1.52
30 [1]	0.5 (bone)	7.55	0.81
30 [1]	2 (bone)	7.67	0.86

[1] $n = 2$, [2] $n = 4$.

An increase in bone seed concentration did not dramatically increase the percent $[Ca^{2+}]$ or [Pi] reduction, but did reduce the theoretical precipitate Ca/P from 1.44 to 0.95. 5 and 7 mM [Pi] solutions seeded with homogeneously nucleated precipitate (ppt) increased the equilibrium pH and reduced the equilibrium $[Ca^{2+}] \times [Pi]$ (Table 8). Figure 3 compares the $[Ca^{2+}]$ and [Pi] changes for the bone mineral- and precipitate-seeded experiments; it highlights the additional $[Ca^{2+}]$ reduction with ppt-seeded experiments, and the small effect of increased bone mineral seed concentration on the equilibrium solution composition.

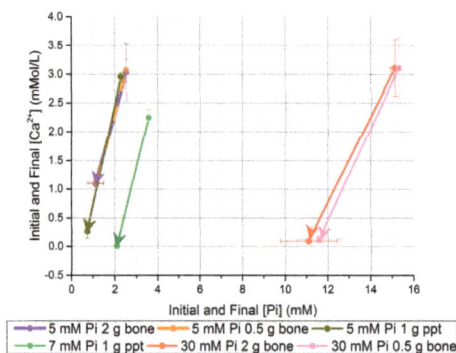

Figure 3. Change in $[Ca^{2+}]$ and [Pi] after five days in seeded batch precipitation (bone and homogeneously nucleated precipitate (ppt)).

The simplified saturation index ($[Ca^{2+}] \times [Pi]$) and final pH for the seeded and unseeded experiments are listed in Table 8.

Table 8. Measured $[Ca^{2+}] \times [Pi]$ values before and after batch precipitation experiments.

[Pi] Solution (mM)	[Seed] (g/L) & Type	$[Ca^{2+}] \times [Pi]$ Initial (mM)^2	$[Ca^{2+}] \times [Pi]$ Final (mM)^2	Final pH
2.5 [1]	0	3.09 ± 0.51	1.90 ± 0.48	7.07 ± 0.19
5 [1]	0	6.36 ± 1.38	1.73 ± 1.40	6.95 ± 0.21
5 [2]	0.50 (bone)	7.73	1.25	7.06
5 [2]	2.0 (bone)	7.65	1.19	7.06
5 [2]	1.0 (ppt)	6.75	0.20	7.56
7 [2]	1.0 (ppt)	7.98	0.02	8.23 ± 0.05
10 [1]	0	13.6 ± 2.64	1.19 ± 0.61	7.21 ± 0.15
20 [1]	0	29.9 ± 6.71	1.35 ± 0.54	7.34 ± 0.48
30 [1]	0	36.1 ± 2.69	0.70 ± 0.11	7.50 ± 0.08
30 [2]	0.50 (bone)	47.5	1.41	7.55
30 [2]	2.0 (bone)	45.9	0.99	7.67

[1] $n = 4$, [2] $n = 2$.

These initial and final $[Ca^{2+}] \times [Pi]$ data plotted with respect to the initial and final [Pi] are summarized in Figure 4a,b.

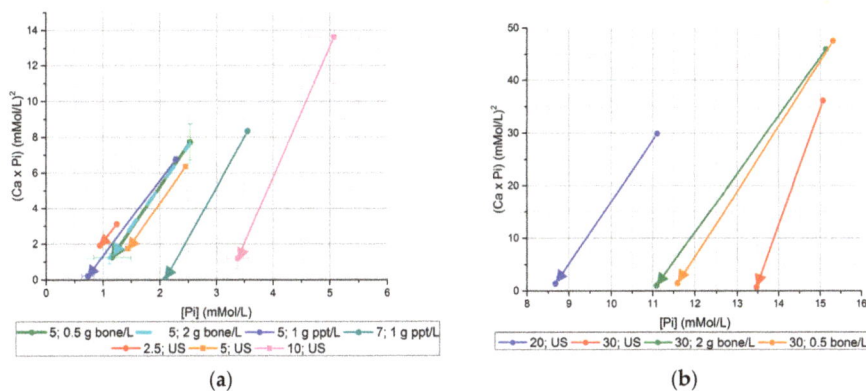

(a)

(b)

Figure 4. Initial and final (Ca × P) values, versus initial and final [Pi] ((**a**) initial [Pi] < 20 mM, (**b**) initial [Pi] ≥ 20 mM). Using seed from previous unseeded experiments generated lower final $[Ca^{2+}] \times [Pi]$ values than bone mineral seed or no seed for equivalent initial [Pi].

All final $[Ca^{2+}] \times [Pi]$ concentrations for bone mineral-seeded and unseeded precipitation experiments were at or higher than 1 mM^2 except for the 30 mM initial [Pi], unseeded experiment. Bone mineral seed lowered the final $[Ca^{2+}] \times [Pi]$ value for the 5 mM [Pi] experiment, but not the 30 mM [Pi] experiment. The lowest final $[Ca^{2+}] \times [Pi]$ values were achieved with the use of precipitate seed from previous unseeded experiments, where the values were 0.2 mM^2 or less for initial concentrations of 5 and 7 mM initial [Pi].

The pH followed an inverse trend of final $[Ca^{2+}] \times [Pi]$, with the highest pH values associated with the 7 mM initial [Pi] group with precipitate seed, and the 30 mM initial [Pi] unseeded precipitations. A plot of the final $[Ca^{2+}] \times [Pi]$ values vs. pH (Figure 5) shows the correlations for the increase in pH with decrease in $[Ca^{2+}] \times [Pi]$. This reflects the increase in $[PO_4^{3-}]$ and $[CO_3^{2-}]$ equilibrium speciation with increasing pH that would be expected to decrease the equilibrium [Pi] above apatite saturation.

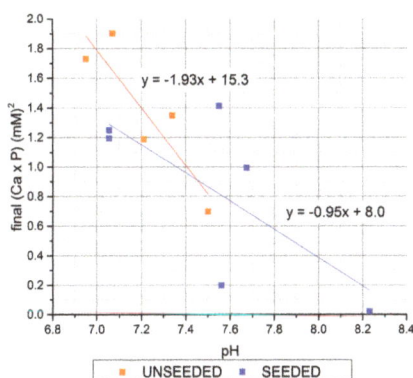

Figure 5. Final (Ca × P) versus final pH (seeded and unseeded).

The different slopes of the linear regression lines for the seeded and unseeded precipitation cases suggests a possible effect of the seed chemistry on the solution chemistry.

3.4. Precipitate Characterization

3.4.1. Scanning Electron Microscopy

Representative images for 5 and 30 mM initial [Pi] (unseeded) and 7 mM (seeded) taken at 5000× magnification are presented in Figure 6. Products produced at intermediate initial [Pi] were similar to the 30 mM results. Additional SEM images of unseeded and precipitate-seeded experiments are presented in Appendix A, Figure A1.

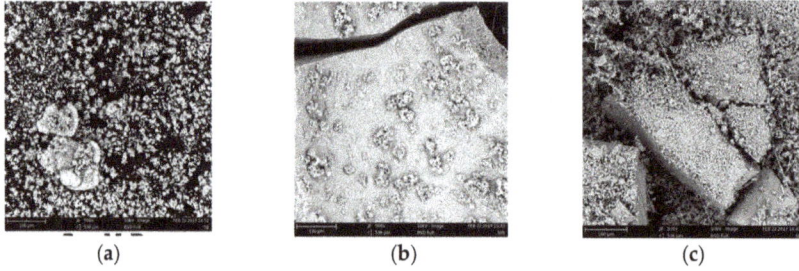

(a) (b) (c)

Figure 6. SEM images of precipitates from unseeded experiments (**a**) 5 mM; (**b**) 30 mM [Pi] and precipitate-seeded experiments (**c**) 7 mM [Pi].

An example of higher-resolution images of the seeded precipitate from the 7 mM [Pi] experiment are shown in Figure 7. As the crystal habit for synthetic carbonate apatite is plate-like [30], the spherical solids, ranging in diameter from approximately 2 to 7 μm, suggest the preservation of the spherical structure characteristic of an amorphous precursor.

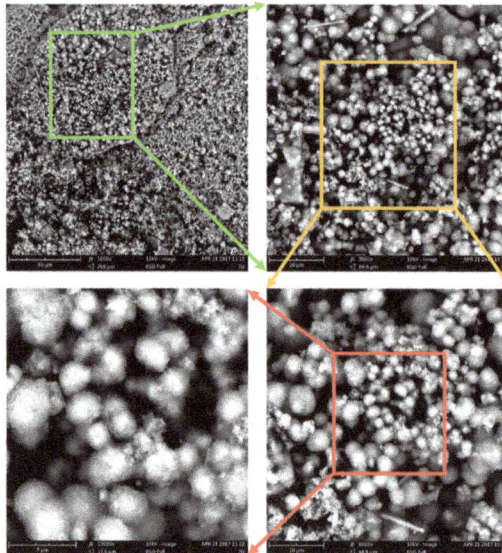

Figure 7. Spherical precipitates imaged with SEM with increasing magnification (1000×, 3000×, 6000×, and 15,000×).

3.4.2. Raman Spectroscopy

Raman shifts of the unseeded precipitates and the product from the seeded, 7 mM [Pi] precipitation are compared in Figure 8a,b. There was no evidence of an amorphous calcium phosphate (ACP) phase, which has been associated with a ν_1 phosphate molecule shift at 951 cm^{-1}, nor a ν_1 carbonate shift at 1081 ± 3 cm^{-1} for ACP precipitating systems that absorbed ambient carbon dioxide [24]. The dominant precipitate Raman shift is for the ν_1 phosphate molecule shift at 960 cm^{-1} of apatite, with shifts similar to the ν_3 phosphate molecule attributed at 1041 ± 3 cm^{-1}, and a ν_1 carbonate shift at 1073 ± 2 cm^{-1} [24]. The intensity of the ν_1 carbonate shift was largest for bone mineral and the 7 mM precipitate with homogeneously nucleated seed, which is supported by their larger measured carbonate weight fractions.

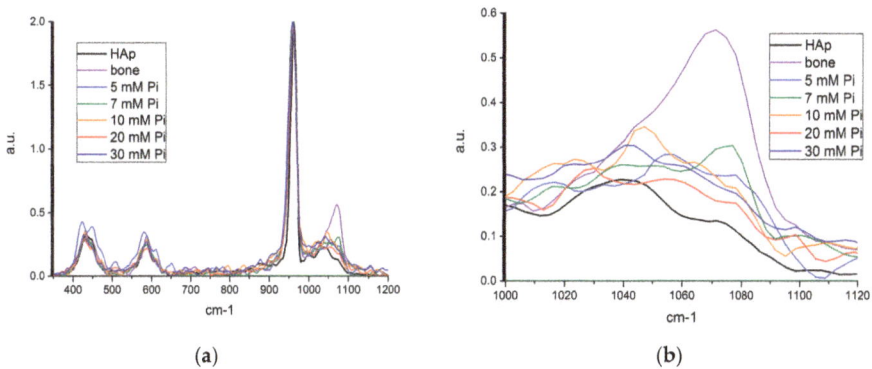

(a)   (b)

Figure 8. Raman spectra for synthetic hydroxyapatite (HAP, black), bone mineral (violet), unseeded precipitate from 5 (blue), 10 mM (orange), 20 mM (red), and 30 mM (navy), and 7 mM seeded precipitate (green) from (**a**) 350–1200 cm^{-1}, and (**b**) 1000–1120 cm^{-1}.

Raman spectroscopy allowed for the identification of the characteristic P-O shift for phosphate in apatite, and supporting evidence for the measured inorganic carbonate fraction (Section 3.4.4). There was no evidence of calcite, with its ν_1 carbonate shift at 1090 cm^{-1}, or another crystalline phosphate mineral phase that is distinguishable from apatite, such as octacalcium phosphate, brushite, or whitlockite.

3.4.3. Powder X-ray Diffraction

Powder X-ray diffraction patterns generated by bone mineral, the 30 mM Pi unseeded experiment, and the seed and precipitate from the 7 mM [Pi] experiment are presented in Figure 9.

These X-ray diffraction patterns are similar in peak position, and are representative of the low intensity and wide shape of biological apatite. This indicates that the precipitates are composed of small polycrystalline domains and/or are poorly crystalline, and/or highly substituted apatite. For orientation, the ICDD powder diffraction file for hydroxyapatite (01-089-4405) is overlaid on the precipitate product data.

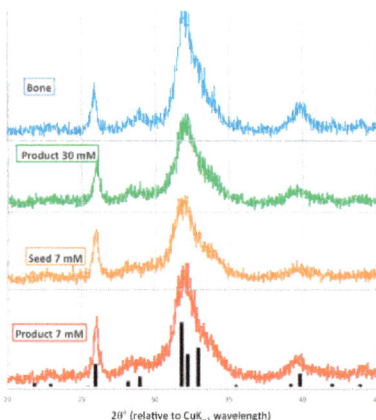

Figure 9. X-ray powder diffraction patterns from bone mineral (blue), 30 mM [Pi] unseeded experiment (green), 7 mM [Pi] seed (orange) and seeded precipitate (red), overlaid with ICDD 01-089-4405 PDF (hydroxyapatite, black).

3.4.4. Inorganic Carbon Coulometry

The measured carbonate weight fraction of the unseeded precipitates and the 7 mM [Pi] seeded precipitate are presented in Table 9.

Table 9. Inorganic carbon coulometry measurement of weight percent carbonate (CO_3^{2-})

[Pi] Solution (mM)	[Seed] g/L	Weight % CO_3^{2-}
5 [1]	0	1.95 ± 0.08
7 [2]	1.0 (ppt)	2.59 ± 0.09
10 [2]	0	1.21 ± 0.43
20 [2]	0	1.13 ± 0.14
30 [1]	0	1.10 ± 0.04

[1] $n = 3$, [2] $n = 4$

There is a trend of decreasing carbonate weight fraction with increasing initial [Pi] for the unseeded experiments. The carbonate content of the seeded product at 7 mM Pi is higher than the unseeded products. The carbonate weight fraction for the unseeded precipitates are statistically significantly different (one-way ANOVA). The carbonate values for the seeded precipitate and the highest Ca-CO_3/Pi ratio (5 mM initial [Pi]) are within the reported carbonate weight fraction of phosphate rock (1.4–6.3 weight % CO_3^{2-} [10]), and are lower than the percent carbonate measured in bone mineral (5–9 weight % CO_3^{2-} [28,29]). These data indicate that carbonate apatite with a carbonate content similar to PR precipitates from mixing solutions with a [Pi] at or lower than 5 mM with saturated Ca-CO_3 solutions in equilibrium with carbon dioxide gas and calcite, without seed, and a solution with 7 mM [Pi] with seed, at room temperature. The carbonate content is less than that for biological apatite. Future work is required to investigate the effect of seed on this carbonate apatite precipitation process.

3.4.5. Solubility

Bone mineral was determined to be more soluble that the homogeneously nucleated precipitates, with an equilibrium [Ca^{2+}] of 0.34 mM, and [Pi] of 0.13 mM. Therefore, samples that contained bone mineral were not included in the comparison of precipitate solubility (Table 10).

Table 10. Measured dissolved $[Ca^{2+}]$, [Pi], and Ca/P in distilled and deionized water compared with Ca/P values predicted from the change in precipitating solution composition.

Initial [Pi] (mM)	[Seed] (g/L)	$[Ca^{2+}]$ (mM)	[Pi] (mM)	Ca/P	Ca/P (Tables 4 and 7)
5 [1]	0	0.16 ± 0.02	0.16 ± 0.02	1.03 ± 0.15	1.47 ± 0.23
5 [2]	1.0	0.24	0.23	1.02	1.74
7 [1]	1.0	0.21 ± 0.01	0.22 ± 0.01	0.97 ± 0.05	1.52 ± 0.07
10 [1]	0	0.18 ± 0.03	0.23 ± 0.04	0.79 ± 0.04	1.46 ± 0.43
20 [1]	0	0.18 ± 0.02	0.51 ± 0.12	0.38 ± 0.14	1.08 ± 0.19
30 [3]	0	0.17 ± 0.06	0.45 ± 0.09	0.39 ± 0.17	0.48 ± 0.18

[1] $n = 4$, [2] $n = 2$, [3] $n = 3$

Precipitates grown from previously homogeneously nucleated seed generated the highest average dissolved $[Ca^{2+}]$. A one-way ANOVA showed that the average dissolved $[Ca^{2+}]$ from the unseeded precipitates did not change significantly with initial [Pi]. Dissolved [Pi] concentrations follow a proportional trend with increasing initial [Pi]. Statistically significant differences between [Pi] means were noted for all groups but not between any two of the 5, 7, and 10 mM Pi groups, and between the 30 and 40 mM Pi groups (Bonferroni test). Ca/P ratios for the unseeded and precipitate-seeded experiments followed an inverse trend with the initial [Pi] from which they were nucleated, and have the same statistically significant differences as the final [Pi]. Within the margins of error, a similar trend is seen with the calculated Ca/P from the $[Ca^{2+}]$ and [Pi] decrease in the precipitation experiments. The average calculated Ca/P values were higher than the measured solubility Ca/P values; this may result from the different solution compositions in which the precipitates were in equilibrium, and the effect of the unmeasured carbonate concentrations.

4. Discussion

The goal of precipitating a carbonate apatite from industrial-scale solutions with uncontrollable Pi concentrations generated by municipal wastewater treatment requires supersaturated conditions for carbonate apatite. This supersaturation condition could be generated by mixing Pi-rich solutions with dissolved calcium carbonate. The calcium concentrations required to precipitate carbonate apatite from a minimum of 5 mM Pi were generated by increasing calcite solubility with dissolved CO_2. Calcium concentrations increased and the pH decreased by contacting increasing percentages of CO_2 gas with synthetic calcite slurried in potable water, up to 6 ± 1 mM $[Ca^{2+}]$. This solution, although unstable when removed from contact CO_2, homogeneously nucleated and precipitated carbonate apatite from synthetic solutions with 5–30 mM Pi when mixed in a 1:1 volume ratio. Limestone and combustion flue gas could provide these reagents on the scale of municipal wastewater treatment plants; the effects of other components in these materials remain to be tested.

Unseeded precipitation tests with the highest initial $[Ca^{2+}] \times$ [Pi] values generally produced the lowest final $[Ca^{2+}] \times$ [Pi] values. The seeded tests did not follow this trend; homogeneously nucleated precipitate used as seed resulted in a lower final $[Ca^{2+}] \times$ [Pi] value than bone mineral-seeded experiments. It is not known if this is due to the different carbonate content of the seed minerals, and/or could be attributed to the incongruent dissolution of apatite minerals [31–33]. Further investigation of the effect of seed and its chemistry is required. Seed dissolution and reprecipitation may also affect the equilibrium $[Ca^{2+}]$ and [Pi]; this possibility requires further precipitation characterization as a function of time. The disparity between the calculated Ca/P precipitate ratios and the measured Ca/P dissolved precipitate ratios for the unseeded and seeded products could be explained by the solution carbonate concentrations; these were not measured, and are the subject of future work. The Ca/P trend is inversely proportional to the initial [Pi], which suggests a lower carbonate fraction in the precipitate that nucleated from higher [Pi]. This was confirmed with decreasing precipitate carbonate content produced from increasing initial [Pi].

pH was not controlled in these precipitation experiments. An inverse correlation between pH and percent Pi removed was observed, which suggests that a carbonate apatite precipitation process for Pi removal could be controlled to meet the effluent [Pi] by increasing the precipitation pH. The different observed relationships between the final pH of unseeded and seeded experiments and percent Pi removed suggest that the seed may have affected the solution chemistry. Dissolution of the known higher carbonate content in bone mineral would also increase pH, $[Ca^{2+}]$, and [Pi]. Future work will include measuring the dissolved carbonate concentrations before and after precipitation.

Precipitates were characterized as carbonate apatite by Raman spectroscopy and X-ray powder diffraction. Inorganic carbon coulometry confirmed the carbonate content of the samples with 5 and 7 mM initial [Pi] to be within reported carbonate bounds of PR. These samples would have had a higher dissolved carbonate to Pi ratio, and this could have contributed to their higher precipitate carbonate content [34].

The spherical shape of the poorly crystalline carbonated apatite represent a possible different precursor pathway than reported by Habraken et al. [35], as they reported that ACP spheres crystallized to hydroxyapatite plates [35]. However, carbonate ions were not present in significant concentrations in their system. In this work, carbonate concentrations ranging from 1.1–1.95 weight percent were measured in unseeded precipitate; the effect of this carbonate fraction on a potential amorphous calcium carbonate phosphate precursor is not known. It is also possible that initial nuclei resulted in a spherical solid formed of small crystal domains due to a fractal growth pattern [36]. The higher weight percent content in the precipitate produced from 7 mM initial [Pi] with seed produced from homogeneous nucleation is interesting, and will be given future attention. It is not known if the seed changed the solution chemistry, and/or if heterogeneous nucleation may favor a different precipitate chemistry.

Citrate-stabilized, spherical ACP particles in phosphate-buffered saline were observed to generate crystalline apatite domains within the ACP spheres [37]. It is possible that the carbonate ions have impacted the restructuring of the initial, spherical amorphous precursor solids as they transform to a poorly crystalline carbonate apatite, without significantly changing morphology. Carbonate apatite precipitated at 60–85 °C temperatures exhibited plate morphology with a *c*-axis length of 291.8 ± 222.0 at lower carbonate weight percent (3.63 wt %), and a reduced *c*-axis length of 36.0 ± 28.4 nm for 17.8 weight percent carbonate, as measured by TEM [34]. This result also indicates a carbonate effect on apatite crystal habit, although the carbonate apatite precipitation conditions for this work were different than presented here. Seeding with a homogeneously nucleated precipitate resulted in spherical carbonate apatite solids with a significantly higher carbonate content (2.59%) when precipitated with a 7 mM [Pi] solution. The effect of seeding with homogeneously nucleated solids on the precipitate characteristics and its carbonate content requires further study.

Zou et al. reported a phase transformation from spherical amorphous calcium carbonate (ACC) to spherical vaterite for smaller (~100 nm) and more soluble ACC particles, and a dissolution-reprecipitation pathway for larger ACC spheres that generated calcite rhombohedral crystals [38]. We propose that our homogeneous precipitates initially nucleated as soluble, amorphous calcium carbonate phosphate spheres that underwent a phase transformation to carbonated apatite without creating the plate-like crystal habit of apatite. The surface of this spherical, highly substituted carbonate apatite precipitate may enable epitaxial growth of carbonated apatite with a higher carbonate fraction, but further experiments that include initial precipitate characterization, and tracking the change in precipitate size in seeded experiments are required to confirm this hypothesis.

5. Conclusions

These precipitation conditions were inspired by the availability of waste phosphate-containing solutions from municipal wastewater treatment plants, carbon dioxide gas, calcium carbonate equilibrium, and the apatite lattice that is tolerant of many substitutions. A saturated solution of potable water, calcite, and carbon dioxide generated a sufficiently high calcium concentration to precipitate carbonate apatite from synthetic phosphate solutions at room temperature that are representative of possible phosphate-containing streams from municipal wastewater treatment plants, with or without seed.

When mixed at a 1:1 volume ratio with a CO_2-calcium carbonate-saturated solution with a 6 ± 1 mM $[Ca^{2+}]$, the minimum [Pi] to nucleate and grow carbonate apatite was 5 mM [Pi]. Homogeneous nucleation was evident for [Pi] at or greater than 10 mM [Pi]. The unseeded precipitation experiment with highest initial carbonate/phosphate concentration, and the experiment seeded with solids from homogenously nucleated precipitation experiments generated spherical precipitates less than 10 μm in size, with some groups producing carbonate fractions similar to phosphate rock. The precipitate weight fraction of carbonate was higher when precipitation occurred in the presence of seed produced from homogeneous nucleation conditions.

The spherical shape of the precipitates and poorly crystalline structure suggest an amorphous precursor that transformed to a poorly crystalline carbonate apatite without significant morphological change. Future work will involve characterizing the first and subsequent solid phases to nucleate in these conditions between initial mixing and the equilibrium state after 5 days, further investigating the effect of seed on this precipitation reaction, and extending these tests to the complex P-containing solutions from municipal wastewater treatment.

Acknowledgments: The Natural Sciences and Research Council of Canada's Discovery Grant program is acknowledge for supporting this research. We thank Professor Fabio Variola for use of his Raman spectrometer, Duane Dukart and Christian Kabbe for guidance and support. Bulat Gabidullin is acknowledged for generating the powder X-ray diffraction data. The reviewers are thanked for their helpful feedback, which improved the manuscript.

Author Contributions: J.R., E.A., and D.S. conceived, designed and performed the precipitation experiments, measured the solution chemistry, and undertook the SEM work. O.M. and L.G. designed and performed the calcium carbonate dissolution experiments. S.O. conceived the project, measured carbonate content with H.M. and J.R., and with J.R., wrote the manuscript.

Conflicts of Interest: The authors declare no conflict of interest.

Appendix A

SEM summary of products at $500\times$ magnification. Top row are products of unseeded experiments, in order of increasing initial [Pi]. The bottom row shows the seed, and precipitates formed from seeded experiments. Seed 2 was formed by mixing a solution of 27 mM Pi with a CO_2-saturated solution of $CaCO_3$, using the same method as the unseeded tests. It was subsequently used as seed for one of the 5 mM Pi tests. The precipitates formed during both 5 mM Pi seeded tests were combined and used as seed ("Seed 5") for the 7 mM Pi tests.

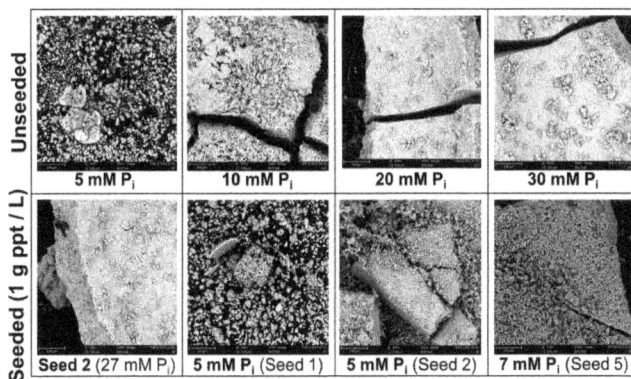

Figure A1. SEM images of precipitates from unseeded and seeded experiments, with the corresponding initial [Pi] used to produce the precipitate when mixed with saturated Ca-CO_3 solutions.

References

1. Kazakov, A. The phosphorite facies and the genesis of phosphorites. *USSR Trans. Sci. Inst. Fert. Insetofung* **1937**, *142*, 95–113.
2. McConnell, D. Precipitation of phosphates in sea water. *Econ. Geol.* **1965**, *60*, 1059–1062. [CrossRef]
3. Schulz, H.N.; Brinkhoff, T.; Ferdelman, T.G.; Marine, M.H.; Teske, A.; Jorgensen, B.B. Dense populations of a giant sulfur bacterium in namibian shelf sediments. *Science* **1999**, *284*, 493–495. [CrossRef] [PubMed]
4. Bruchert, V.; Jorgensen, B.B.; Neumann, K.; Riechmann, D.; Schlosser, M.; Schulz, H. Regulation of bacterial sulfate reduction and hydrogen sulfide fluxes in the central namibian coastal upwelling zone. *Geochim. Cosmochim. Acta* **2003**, *67*, 4505–4518. [CrossRef]
5. Goldhammer, T.; Bruchert, V.; Ferdelman, T.G.; Zabel, M. Microbial sequestration of phosphorus in anoxic upwelling sediments. *Nat. Geosci.* **2010**, *3*, 557–561. [CrossRef]
6. Schulz, H.N.; Schulz, H.D. Large sulfur bacteria and the formation of phosphorite. *Science* **2005**, *307*, 416–418. [CrossRef] [PubMed]
7. Conkright, M.E.; Gregg, W.W.; Levitus, S. Seasonal cycle of phosphate in the open ocean. *Deep Sea Res. Part I Oceanogr. Res. Pap.* **2000**, *47*, 159–175. [CrossRef]
8. Sheldon, R.P. Ancient marine phosphorites. *Annu. Rev. Earth Planet. Sci.* **1981**, *9*, 251–284. [CrossRef]
9. Dawson, C.J.; Hilton, J. Fertiliser availability in a resource-limited world: Production and recycling of nitrogen and phosphorus. *Food Policy* **2011**, *36*, S14–S22. [CrossRef]
10. McArthur, J.M. Francolite geochemistry—Compositional controls during formation, diagenesis, metamorphism and weathering. *Geochim. Cosmochim. Acta* **1985**, *49*, 23–35. [CrossRef]
11. *World Fertilizer Trends and Outlook to 2019*; Food and Agriculture Organization of the United Nations: Rome, Italy, 2016. Available online: www.fao.org/3/a-i5627e.pdf (accessed on 19 June 2017).
12. Cordell, D.; Neset, T.S.S. Phosphorus vulnerability: A qualitative framework for assessing the vulnerability of national and regional food systems to the multi-dimensional stressors of phosphorus scarcity. *Glob. Environ. Chang.* **2014**, *24*, 108–122. [CrossRef]
13. Kuba, T.; Smolders, G.; van Loosdrecht, M.C.M.; Heijnen, J.J. Biological phosphorus removal from wastewater by anaerobic-anoxic sequencing batch reactor. *Water Sci. Technol.* **1993**, *27*, 241–252.
14. Ulrich, A. Taking stock: Phosphorus supply from natural and anthropogenic pools in the 21st century. *Sci. Total Environ.* **2016**, *542*, 1005–1007. [CrossRef] [PubMed]
15. Morse, G.K.; Brett, S.W.; Guy, J.A.; Lester, J.N. Review: Phosphorus removal and recovery technologies. *Sci. Total Environ.* **1998**, *212*, 69–81. [CrossRef]
16. *Phosphorus and Nitrogen Removal from Municipal Wastewater: Principles and Practice*, 2nd ed.; Sedlak, R.I., Ed.; Lewis Publishers: Washington, DC, USA, 1991.
17. Minh, D.P.; Lyczko, N.; Sebei, H.; Nzihou, A.; Sharrock, P. Synthesis of calcium hydroxyapatite from calcium carbonate and different orthophosphate sources: A comparative study. *Mater. Sci. Eng. B* **2012**, *177*, 1080–1089. [CrossRef]
18. McKendry, P. Energy production from biomass (part 2): Conversion technologies. *Bioresour. Technol.* **2002**, *83*, 47–54. [CrossRef]
19. Kashchiev, D.; van Rosmalen, G.M. Review: Nucleation in solutions revisited. *Cryst. Res. Technol.* **2003**, *38*, 555–574. [CrossRef]
20. Songolzadeh, M.; Soleimani, M.; Takht Ravanchi, M.; Songolzadeh, R. Carbon dioxide separation from flue gases: A technological review emphasizing reduction in greenhouse gas emissions. *Sci. World J.* **2014**, *2014*, 34. [CrossRef] [PubMed]
21. Kuo, J.; Dow, J. Biogas Production from Anaerobic Digestion of Food Waste and Relevant Air Quality Implications. *J. Air. Waste Manag. Assoc.* **2017**. [CrossRef] [PubMed]
22. Schwarzenbach, G. The complexones and their analytical application. *Analyst* **1955**, *80*, 713–729. [CrossRef]
23. Robinson, R.; Roughan, M.E.; Wagstaff, D. Measuring inorganic phosphate without using a reducing agent. *Ann. Clin. Biochem.* **1971**, *8*, 168–170. [CrossRef]
24. Kazanci, M.; Fratzl, P.; Klaushofer, K.; Paschalis, E.P. Complementary information on in vitro conversion of amorphous (precursor) calcium phosphate to hydroxyapatite from raman microscopy and wide-angle X-ray scattering. *Calcif. Tissue Int.* **2006**, *79*, 354–359. [CrossRef] [PubMed]

25. Niu, X.; Chen, S.; Tian, F.; Wang, L.; Feng, Q.; Fan, Y. Hydrolytic conversion of amorphous calcium phosphate into apatite accompanied by sustained calcium and orthophosphate ions release. *Mater. Sci. Eng. C Mater. Biol. Appl.* **2017**, *70*, 1120–1124. [CrossRef] [PubMed]

26. Morse, J.W. Dissolution kinetics of calcium carbonate in sea water, iii. A new method for the study of carbonate reaction kinetics. *Am. J. Sci.* **1974**, *274*, 97–107. [CrossRef]

27. Fleisch, H. Role of nucleation and inhibition in calcification. *Clin. Orthop. Relat. Res.* **1964**, *32*, 170–180. [CrossRef] [PubMed]

28. Penel, G.; Leroy, G.; Rey, C.; Bres, E. Microraman spectral study of the po4 and co3 vibrational modes in synthetic and biological apatites. *Calcif. Tissue Int.* **1998**, *63*, 475–481. [CrossRef] [PubMed]

29. Kuhn, L.T.; Grynpas, M.D.; Rey, C.C.; Wu, Y.; Ackerman, J.L.; Glimcher, M.J. A comparison of the physical and chemical differences between cancellous and cortical bovine bone mineral at two ages. *Calcif. Tissue Int.* **2008**, *83*, 146–154. [CrossRef] [PubMed]

30. Moradian-Oldak, J.; Weiner, S.; Addadi, L.; Landis, W.J.; Traub, W. Electron imaging and diffraction study of individual crystals of bone, mineralized tendon and synthetic carbonate apatite. *Connect. Tissue Res.* **1991**, *25*, 219–228. [CrossRef] [PubMed]

31. Jahnke, R.A. The synthesis and solubility of carbonate fluorapatite. *Am. J. Sci.* **1984**, *284*, 58–78. [CrossRef]

32. Perrone, J.; Fourest, B.; Giffaut, E. Surface characterization of synthetic and mineral carbonate fluorapatites. *J. Colloid Interface Sci.* **2002**, *249*, 441–452. [CrossRef] [PubMed]

33. Tang, R.; Henneman, Z.J.; Nancollas, G.H. Constant composition kinetics study of carbonated apatite dissolution. *J. Cryst. Growth* **2003**, *249*, 614. [CrossRef]

34. Deymier, A.C.; Nair, A.K.; Depalle, B.; Qin, Z.; Arcot, K.; Drouet, C.; Yoder, C.H.; Buehler, M.J.; Thomopoulos, S.; Genin, G.M.; et al. Protein-free formation of bone-like apatite: New insights into the key role of carbonation. *Biomaterials* **2017**, *127*, 75–88. [CrossRef] [PubMed]

35. Habraken, W.J.E.M.; Tao, J.; Brylka, L.J.; Friedrich, H.; Bertinetti, L.; Schenk, A.S.; Verch, A.; Dmitrovic, V.; Bomans, P.H.H.; Frederik, P.M.; et al. Ion-association complexes unite classical and non-classical theories for the biomimetic nucleation of calcium phosphate. *Nat. Commun.* **2013**, *4*, 1507. [CrossRef] [PubMed]

36. Simon, P.; Schwarz, U.; Kniep, R. Hierarchical architecture and real structure in a biomimetic nano-composite of fluorapatite with gelatine: A model system for steps in dentino- and osteogenesis? *J. Mater. Chem.* **2005**, *15*, 4992–4996. [CrossRef]

37. Chatzipanagis, K.; Iafisco, M.; Roncal-Herrero, T.; Bilton, M.; Tampieri, A.; Kroger, R.; Delgado-Lopez, J.M. Crystallization of citrate-stabilized amorphous calcium phosphate to nanocrystalline apatite: A surface-mediated transformation. *CrystEngComm* **2016**, *18*, 3170–3173. [CrossRef]

38. Zou, Z.; Bertinetti, L.; Politi, Y.; Jensen, A.C.S.; Weiner, S.; Addadi, L.; Fratzl, P.; Habraken, W.J.E.M. Opposite particle size effect on amorphous calcium carbonate crystallization in water and during heating in air. *Chem. Mater.* **2015**, *27*, 4237–4246. [CrossRef]

minerals

MDPI

Article

Fabrication of Single-Crystalline Calcite Needle-Like Particles Using the Aragonite–Calcite Phase Transition

Yuki Kezuka *, Kosuke Kawai, Kenichiro Eguchi and Masahiko Tajika

Shiraishi Central Laboratories Co. Ltd., 4-78 Motohama-cho, Amagasaki, Hyogo 660-0085, Japan;
kawai_kosuke@shiraishi.co.jp (K.K.); eguchi_kenichiro@shiraishi.co.jp (K.E.);
tajika_masahiko@shiraishi.co.jp (M.T.)
* Correspondence: kezuka_yuki@shiraishi.co.jp; Tel.: +81-6-6417-3130

Received: 24 July 2017; Accepted: 30 July 2017; Published: 1 August 2017

Abstract: Calcium carbonate ($CaCO_3$) occurs in two major polymorphs: rhombohedral calcite and orthorhombic aragonite, the latter is thermodynamically metastable. In this study, we first prepared aragonite needle-like particles by introducing CO_2-containing gas into $Ca(OH)_2$ aqueous slurry. Then, the resulted aragonite particles were heat treated at 500 °C for 1 h, in order to induce the aragonite–calcite phase transition. Particle structures before and after the heat treatment were characterized mainly by powder X-ray diffractometry (XRD), field emission-scanning electron microscopy (FE-SEM) and transmission electron microscopy (TEM). We found that single-crystalline calcite needle-like particles with zigzag surface structures can be fabricated using the phase transition.

Keywords: calcium carbonate; phase transition; calcite; needle-like particle; transmission electron microscopy (TEM)

1. Introduction

In order to design composite materials with desired properties, choosing suitable fillers is crucially important. For instance, mechanical properties of composite materials can be affected by particle morphology and size of fillers [1–5]. In particular, anisotropic materials such as needle-like or fiber-like particles are known to improve mechanical strength of composites much more effectively than spherical particles, because of increase in filler entanglement density. Among a wide variety of fillers, calcium carbonate ($CaCO_3$) is one of the most used fillers in the industry. Calcium carbonate naturally occurs abundantly in the forms of limestone, marble, stalactites, shells, corals, etc. Both ground calcium carbonate (GCC) and precipitated calcium carbonate (PCC) have been used in many industrial fields [6]. The former is obtained by mechanical grinding of limestone and the latter is synthesized mainly using the following reaction:

$$Ca(OH)_2 + CO_2 \rightarrow CaCO_3 + H_2O \tag{1}$$

in recent years, the use of PCC has received much attention, because the particle shape/size of PCCs are relatively uniform and can be controlled by several synthetic conditions [7–10]. For instance, previous studies have shown that dimensional stability and impact strength of plastic composite materials [1–5], mechanical properties of rubber materials [11,12], rheological properties of sealants [13] and gloss characteristics of paper [14,15] can be controlled by changing shape/size of PCCs.

Anhydrous calcium carbonate has three crystalline polymorphs: calcite, aragonite and vaterite. Precipitation in aqueous systems commonly leads to calcite rhombic particles, aragonite needle-like particles and vaterite polycrystalline spherical particles [16–21]. Among the three polymorphs, calcite is the thermodynamically most stable, and vaterite is the least stable [18]. It is reported that metastable aragonite minerals transform into the calcite from 400 to 450 °C, under atmospheric conditions [22,23].

The aragonite–calcite phase transition temperature depends on the origin, crystallinity, calcite content, impurity (e.g., Sr^{2+}) content and water content of aragonite [23–26]. For instance, aragonite phases in biominerals can transform into calcite phases below 350 °C [23]. Here, there are few publications that shed light on the particle morphology changes during the phase transitions. Original morphologies may remain unchanged after the phase transitions.

Synthesis of aragonite needle-like particles, or whiskers, has been studied by many researchers [19–21]. Mg^{2+} ions are known to be one of the most effective impurity ions for the promotion of aragonite formation, because they inhibit the nucleation and crystal growth of calcite [27] and the phase transition from aragonite to calcite [28]. Furthermore, Ota et al. reported that aragonite needle-like particles can be formed efficiently by blowing CO_2-containing gas into $Ca(OH)_2$ suspended $MgCl_2$ aqueous solution [19]. They also mentioned that $MgCl_2$ aqueous solution is reusable for another carbonation process because Mg^{2+} ions and Cl^- ions will not be incorporated into aragonite particles.

In the present study, aragonite needle-like particles were prepared by carbonation reaction of $Ca(OH)_2$ aqueous slurry using CO_2 gas bubbling method, and transformed into the calcite phase by heating at high temperatures. Crystallographic changes and particle morphology changes before and after the heat treatment were investigated mainly by powder X-ray diffractometry (XRD), field emission-scanning electron microscopy (FE-SEM), transmission electron microscopy (TEM) and selected area electron diffraction (SAED) analysis. The main aim of this work is to fabricate and characterize single-crystalline calcite needle-like particles. The fabrication of thermodynamically stable calcite particles with a high aspect ratio may open up new possibilities of PCC as a filling material.

2. Materials and Methods

2.1. Chemicals

To prepare aragonite needle-like particles, chemicals were obtained from manufacturers and used without any purification procedures. $Ca(OH)_2$ aqueous slurry (Taiyo Kagaku Kogyo Kaisha Ltd., Tsukumi, Japan; purity ~99.5%) and $MgCl_2 \cdot 6H_2O$ wafer (Ako Kasei Co., Ltd., Ako, Japan; purity; ~99.8%) were employed as raw materials. CO_2 gas and clean-air was purchased from Showa Denko Gas Products Co., Ltd (Kawasaki, Japan). Impurity content of the starting material, $Ca(OH)_2$, was determined by inductivity coupled plasma-atomic emission spectroscopy (ICP-AES) in an iCE 3300-Uni (Thermo Fisher Scientific Inc., Waltham, MA, USA). Sample preparation for ICP-AES measurement was carried out by the following steps: (1) dissolution of 0.5 g of the powder in 2 mL of nitric acid, (2) dilution up to 100 mL with distilled water, and (3) heating at 180 °C for 1 h.

2.2. Synthetic Procedures and Heat Treatments

Synthetic procedures were determined by following the well-established methods in the literature [19–21]. Aqueous suspension with 1.5 wt % $Ca(OH)_2$ and 4.3 wt % $MgCl_2$ (the molar ratio Ca/Mg = 1/2.3) were prepared by dilution and agitation. The suspension was poured into the reaction tank made of stainless steel and maintained at 70 °C. A CO_2 (30%)-clean air (70%) mixed gas was introduced into the tank from the bottom until pH reached 6.7, at an approximate rate of 900 mL/min for 100 g of $Ca(OH)_2$. During the reaction, the suspension was mechanically stirred at an approximate constant rate of 350 rpm, and pH of the suspension was monitored using a handheld electrical pH meter (DKK-TOA Co., WM-32EP, Tokyo, Japan). After the carbonation, the suspension was dehydrated and dried.

About 5 g of thus prepared powder was put into a crucible and placed in an electric furnace. The powder was then heat treated at 500 °C for 1 h in air, so as to induce the aragonite–calcite phase transition. Then, the sample was slowly furnace-cooled to room temperature.

2.3. Characterization

The as-prepared calcium carbonate dry powder was subjected to the simultaneous thermogravimetry-differential thermal analysis (TG-DTA) (Thermo plus EVO2, Rigaku Co., Ltd., Tokyo, Japan). Here, approximately 20.0 mg powder was placed in a platinum pan and heated at a rate of 20 °C/min up to 1000 °C under the clean-air flow (500 mL/min).

The crystallographic phases were identified by powder XRD (MultiFlex, Rigaku Co. Ltd., Tokyo, Japan) with Cu-K$_\alpha$ line (X-ray wavelength λ = 0.154 nm) at 40 kV and 40 mA. The XRD patterns were collected in the range from $2\theta = 20°$ to $50°$, at a counting rate of $2\theta = 0.02°/s$. The polymorphic ratio (aragonite content ratio) in the powder sample was estimated from the peak intensity ratio, using the following equation:

$$F_a = \frac{(I_{a111} + I_{a221})}{(I_{a111} + I_{a221} + 0.5 \times I_{c104})} \times 100 \tag{2}$$

where F_a is the fraction (wt %) of aragonite and I_{a111}, I_{a221} and I_{c104} are the intensities of the aragonite (111) reflection peak at 26.3°, the aragonite (221) reflection peak at 45.9° and the calcite (104) reflection peak at 29.5°, respectively [20,29].

The particle morphologies were observed by FE-SEM (JSM-6330F, JEOL Co. Ltd., Tokyo, Japan) equipped with a field emission electron gun operated at 3.0 kV and TEM (JEM-2100HR, JEOL Co. Ltd., Tokyo, Japan) equipped with a LaB$_6$ electron gun operated at 200 kV. Furthermore, SAED patterns were taken in TEM, in order to analyze the crystallographic structure changes. Here, specimens for SEM observations were prepared by mounting powder samples on an electrically conductive tape and coated with gold using a FINE COATER JFC-1200 (JEOL Co. Ltd., Tokyo, Japan) to impart conductivity. Specimens for TEM observations were prepared by dispersing the sample particles into ethanol to the concentration of 0.8 wt %, dropping the suspension on carbon/collodion-coated copper grids, removing excess ethanol and drying.

The Brunauer-Emmett-Teller (BET) one-point method was applied to measure the specific surface area (SSA) of the samples. Measurements were made by nitrogen gas adsorption to the particle surfaces at liquid nitrogen temperature [30,31] using a Macsorb HM Model-1208 (Mountech Co. Ltd., Tokyo, Japan). The powder samples were pre-heated at 110 °C for 1 h in vacuum and degassed at 110 °C for 20 min in vacuum, respectively.

3. Results and Discussion

3.1. Elemental Analysis and Carbonation Reaction

The impurity concentrations of the starting material, Ca(OH)$_2$, determined by the ICP-AES technique are listed in Table 1. Figure 1 shows the pH change in the slurry as the function of carbonation reaction time. The neutralization reaction seems to be completed within 350 min under the present experimental system. Ota et al. pointed out that keeping the pH value of the slurry < 9 during the carbonation reaction was important for efficient formation of aragonite particles [19]. The present system fulfilled that criteria; thus, aragonite particles should be favorably produced. This will be confirmed later. Gas efficiency can be estimated by comparing the total volume of CO$_2$ introduced to the slurry and the amount of resulted calcium carbonate particles. Assuming the completion of the neutralization reaction at 6.7 in pH, which correspond to the reaction time of 350 min, the gas efficiency was roughly estimated to be 32%.

Table 1. Impurity concentrations, in µg/g (ppm), in the Ca(OH)$_2$ used in this study measured by ICP-AES.

Mg	Sr	Si	P	Fe	Na	Al	Mn
2900	360	450	55	96	65	190	10

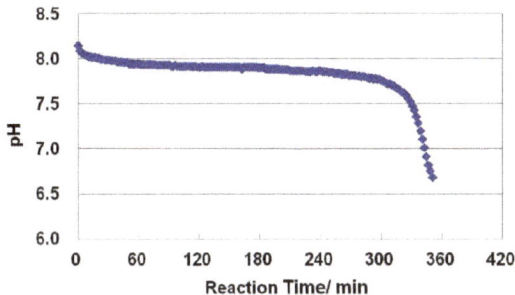

Figure 1. The change in pH of the slurry during the carbonation of Ca(OH)$_2$ at 70 °C with the approximate flow rate of 900 mL/min of CO$_2$ (30%)-clean air (70%) mixed gas for 100 g of Ca(OH)$_2$, as the function of reaction time.

3.2. Thermal Analysis

Figure 2 shows simultaneously taken TG-DTA curves of the as-prepared calcium carbonate dry powder synthesized in this study. An initial weight loss of 0.5 wt % until the temperature reaches 372 °C is mainly attributed to desorption of water and organic species physically absorbed on the particle surface. Then, a small endothermic peak between 372 °C and 436 °C, accompanied with a weight loss of 0.8%, can be confirmed. This is considered to be the sum of the removal of water from residual Ca(OH)$_2$, according to the following equation:

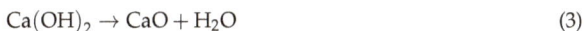

$$Ca(OH)_2 \rightarrow CaO + H_2O \tag{3}$$

and the aragonite–calcite phase transition. Phase transition temperature from aragonite to calcite varies with states of aragonite; mineral origin aragonite tends to transform at higher temperatures (<450 °C) compared to aragonitic biominerals such as shells and coral (<350 °C) [23]. Thereafter, a larger endothermic peak between 552 °C and 806 °C, accompanied with the main weight loss of 42.4%, can be seen. This is assigned to the elimination of carbon dioxide along with the carbonate decomposition according to the following equation:

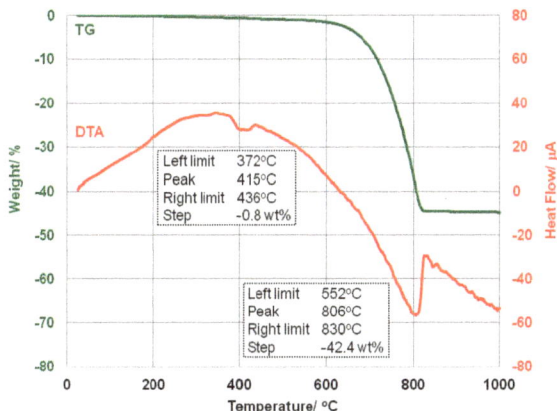

$$CaCO_3 \rightarrow CaO + CO_2 \tag{4}$$

Figure 2. Simultaneous TG (green)-DTA (red) curves of the as-prepared calcium carbonate powder.

Further decrease in the sample weight is negligible until the temperature reaches 1000 °C. The heat treatment temperature chosen in this study, 500 °C, is located in the middle of the two endothermic peaks so that it can be anticipated that the phase transition would occur, but the decarbonation will not occur during the heat treatment.

3.3. Identification of Crystallographic Phases

Figure 3 shows the XRD patterns of the powders before (blue) and after (red) the heat treatment at 500 °C for 1 h. The major peaks for aragonite and calcite, used in Equation (2), are explicitly indicated in the figure. For the XRD pattern of the powder before the heat treatment (blue), all of the peaks were indexed as either orthorhombic aragonite phase or rhombohedral calcite phase. According to Equation (2), the aragonite content in the as-prepared powder was calculated approximately to be 96 wt %. The presence of unreacted $Ca(OH)_2$ was not detected within the sensitivity of the XRD analysis. This indicates that only a trace amount of $Ca(OH)_2$ remained in the powder, if any. On the other hand, for the XRD pattern of the powder after the heat treatment (red), the presence of aragonite phase is not confirmed. This indicates that all the aragonite particles transformed into calcite phase due to the heat treatment. The Scherer equation [32] was not applicable to calculate the crystallite sizes of the aragonite and calcite particles from the XRD patterns, because the peaks were highly sharp and not enough peak broadenings were observed. This suggests that both particles were greater than 100 nm.

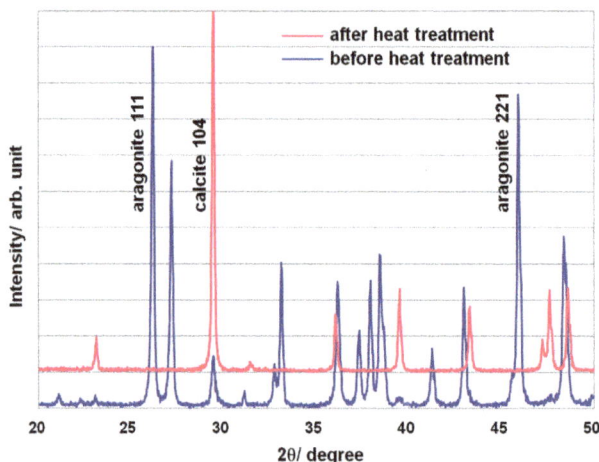

Figure 3. Powder XRD patterns of the calcium carbonate particles before (blue) and after (pink) the heat treatment.

3.4. Particle Characterization

Figure 4 shows low- and high-magnification SEM images for aragonite powder (Figure 4a,c) and for calcite powder (Figure 4b,d). It is clear that the carbonation reaction produced aragonite needle-like particles. These particles have sizes of 3–10 μm in length and 130–280 nm in width, i.e., the aspect ratio ranged from 10 to 30. On the other hand, it was confirmed that the calcite particles almost preserved the needle-like morphology of the aragonite particles. Therefore, it is concluded that the calcite needle-like particles are successfully fabricated via the phase transition of aragonite due to heating at 500 °C for 1 h. Here, it can be seen in Figure 4d that calcite needle-like particles have rougher surfaces compared to aragonite needle-like particles. In addition, some of the calcite needle-like particles are slightly winding. This might indicate that the particle morphology changes has not thoroughly completed for all the particles by the heat treatment conditions employed in this study, even after the aragonite–calcite phase transition has completed. The densities of aragonite and calcite are 2.93 g/mm^3 and 2.71 g/mm^3, respectively. Therefore, volume expansion of about 8%, which corresponds to the length and width expansion of about 2%, should occur after the phase transition, whereas, these are not confirmed by the SEM observations in this study, because powder samples were directly observed and most of the needle-like particles in the images are being inclined.

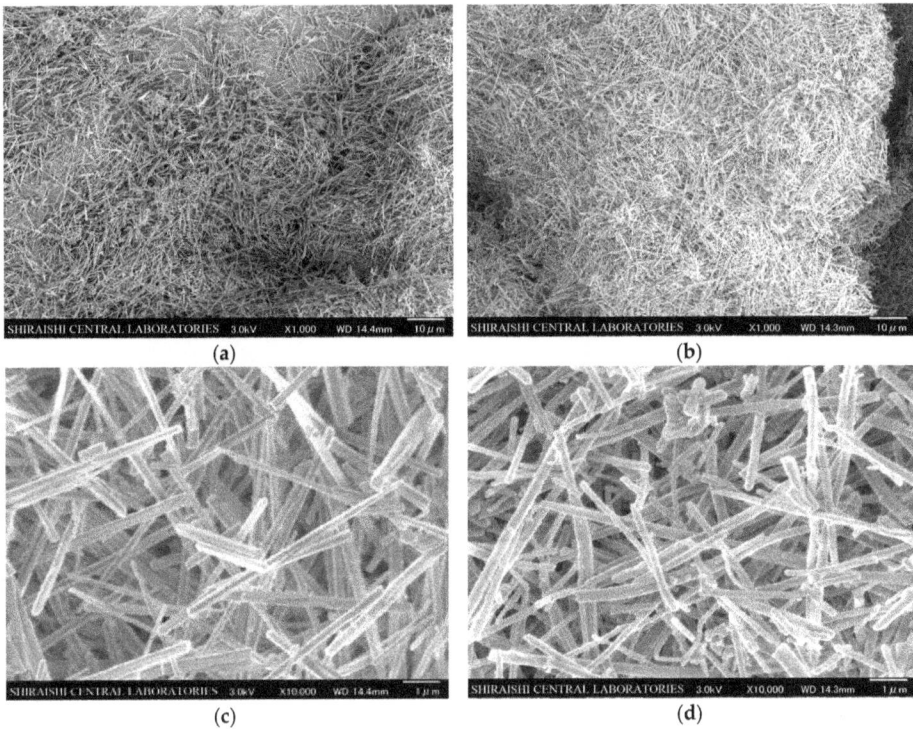

Figure 4. Low- and high-magnification SEM images for aragonite powder (**a,c**) and for calcite powder (**b,d**). Scale-bars are 10 μm for low-magnification and 1 μm for high-magnification observations. The needle-like morphology of aragonite particles was maintained even after the heat treatment, to form calcite needle-like particles.

BET-SSA of the aragonite needle-like particles and the calcite needle-like particles are measured to be 6.6 m^2/g and 6.0 m^2/g, respectively. These small values might indicate that both particles are solid crystals without pores.

In order to clarify the particle structures in more detail, TEM observations were carried out for both samples. Figure 5a shows a bright-field TEM image of an aragonite particle. The insert is a SAED pattern taken along the [1$\bar{1}$0] axis of the aragonite structure. A single-crystalline aragonite needle-like particle is observed. The long axis of the aragonite particle is along the *c*-axis, as reported in the literature [19]. Figure 5b shows a bright-field TEM image of a calcite particle after the heat treatment. A needle-like particle with zigzag surface structure is observed. Judging only from the TEM image, the particle looks superficially polycrystalline. Each surface step is parallel to one another, and the length between each surface steps is approximately 10–40 nm. The insert is a SAED pattern taken along the [241] axis of the calcite structure. Here, splitting of the diffraction spots is not observed in the SAED pattern. This means that the crystallographic orientation differences between neighboring nano-blocks are negligibly small and each calcite needle-like particle is almost single-crystalline.

Thus, it is deduced that the darker contrasts in between the neighboring nano-blocks are caused by the surface steps. It is concluded that a fabrication method of single-crystalline calcite particles with a high aspect ratio using the aragonite–calcite phase transition is established. The use of the calcite needle-like particles as fillers may be beneficial for strengthening composites, because the filler-matrix contacting area per filler length becomes larger by the zigzag surface structures.

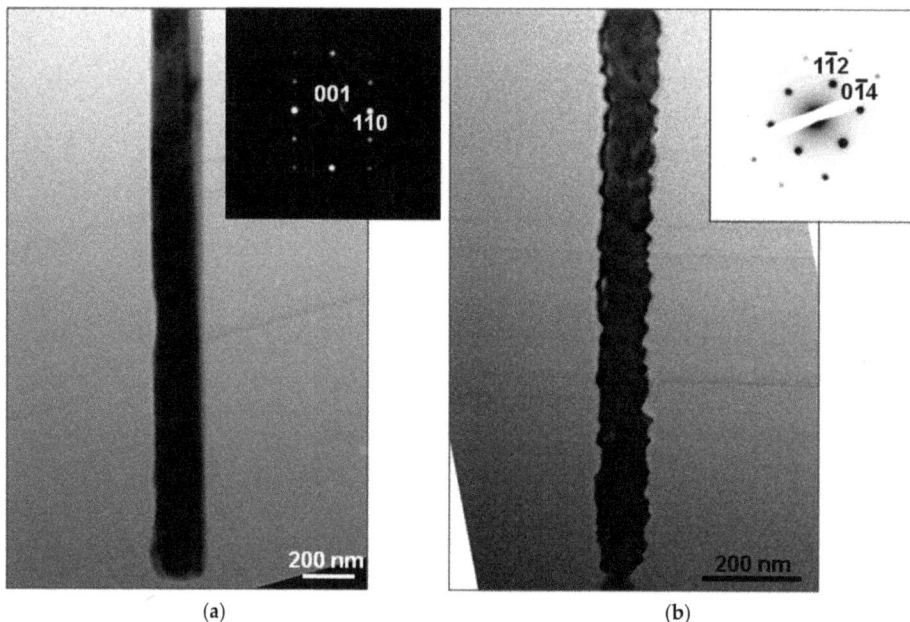

Figure 5. Bright-field TEM images of (**a**) an aragonite needle-like particle and (**b**) a calcite needle-like particle with corresponding SAED patterns. SAED patterns indicate that both particles are single-crystalline.

4. Conclusions

Aragonite needle-like particles were synthesized by bubbling CO_2-containing gas into $Ca(OH)_2$ aqueous slurry with the presence of enough amount of $MgCl_2$. It was elucidated that calcite needle-like particles can be fabricated by calcining aragonite needle-like particles at 500 °C for 1 h in air. By TEM observations, it was found that the fabricated calcite needle-like particles have zigzag surface structures. SAED analyses revealed that neighboring nano-blocks are highly mono-oriented to one another to form almost single-crystalline particles. This fabrication method of thermodynamically stable calcite needle-like particles may broaden the applications of calcium carbonate as a filling material.

Acknowledgments: We thank Akiyoshi Osaka, Yuichi Ikuhara, Eita Tochigi, and two anonymous reviewers for useful discussion and thoughtful comments.

Author Contributions: Yuki Kezuka, Kosuke Kawai, Kenichiro Eguchi, and Masahiko Tajika conceived and designed the experiments; Yuki Kezuka, Kosuke Kawai, and Kenichiro Eguchi performed the experiments, analyzed the data, and contributed reagents/materials/analysis tools; Yuki Kezuka wrote the paper.

Conflicts of Interest: The authors declare no conflicts of interest.

References

1. Eirich, F.R. *Handbook of Fillers and Reinforcements for Plastics*; Katz, H.S., Milewski, J.V., Eds.; Litton Educational Publishing, Inc.: New York, NY, USA, 1978.

2. Bartczak, Z.; Argon, A.S.; Cohen, R.E.; Weinberg, M. Toughness mechanism in semi-crystalline polymer blends: II. High-density polyethylene toughened with calcium carbonate filler particles. *Polymer* **1999**, *40*, 2347–2365. [CrossRef]
3. Da Silva, A.L.N.; Rocha, M.C.G.; Moraes, M.A.R.; Valente, C.A.R.; Coutinho, F.M.B. Mechanical and rheological properties of composites based on polyolefin and mineral additives. *Polym. Test.* **2002**, *21*, 57–60. [CrossRef]
4. Thio, Y.S.; Argon, A.S.; Cohen, R.E.; Weinberg, M. Toughening of isotactic polypropylene with $CaCO_3$ particles. *Polymer* **2002**, *43*, 3661–3674. [CrossRef]
5. Lama, T.D.; Hoang, T.V.; Quang, D.T.; Kim, J.S. Effect of nanosized and surface-modified precipitated calcium carbonate on properties of CaCO3/polypropylene nanocomposites. *Mater. Sci. Eng. A* **2009**, *501*, 87–93. [CrossRef]
6. Roskill Information Services Ltd. *Ground and Precipitated Calcium Carbonate: Global Industry Markets and Outlook*, 1st ed.; Roskill Information Services Ltd.: London, UK, 2012; pp. 270–359.
7. Shiraishi, T. Colloidal Calcium Carbonate and Method of Producing the Same. U.S. Patent No. 1,654,099, 27 December 1927.
8. Shiraishi, T. Method of Manufacturing Colloidal Carbonate of Alkali Earths. U.S. Patent No. 1,863,945, 21 June 1932.
9. Wray, J.L.; Daniels, F. Precipitation of Calcite and Aragonite. *J. Am. Chem. Soc.* **1957**, *79*, 2031–2034. [CrossRef]
10. Yamada, H.; Hara, N. Formation Process of Colloidal Calcium Carbonate in the Reaction of the System $Ca(OH)_2$-H_2O-CO_2. *Gypsum Lime* **1985**, *194*, 3–12.
11. Payne, A.R. Dynamic Properties of Filler-Loaded Rubbers. In *Reinforcement of Elastomers*; Kraus, G., Ed.; John Wiley & Sons, Inc.: New York, NY, USA, 1965; pp. 69–124.
12. Fang, Q.; Song, B.; Tee, T.; Sin, L.T.; Hui, D.; Bee, S. Investigation of dynamic characteristics of nano-size calcium carbonate added in natural rubber vulcanizate. *Compos. Part B* **2014**, *60*, 561–567. [CrossRef]
13. Damusis, A. Pigments in Sealants. In *Sealants*; Damusis, A., Ed.; Reinhold Publishing Co.: New York, NY, USA, 1967; pp. 52–91.
14. Hagemeyer, R.W. *Pigments for Paper*, 1st ed.; O'Shea, J.E., Ed.; Tappi Press: Atlanta, GA, USA, 1997.
15. Kumar, N.; Bhardwaj, N.K.; Chakrabarti, S.K. Influence of pigment blends of different shapes and size distributions on coated paper properties. *J. Coat. Technol. Res.* **2011**, *8*, 605–611. [CrossRef]
16. García-Carmona, J.; Morales, J.G.; Clemente, R.R. Morphological control of precipitated calcite obtained by adjusting the electrical conductivity in the $Ca(OH)_2$–H_2O–CO_2 system. *J. Cryst. Growth* **2003**, *249*, 561–571. [CrossRef]
17. Han, Y.S.; Hadiko, G.; Fuji, M.; Takahashi, M. Factors affecting the phase and morphology of $CaCO_3$ prepared by a bubbling method. *J. Eur. Ceram. Soc.* **2006**, *26*, 843–847. [CrossRef]
18. Somani, R.S.; Patel, K.S.; Mehta, A.R.; Jasra, R.V. Examination of the Polymorphs and Particle Size of Calcium Carbonate Precipitated Using Still Effluent (i.e., $CaCl_2$ + NaCl Solution) of Soda Ash Manufacturing Process. *Ind. Eng. Chem. Res.* **2006**, *45*, 5223–5230. [CrossRef]
19. Ota, Y.; Inui, S.; Iwashita, T.; Kasuga, T.; Abe, Y. Preparation of Aragonite whiskers. *J. Am. Ceram. Soc.* **1995**, *78*, 1983–1984. [CrossRef]
20. Park, W.K.; Ko, S.J.; Lee, S.W.; Cho, K.H.; Ahn, J.W.; Han, C. Effects of magnesium chloride and organic additives on the synthesis of aragonite precipitated calcium carbonate. *J. Cryst. Growth* **2008**, *310*, 2593–2601. [CrossRef]
21. Hu, Z.; Shao, M.; Cai, Q.; Ding, S.; Zhong, C.; Wei, X.; Deng, Y. Synthesis of needle-like aragonite from limestone in the presence of magnesium chloride. *J. Mater. Process. Technol.* **2009**, *209*, 1607–1611. [CrossRef]
22. Faust, G.T. Differentiation of Aragonite from Calcite by Differential Thermal Analysis. *Science* **1949**, *110*, 402–403. [CrossRef] [PubMed]
23. Omari, H. Thermal stability of calcium carbonate gallstone crystal. *J. Jpn. Biliary Assoc.* **1989**, *3*, 109–117.
24. Rao, M.S.; Yoganarasimhan, S.R. Preparation of pure aragonite and its transformation to calcite. *Am. Mineral.* **1965**, *50*, 1489–1493.
25. Rao, G.V.S.; Natarajan, M.; Rao, C.N.R. Effect of Impurities on the Phase Transformations and Decomposition of $CaCO_3$. *J. Am. Ceram. Soc.* **1968**, *51*, 179–181. [CrossRef]
26. Yoshioka, S.; Kitano, Y. Transformation of aragonite to calcite through heating. *Geochem. J.* **1985**, *19*, 245–249. [CrossRef]

27. Beruto, D.; Giordani, M. Calcite and aragonite formation from aqueous calcium hydrogencarbonate solutions: Effect of induced electromagnetic field on the activity of $CaCO_3$ Nuclei Precursors. *J. Chem. Soc. Faraday Trans.* **1993**, *89*, 2457–24666. [CrossRef]

28. Bischoff, J.L.; Fyfe, W.S. Catalysis, inhibition, and the calcite-aragonite problem; Part 1, The aragonite–calcite transformation. *Am. J. Sci.* **1968**, *266*, 65–79. [CrossRef]

29. Wada, N.; Okazaki, M.; Tachikawa, S. Effects of calcium-binding polysaccharides from calcareous algae on calcium carbonate polymorphs under conditions of double diffusion. *J. Cryst. Growth* **1993**, *132*, 115–121. [CrossRef]

30. Brunauer, S.; Emmett, P.H.; Teller, E. Adsorption of gases in multimolecular layers. *J. Am. Chem. Soc.* **1938**, *60*, 309–319. [CrossRef]

31. Mikhail, R.S.; Brunauer, S. Surface area measurements by nitrogen and argon adsorption. *J. Colloid Interface Sci.* **1975**, *52*, 572–577. [CrossRef]

32. Scherrer, P. Estimation of the size and internal structure of colloidal particles by means of röntgen. *Nachr. Ges. Wiss. Göttingen* **1918**, *2*, 96–100.

minerals

MDPI

Article

Crystallization of Jarosite with Variable Al^{3+} Content: The Transition to Alunite

Franca Jones

Curtin Institute of Functional Molecules and Interfaces, Department of Chemistry, Curtin University, GPO Box U1987, Perth WA 6845, Australia; F.Jones@curtin.edu.au; Tel.: +61-8-9266-7677; Fax: +61-8-9266-4699

Academic Editor: Denis Gebauer
Received: 28 March 2017; Accepted: 29 May 2017; Published: 1 June 2017

Abstract: This study focused on the formation of the jarosite-alunite solid solution at relatively low temperature, 90 °C. It was found that the transition from jarosite to alunite results in significant changes in the powder X-ray diffraction pattern, the infrared spectrum and thermal behavior when the degree of substitution reached $\geq$50%. The initial Al/(Al + Fe) in solution, however, required to achieve these substitution levels in the solid is $\geq$90%. The morphology shows that the faceted jarosite form goes through an intergrown transition to a spherical morphology of pure alunite. This morphology has not been previously observed for alunite and most likely reflects the formation temperature. Rietveld analysis shows that the *a* lattice parameter obeys Vegard's Rule while the *c* lattice parameter behavior is more complex. Empirical modelling of the incorporation of Fe into alunite supports the general trends found in the X-ray diffraction data for the behaviour of the *a*-axis with Al/Fe content. The dehydration of the Al^{3+} ion could be a significant contribution to the activation energy barrier to alunite formation as found for other minerals. Finally, dynamic light scattering showed that the nucleation behavior for jarosite and Fe-containing alunite are significantly different. Alunite appears to nucleate continuously rather than in a single nucleation event.

Keywords: crystal growth; nucleation; infrared spectroscopy; XRD data; jarosite; alunite

1. Introduction

The mixed iron sulphate mineral jarosite [KFe$_3$(SO$_4$)$_2$(OH)$_6$] forms under acidic, wet conditions such as acid sulphate soils [1], acid mine wastes [2], saline lakes [3] and hypogene systems [4]. Interest in jarosite increased after it was found on Mars in 2004 and confirmed that the planet had water at some point in its history [5]. In terrestrial environments, jarosite is also of interest. Acid-mine drainage is a serious environmental concern with jarosite a major mineral constituent of these systems [6,7]. The formation of jarosite in acid-mine drainage systems can be beneficial in that heavy metals can be incorporated into the structure, along with consumption of some of the acid [7,8], and thereby limit environmental harm [9,10]. Jarosite is also produced in some industrial hydrometallurgical operations (such as zinc processing) to remove unwanted iron and improve metal concentrates [11,12]. Understanding the formation of jarosite (and alunite) in the environment is important, therefore, for these reasons.

Alunite is isostructural with jarosite, where the Fe^{3+} is replaced with Al^{3+} [13–17]. Recently, it has also been found on Mars [18]. In nature, jarosite is said to be rarely found as the pure Fe end member and has some Al substitution [19,20]. In nature, however, many other substitutions are likely to be observed in the jarosite/alunite structures in addition to the Fe/Al substitution, as found by Scott [21,22]. For this reason, synthetic samples are often used to determine relationships free from other substitutional interferances [13–15]. Alunite is also said to be found in nature with significant Fe substitution [13–15], although studies on natural K-alunite/jarosite samples are rare in the literature.

However, the work of Brophy et al. [23] makes the comment that "alunite show little, if any, Fe^{3+} even when the mineral occurs in an iron rich environment". Despite this, Brophy et al. [23] were able to synthesise a variety of different compositions spanning the solid solution. Therefore, the lack of natural intermediates more likely reflects formation conditions. The work of Basciano and Peterson [24] mentions that, in many samples, vacancies can exist in both the A position (where the monovalent cation is found) and the octahedral B position (occupied by the Al or Fe ion in this case). Vacancies can result in hydronium ion substitution elsewhere in the structure for charge balancing reasons. However, Nielson et al. [25] found that the presence of hydronium ion is not common in alunite. In the alunite system, charge balancing probably occurs by having vacancies in the B site. Lower temperatures are also said to favour B site vacancies according to Scarlett et al. [26]. In addition, vacancies in the B site can cause significant *c* lattice parameter reductions.

Other comprehensive investigations of the jarosite-alunite solid solution are from the group of Navrotsky [13,14]. These investigations mainly focus on the thermodynamics [14] of formation and in particular the difference between Na and K containing jarosites [13]. There is a lack of information in these manuscripts on how the infrared spectroscopy or thermal behaviour alters for the entire solid solution from 0 to 1, but there is previously published infrared spectroscopy information on jarosite and alunite [9,15,27,28] looking at the formation of the Fe and Al end members of the jarosite family to determine their spectroscopic characteristics [9]. Similarly, the work of Rudolph et al. [29] looked at the formation of synthetic alunites and their characterisation by diffraction and thermal techniques. Finally, the work of Grube [15] looked at the impact of reaction times on alunite formation at 190 °C showing that longer times increases the potassium content.

This work aims to expand on these previous studies by investigating the formation of jarosite and alunite across the entire compositional range. The solids are investigated using X-ray diffraction, infrared spectroscopy, microscopy and thermogravimetry. The results from this set of data will clarify the influence of Fe and Al content on the properties of the particles, from morphology to spectroscopic and thermal behaviour. In particular, these solids are formed at relatively low temperatures (90 °C) in comparison to other studies (>100 °C) and can, therefore, serve as a starting point to understanding the role of formation temperatures on jarosite/alunite properties. The formation temperature of jarosite/alunite is important for many reasons. Geologically, the formation temperature is of fundamental importance, and, in acid-mine drainage systems, the formation temperature is closer to ambient conditions. In order to fully predict jarosite-alunite formation and the particle properties expected, research into lower formation temperatures is required. In addition, in many terrestrial environments, living organisms are able to interact with the growing mineral. Thus, this work is to determine the properties of the particles in the absence of organic matter so that how the solids differ in the presence of living organisms [30] can be properly assessed. These differences (if they exist) may be used as a proxy for determining whether the solids have had contact with living forms during their formation and could help identify when the sample is merely contaminated with organic matter. To complement the experimental data, the incorporation of Fe into alunite using empirical modelling methods has been undertaken to gain further insights into the mechanistic drivers for the solid solution formation. Finally, nucleation kinetics are presented in the presence and absence of Al^{3+} ions to determine whether the impurity cations significantly alter this.

2. Materials and Methods

2.1. Materials

Ferric sulphate $Fe_2(SO_4)_3 \cdot xH_2O$ was obtained from Chem Supply, Gillman, Australia (*x* was found to be ~9) and potassium nitrate was analytical reagent (AR) grade from Ajax Chemicals, Australia. To alter pH, concentrated sulphuric acid (H_2SO_4) >95% from Ajax Chemicals or potassium hydroxide (KOH), AR grade from BDH was used. Aluminium sulphate $18H_2O$ (99% purity from BDH) was used as received and digestion of solids involved the use of concentrated hydrochloric acid (32%, AR grade from Ajax).

2.2. Jarosite Formation

Potassium jarosite was prepared according to the methods described in [4,30,31]. This involved dissolving $Fe_2(SO_4)_3 \cdot xH_2O$ (6.4 g) and KNO_3 (24 g) into 800 g de-ionised water in a clean, glass bottle. Concentrated H_2SO_4 (0.8 mL) is added to the solution. The bottle is capped and the resulting clear solution is then heated to 90 °C for 24 h. After the allocated time, any solids formed are collected by filtering, washing with de-ionised water three times and drying in a desiccator.

Al^{3+} was added as $Al_2(SO_4)_3 \cdot 18H_2O$ solid when present, prior to the addition of water, and dissolved along with the other solids (potassium nitrate and, when present, iron sulphate). The amount of aluminium sulphate was varied from 0 to 0.22 M in the solution.

2.3. Characterisation of Solids

2.3.1. Vibrational Spectroscopy

Infrared spectroscopy is a well-known method to characterize mineral forms as the spectra are phase-specific. The solids for Fourier transform infrared spectroscopy (FTIR) were placed onto a diamond attenuated total reflection (ATR) accessory of a Nicolet iS50 FTIR Spectrometer. A background spectrum was collected before each new scan. Data were collected from 400 to 4000 cm^{-1} with a spectral resolution of 4 cm^{-1} in absorbance mode. All spectra were corrected using the ATR correction mode of the instrument using the supplied OMNIC® software.

2.3.2. Scanning Electron Microscopy (SEM)

The washed and dried solids from crystallization experiments were placed on carbon coated SEM stubs and placed in a dessicator to let the stubs dry. They were then sputter-coated with gold or platinum prior to viewing on an Evo Zeiss (Oberkochen, Germany) instrument. The images were collected at a working distance of 10 mm and a voltage of 15 kV unless otherwise stated.

2.3.3. X-ray Diffraction (XRD)

Wide angle, powder XRD was obtained using a D8 Advance (Bruker, Billerica, MA, USA) diffractometer. The solids were dispersed in ethanol and cast onto low-background silicon holders. The XRD pattern was collected using Cu K$_\alpha$ radiation at a 2theta range of 10–50° and 10–120°. The step size was 0.001 for the small angle scan and 0.02 for the larger angle scan using a divergence slit of 0.3°. The sample holder was spun at 30 rpm. Rietveld analysis was performed using the Topas® software. Appropriate cif files for alunite (ICSD#12106)) and jarosite (ICSD#34344) were obtained and the crystal structures were refined against these. Goodness of fit (GOF) values ranged from 1.81 to 2.96 (fitting statistics can be found in the Supplementary Materials, Table S1). A Rietveld refined structure with a GOF of 2.8 is shown in the Supplementary Materials (Figure S1).

2.3.4. Thermogravimetric Analysis (TGA)

TGA was conducted on a TA Instruments SDT 2960, capable of simultaneous DSC-TGA measurements. The thermograms were obtained from ambient to 800 °C at 5 °C per minute in air at a flow rate of 40 mL/min. Approximately 15 mg of sample was heated in a platinum pan for each measurement. The temperature of the instrument was calibrated against the melting points of indium, zinc, tin, silver and gold. The balance was calibrated over the temperature range with standard alumina weights as provided by the vendor. The reproducibility in the mass loss measured was previously found to be ~3% [30].

2.3.5. Elemental Analysis

The solids obtained from synthesis were digested in concentrated HCl and were analyzed for Fe, Al, K and S content using an Agilent 720 inductively coupled plasma–atomic emission spectroscopy (ICP-AES). Analyses were conducted by the Marine and Freshwater Laboratory at Murdoch University.

2.3.6. Dynamic Light Scattering (DLS)

A Nanosizer ZS (Malvern, UK) was utilised to measure the DLS behaviour of solutions in the absence and presence of Al^{3+} (0.04 M, mole ratio Al:Fe 0.98) cations using a quartz cuvette. This instrument can operate at high temperatures, and so the particle size and particle counts were obtained at 90 °C. Therefore, the concentrations of iron sulphate, potassium nitrate and sulphuric acid and temperature were as per the crystallization experiments for pure jarosite and jarosite I with the pH of the solutions adjusted to pH = 2.1 to allow for comparison with the other experimental results. Both samples were filtered through a Supor 0.2 μm membrane prior to the analysis. The DLS can give two pieces of information: the particle counts and the particle size. The particle counts should be low until nucleation occurs and then will increase until a steady state is reached (in a batch system). Thus, the particle counts were used to determine nucleation rates [32].

2.4. Molecular Modelling

The modelling of cation incorporation was performed by using the previously reported empirical models for alunite [33] and the empirical methods described in [19,33,34] as a starting point. The only variables altered were the A values of the Buckingham Al and Fe-O_{oxide} (Al/Fe-O1) and Al and Fe-$O_{hydroxide}$ (Al/Fe-O2) potentials in order to try and improve the fit, particularly for the *c* lattice parameter. Our previous sulphate potentials, unaltered, were used to model the sulphate molecule [35]. The results of the fitting procedure, along with the experimental values and all potential values are listed in the supplementary information (Tables S2 and S3). The visualization software GDIS [36] was used to construct the simulation cells and GULP [37] was used to minimize their energy. Firstly, a 2 × 2 × 1 supercell of alunite was constructed to give a supercell with similar *a*, *b* and *c* values. Differing degrees of the Al^{3+} ions were substituted for Fe^{3+} until all were substituted (36 ions in total). The starting configuration was minimized prior to the substitution of the different ions.

The equation describing the substitution of aluminium ions with iron ions is:

$$K_{12}Al_{36}(SO_4)_{24}(OH)_{72} + nFe^{3+}{}_{(aq)} \rightarrow K_{12}Fe_nAl_{(36-n)}(SO_4)_{24}(OH)_{72} + nAl^{3+}{}_{(aq)}, \tag{1}$$

where *n* is the number of iron cations susbstituting in the supercell. The energy to replace aluminium ions, the Replacement Energy, E_r, is calculated according to:

$$E_r = (E_{final} + nE_{hydration\ Al}) - (E_{initial} + nE_{hydration\ Fe}), \tag{2}$$

where $E_{initial}$ is the starting energy of the pure alunite, E_{final} is the energy with the iron ion present and *n* is defined as per Equation (1). The hydration energies of the ions, $E_{hydration}$, are the experimental values taken from [38].

3. Results

The jarosite samples discussed herein and the corresponding initial Fe/Al molar ratios can be found in Table 1.

Table 1. List of all samples prepared and the initial molar ratios used to obtain the solids.

Sample Name	Fe Moles ($\times 10^{-3}$)	Al Moles ($\times 10^{-3}$)	Fe/Al Molar Ratio	Initial Al/(Al + Fe)
Jarosite A	2.85	0	-	0
Jarosite B	1.85	0.78	2.37	0.30
Jarosite C	1.28	1.65	0.77	0.56
Jarosite D	0.69	2.28	0.30	0.77
Jarosite E (no solids obtained)	0	3.09	0	1.00
Jarosite F	0.74	6.53	0.11	0.90
Jarosite H	2.22	44.0	0.05	0.95
Jarosite I	0.89	44.0	0.02	0.98
Jarosite J	0.04	44.0	0.01	1.00
Jarosite K	0.02	44.0	0.005	1.00
(Jarosite G) Alunite	0	44.0	0	1.00

3.1. Morphology

The jarosite particles formed (Figure 1) in the absence of Al^{3+} ions were similar to those seen by Sasaki [12,39,40]. As the ratio of Fe/Al decreases, the particle morphology did not alter significantly, but the degree of interparticle growth appeared to increase. It also appears that the individual particle size decreased as the ratio of Fe/Al decreased, although this was difficult to quantify due to the degree of intergrowth occurring.

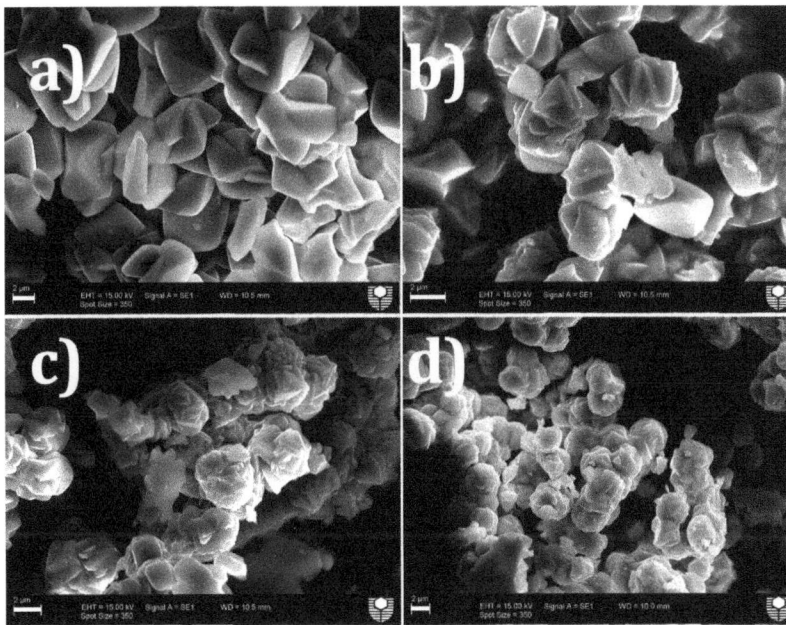

Figure 1. SEM images of jarosite particles formed for sample (**a**) jarosite A (control); (**b**) jarosite C; (**c**) jarosite D; and (**d**) jarosite F (all scale bars 2 μm).

As the ratio of Fe/Al decreased further (<0.11), the morphology of the particles changed from faceted to spherical and remained so with increasing aluminium content (Figure 2 and Supplementary Materials Figure S2). It can be concluded from these images that single-phase solids are formed as there appears to be only one population of particles and not a mixed population as might be expected from mixed phase solids. Furthermore, the alunite particles formed are not faceted as seen for jarosite

(Figure 1a) or for alunite formed at higher temperatures [29,33]. However, very few images of alunite are available in the literature for comparison.

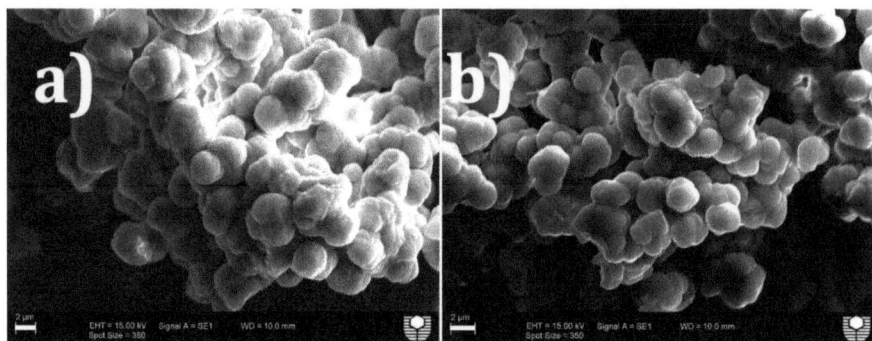

Figure 2. SEM images of jarosite particles formed for sample (**a**) jarosite H, and, (**b**) alunite (all scale bars 2 μm).

3.2. Elemental Analysis

The solids were digested in concentrated acid and sent for ICP analysis to determine their stoichiometry based on elemental analysis. The formula was determined by assuming the sulphate content and determining the number of moles of the K, Al, and Fe relative to sulphate from the ICP data. Finally, the hydronium, water and hydroxide content were determined by charge neutralization considerations as given by $(H_3O)_{1-x}K_xFe_yAl_z(SO_4)_2(OH)_{6-3[y+z]}(H_2O)_{3(y+z)}$ [14]. The calculated compositions are shown in Table 2.

Table 2. Moles and elemental composition of solids formed.

Sample Name	Fe	Al	K	S	OH$^-$	H$_2$O	H$_3$O$^+$
Jarosite A	2.61	0.00	0.88	2	5.91	0.09	0.12
Jarosite B	2.65	0.02	0.90	2	5.89	0.11	0.10
Jarosite C	2.44	0.06	0.87	2	5.63	0.37	0.13
Jarosite D	2.33	0.15	0.84	2	5.82	0.18	0.16
Jarosite F	2.02	0.44	0.91	2	5.18	0.82	0.09
Jarosite H	1.28	1.09	0.89	2	5.02	0.98	0.11
Jarosite I	0.80	1.55	0.86	2	5.23	0.77	0.14
Jarosite J	0.27	2.02	0.92	2	4.52	1.48	0.08
Jarosite K	0.33	1.90	0.82	2	5.16	0.84	0.18
Alunite	0.00	2.23	0.95	2	4.11	1.89	0.06

The values for Fe, K and Al were all within expected ranges [9,14]. These results showed that Al substitution above 20% did not occur until the initial solution Al/(Al + Fe) content had increased above 0.9. The initial Al/(Al + Fe) content plotted against the final Al/(Al + Fe) content found in the solid is given in Figure 3. The non-ideal stoichiometry suggests vacancies are present and the % vacancy based on the ideal 3:2 Fe (and/or Al):S ratio is shown in the Supplementary Materials (Figure S3a).

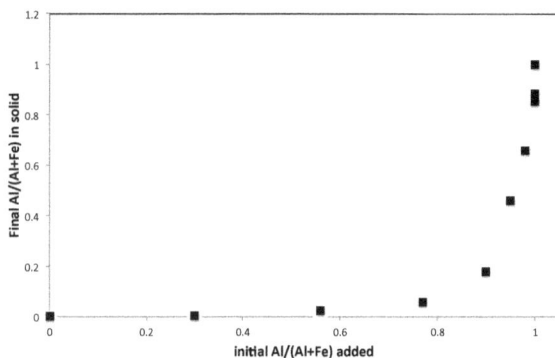

Figure 3. Aluminium ion content in the initial solution versus that found in the solid on collection of the product.

3.3. XRD

Figure 4 shows the XRD patterns for the two end members: jarosite and alunite. The XRD patterns were indexed to reference patterns (jarosite PDF: 01 076 7597 and alunite PDF: 00 047 1884), confirming their mineralogy at the resolution of the XRD (see Supplementary Materials, Figure S4).

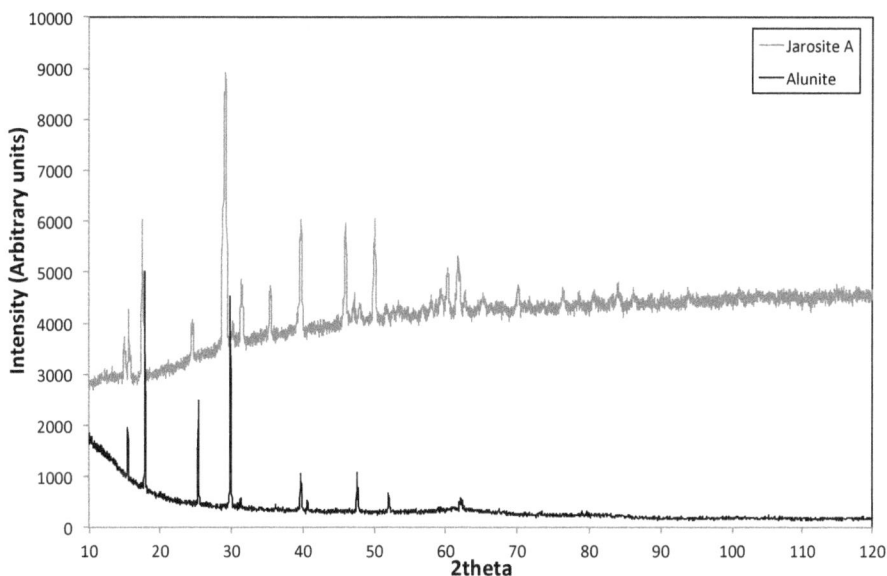

Figure 4. XRD pattern of end member solids, jarosite and alunite.

No new reflections were observed in the pattern (Figure 5) with the introduction of aluminium ions to the solution, but there were 2theta shifts in the reflection positions, consistent with lattice parameter changes and Al^{3+} substitution for Fe^{3+}. Between jarosite H and I, the pattern clearly became alunite-like (Figure 6). This corresponds to a final $Al/(Al + Fe)$ content in the solids of ≥ 0.5; thus, perhaps this transition is not unexpected.

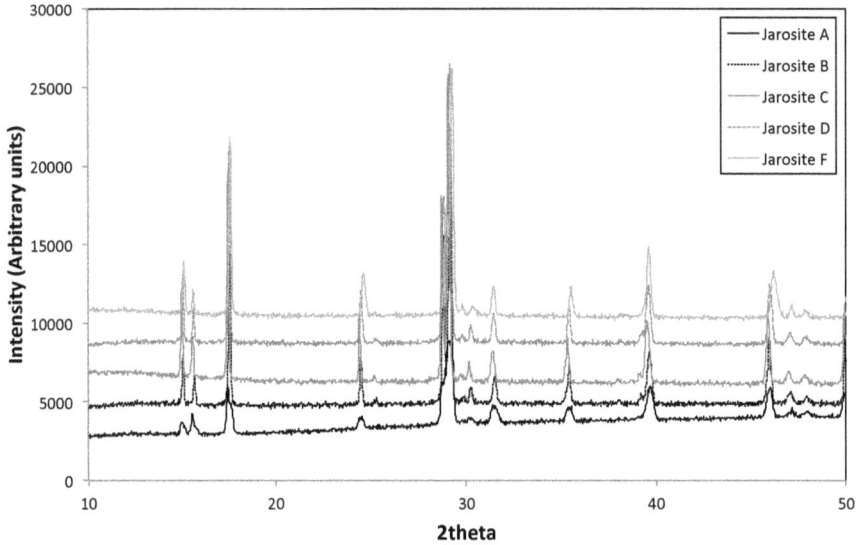

Figure 5. XRD pattern of solids obtained on initial addition of up to 0.9 mole fraction Al. Patterns have been offset in the *y*-direction for clarity.

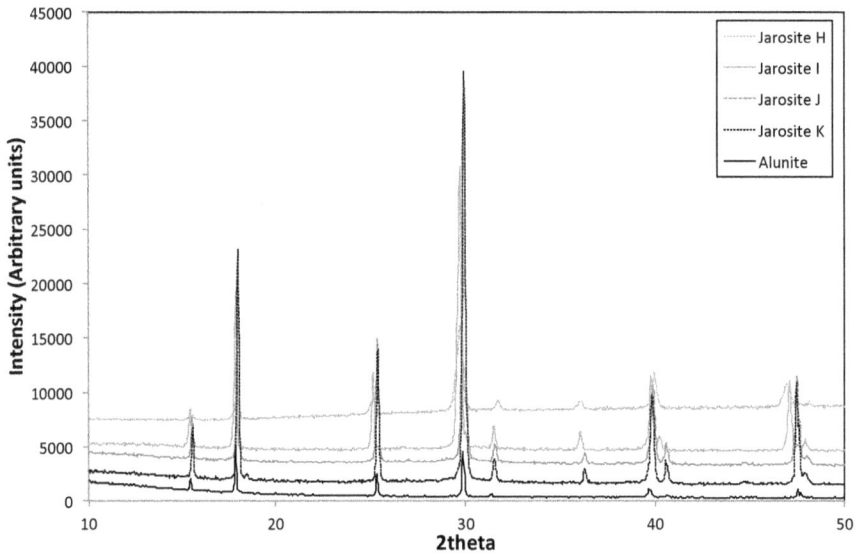

Figure 6. XRD pattern of solids obtained on initial addition of up to 0.9–1.0 mole fraction Al. Patterns have been offset in the *y*-direction for clarity.

All of the XRD patterns obtained at 10°–120° 2theta were subsequently used for Rietveld refinement, whereby the patterns were refined as if they were jarosite or alunite. The average lattice parameters obtained for the jarosite and alunite refinements are listed in Table 3, although it must be stressed that the data were collected on low background holders (due to the mg quantities of solids formed) and as such are not ideal for Rietveld analysis.

Table 3. Rietveld refinement for samples prepared in this study *.

Sample Name	Alunite	Jarosite
Jarosite A		a = 7.306 ± 0.004; c = 17.100 ± 0.008
Jarosite B		a = 7.3139 ± 0.0005; c = 17.116 ± 0.001
Jarosite C		a = 7.3068 ± 0.0003; c = 17.1157 ± 0.0008
Jarosite D		a = 7.2994 ± 0.0003; c = 17.1258 ± 0.0008
Jarosite F		a = 7.2614 ± 0.0003; c = 17.125 ± 0.001
Jarosite H	a = 7.173 ± 0.003; c = 17.104 ± 0.007	
Jarosite I	a = 7.1193 ± 0.0001; c = 17.1205 ± 0.0003	
Jarosite J	a = 7.055 ± 0.0005; c = 17.088 ± 0.001	
Jarosite K	a = 7.0579 ± 0.0003; c = 17.1210 ± 0.0008	
(Jarosite G) Alunite	a = 7.0172 ± 0.0003; c = 17.1204 ± 0.0008	

* The supplementary information gives all the fitting statistics, also structures used are for lowest GOF found for samples with both alunite and jarosite structure possibilities.

The alunite sample in Table 3 shows a *c* lattice parameter smaller than reported in [41]. The sample composition in [41] lists that Al, Na, K and S were present and the occupancy of Al is given as 0.989; thus, it is assumed that the difference between this result and [41] are due to the different chemistries and, possibly, the formation temperature of this sample. The data from Table 3 and the final Al/(Al + Fe) content in the solid as determined from the ICP analysis was used to construct Figure 7. This shows that the *a* lattice parameter has a linear correlation with the Al/(Al + Fe) content and therefore follows Vegard's Rule [42]. The *c* lattice parameter does not appear to be altered in any systematic manner with Al/(Al + Fe) content. Both increasing Al content and vacancies would be expected to contract the *c*-axis, but this is not observed (see Supplementary Materials Figure S3a for vacancies expected). Thus, the *c* lattice parameter is not found to behave as expected but does generally agree with Brophy et al. [23], who also saw little change in the *c*-axis with Al content for the series II samples. Increasing K$^+$ content over long periods of days is seen to increase the *c* lattice parameter according to Grube and Nielsen [15] and the scatter in these values may reflect the variable K$^+$ content.

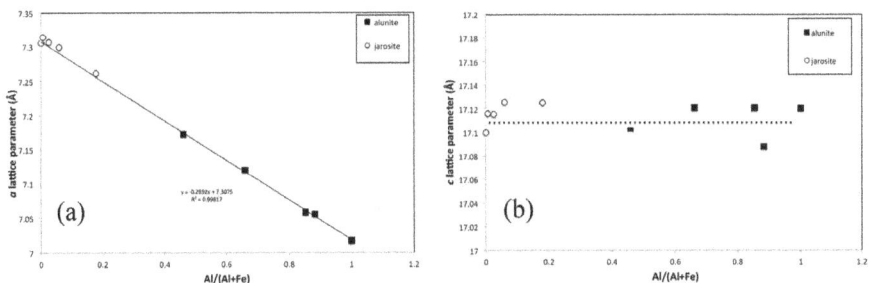

Figure 7. Lattice parameters (**a**) *a* and (**b**) *c* obtained from Rietveld analysis for alunite and jarosite versus Al content in solid.

3.4. Vibrational Sepctroscopy

In Figure 8 below, the control jarosite and alunite infrared spectra are shown. The peaks compare well with literature [9,15,27,28], where the sulphate bands are found between 940 and 1300 cm^{-1}. The main differences between the alunite to jarosite spectrum are that the sulphate band is shifted to higher wavenumbers, as are the bands in the 400–700 cm^{-1} region. This trend is also observed for the water stretch region (see Supplementary Materials, Figure S5).

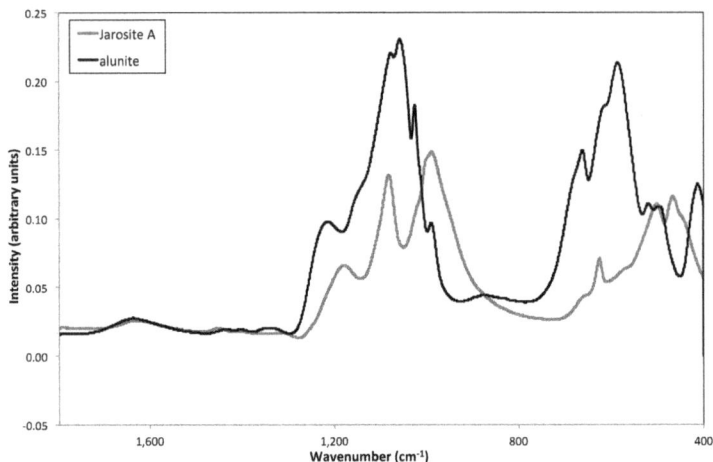

Figure 8. FTIR spectrum of end member solids (jarosite and alunite).

As the initial Al/(Al + Fe) content increased to 0.90, there did not appear to be any significant differences in the infrared spectra of the samples other than a broadening of the features for the jarosite F sample (Figure 9). The 1175 cm^{-1} band in jarosite A did shift to higher wavenumbers (to 1180 cm^{-1}) by jarosite F, but, given the 4 cm^{-1} resolution, this may not be significant.

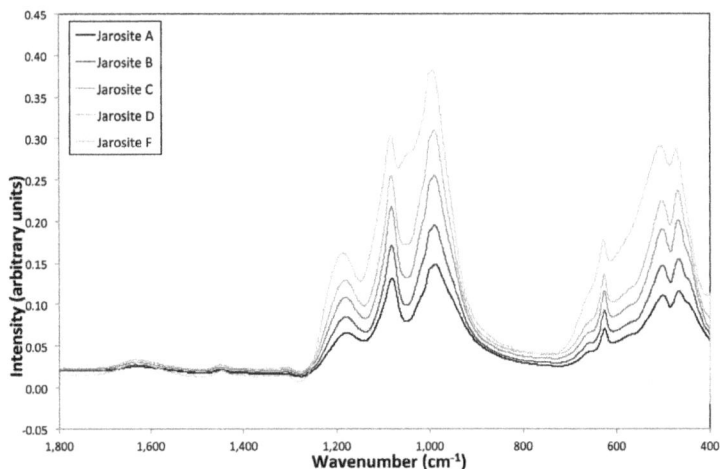

Figure 9. FTIR spectra of solids formed on addition of up to 0.9 mole fraction Al.

When the initial Al/(Al + Fe) content increased further, there was a clear change in the FTIR spectra from jarosite-like to alunite-like. The transition appeared to occur between jarosite F and jarosite J as found for the XRD patterns (Figure 10a). The broadening of the sulphate band observed for jarosite F and H decreased as the alunite bands became more distinct. Similarly, there was a shift in the OH stretching band (~3400 cm^{-1}) to higher wavenumbers as the structure became more and more alunite-like (Figure 10b). The infrared spectrum of jarosite J was identical to that of pure alunite. The ICP results shows <10% Fe is present in this sample, thus an alunite-like spectrum is not surprising.

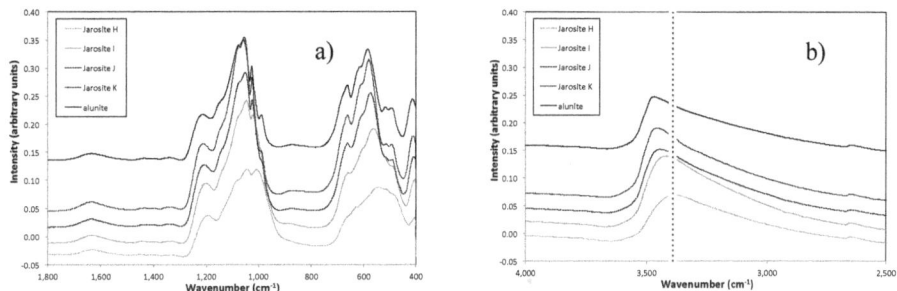

Figure 10. FTIR spectra of solids formed at on addition of 0.9–1.0 mole fraction Al. (**a**) 400–1800 cm^{-1} region and (**b**) 2500–4000 cm^{-1} region. Spectra have been offset in the *y*-direction for clarity.

3.5. Molecular Modelling

The molecular modelling involved taking the pure alunite structure and the empirical potentials as determined by [33] and incorporating more and more iron until a pure jarosite structure was formed under constant pressure and temperature conditions (see Supplementary Materials for potential parameters and comparison to pure end members). The replacement energy and the lattice parameters after minimization were then plotted versus the aluminium molar fraction. The replacement energy was normalized for the number of iron cations replaced and the results presented in Figure 11. It can be seen that the replacement energy for the pure jarosite (Al mole fraction = 0) is the lowest of all, although the value appears to be plateauing to a steady value. This is found despite the fact that the overall energy for the pure alunite structure is lower than the pure jarosite structure, as would be expected from the differences in formation energy [14].

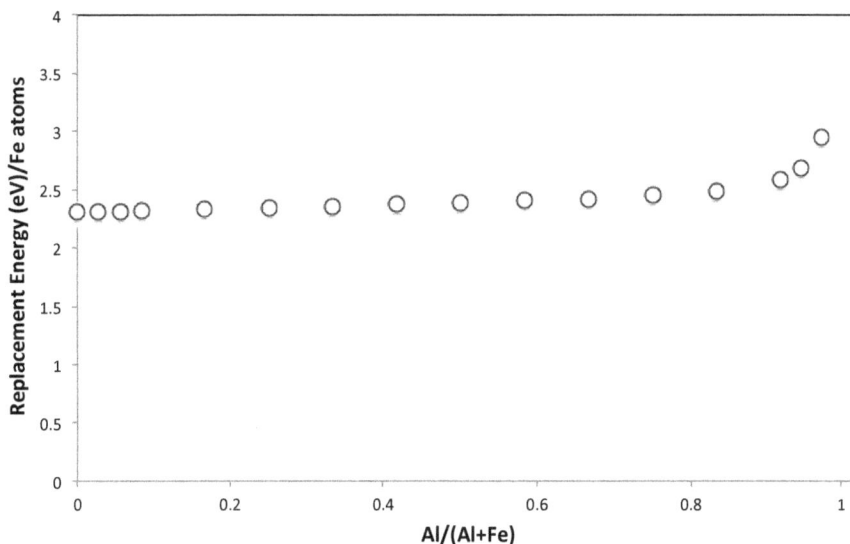

Figure 11. Replacement energy (eV)/Fe ions substituted versus Al mole fraction.

Looking at the lattice parameters and their variation with Al mole fraction, there is a significant similarity in the trends observed between the real data (see Figure 7 above) for the *a*-axis and those from modelling (Figure 12). The most significant difference between the experimental and modelling results appears to be in the *c* lattice parameter behaviour. The *c* lattice parameter for the experimental samples vary from ~17.08–17.12 Å (and no trend is observed with Al content), while the modelled values vary from ~16.9 to ~17.3 Å (and decrease with Al content). The model therefore predicts a greater expansion than found in real samples. The substitution of Al for Fe (ionic size considerations) is seen to contract the *c*-axis if all other impacts are ignored (Figure 12b). However, the real samples show no impact with Al content on the *c*-axis, and this does not correlate with vacancies in the structures (Supplementary Materials Figure S3a). The expected contraction in the *c*-axis must be being offset by other mechanisms, and one possibility is the different K^+ content, which is known to affect the *c*-axis [15]. In empirical modelling, an error in the modelled lattice parameters of <5% is acceptable. However, this is different to errors within the model. Thus, having established the model, the expected changes in the lattice parameter with iron or aluminium ion substitution should replicate experimental trends. For the *a*-axis relationship with Al content, this is clearly the case; a slope of 0.32 from the model versus 0.29 from the XRD refinement. This suggests that the *a*-axis length is indeed dominated by the impact of the Al/Fe content. The significant difference in the *c*-axis between model and experimental results suggest, then, that the Al/Fe content is not the main determinant.

In the modelling, there are also small differences observed between the *a* and *b* lattice parameter values for the substituted solids, suggesting that these solids have structures with a lower symmetry than the pure end members. As the model approaches the pure end members, the *a* and *b* lattice parameters become identical.

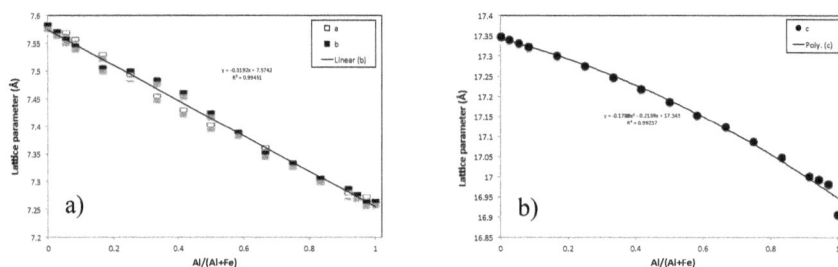

Figure 12. (a) *a* and *b* lattice parameter and (b) *c* lattice parameter versus Al mole fraction.

3.6. Thermal Behaviour

When the intial Al/(Al + Fe) content was below 0.90, the TGA mass loss showed a steady movement of the weight loss regime towards higher temperatures (Figure 13a). At higher Al contents, the weight loss regime also tended towards higher temperatures, but the trend was harder to discern (Figure 13c). The differential (DTA) curves showed this more markedly, particularly for the exotherms occurring at ~200 and 300 °C (Figure 13b). The weight loss up to 300 °C is due to "additional" water as attributed by Kubisz [43]. The 'additional water' content was also calculated for the samples in this work (see supplementary information). The weight loss from 300 °C to 560 °C is considered to be due to the loss of structural water and the conversion to K, Al/Fe sulphate (yavapaiite) [13,44], and, in the case of pure jarosite, conversion to hematite. However, these trends were less obvious in the high Al solids due to the broadening of the low temperature transitions (see Figure 13c,d). The endotherm just below 600 °C was similar to that observed by Drouet and Novrotsky [13]. This endotherm is considered to be due to the formation of hematite or corundum [13] and also increases to higher temperatures as more Al is substituted into the structure. The ~700 °C exotherm is associated with de-sulphurization. Although this can occur at a range of temperatures [13], it is clear that this transition tends to increase in temperature with increasing Al content.

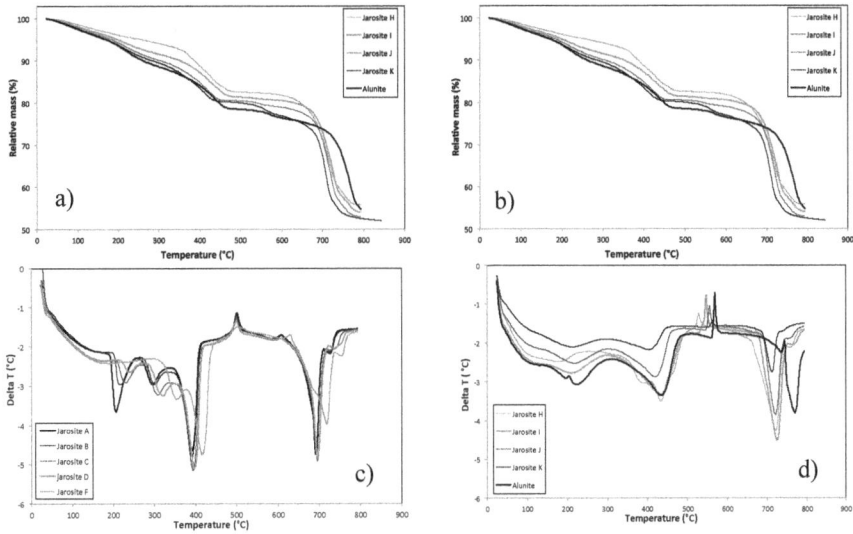

Figure 13. TGA results of mass loss (Relative mass, %) versus temperature (°C) (**a,c**) and DTA results, Delta T in °C (**b,d**) for addition of up to 0.9 Al mole fraction (**a,b**) and 0.9–1.0 mole fraction Al (**c,d**).

3.7. Nucleation Behaviour

The formation of pure jarosite and a sample with the composition of jarosite I at pH 2.1 were assessed for their nucleation behavior. DLS was used to monitor the particle counts (given as kilocounts per second) in this batch system. The expected behavior is that the counts are low until nucleation occurs, at which point the counts will increase and then remain steady. However, if aggregation or growth processes occur, the counts will drop due to sedimentation of the particles. In addition, if nucleation is not a single event, the counts will continue to increase over time. In the case of pure jarosite (see Figure 14), nucleation did indeed appear to be an individual event at ~20 min with subsequent sedimentation of particles due to aggregation or growth. In contrast, the Fe-substituted alunite sample (jarosite I) showed an increase in particles numbers at ~20 min but also an additional further steady increase in particle numbers that continued after 100 min. The number of particles counted (or nuclei formed) was lower for alunite than jarosite, although this would be expected because the more thermodynamically stable solid usually has the higher surface free energy. Although further work is required, these preliminary results imply that the nucleation of alunite is more complex than that of jarosite and was continuous.

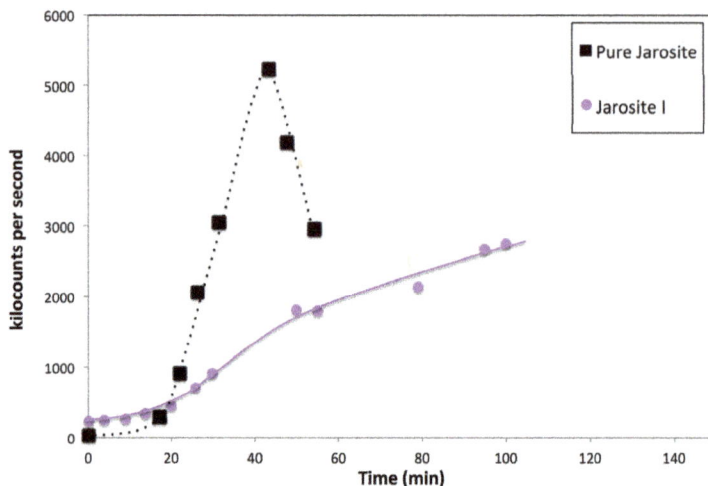

Figure 14. DLS results of particles counted (kilocounts per second) versus time for pure jarosite and jarosite I.

4. Discussion

Morphology results showed that increasing the Al content initially formed smaller, more intergrown particles, which eventually became spherical as the Al mole fraction reached 1. This was supported by the FTIR results that showed that as the Al content increased, the sulphate band broadened and then became alunite-like. A spherical morphology has not previously been mentioned for alunite (as far as the author is aware) and suggests that temperature is a key determinant in the morphology observed for both jarosite and alunite. Thus, at temperatures below 100 °C, alunite formation will not result in solids with distinct crystallographic faces, and this spherical morphology would be expected. As such, the morphology of the solid alunite formed may give clues as to the formation temperature. This morphology may also have consequences for dissolution by virtue of the smaller size and therefore increased surface area.

The added $[Al^{3+}]$ versus incorporated Al^{3+} (Figure 3) was very similar to Brophy et al. [23], whereby only above an initial Al/(Al + Fe) content of 0.7 did the product begin to show >20% Al substitution. The difference to the results presented here are undoubtedly due to the differences in temperature used to prepare the samples (105 °C in [23] and 90 °C here). In addition, this was not an expected result if considering the alunite has been found to have the lower formation enthalpy. This would suggest that alunite would be less soluble than jarosite and should readily form before any jarosite. Brophy et al. [23] proposed the hydrolysis of the iron and aluminium ions as being the critical factor in determining whether the iron or aluminium ion would be prefered in the structure. This hypothesis is supported by our modelling results, where replacing Al^{3+} cations with Fe^{3+} has a lower replacement energy as more substitution occurs. Looking at the replacement energy calculation, every time an Al^{3+} cation was replaced with an Fe^{3+} cation, there was a gain in Al^{3+} hydration energy. Thus, one possible activation energy barrier in the formation of alunite is the dehydration of the Al^{3+} cation. Cation dehydration being a rate determining step has been found for other systems (barite, for example [45]).

The thermal data showed that alunite contained more "additional" water and had higher decomposition temperatures, especially when no iron was present. This was supported by the ICP data, where the alunite samples were calculated to have higher water contents than the jarosite samples. However, within the alunite samples, lower Fe contents did not always correlate with higher water contents. Overall, however, the data support that the higher hydration energy of Al is a key factor in the formation of alunite type structures.

XRD data confirmed that 50% substitution was required to see an intermediate spectrum having both alunite and jarosite like properties, and Rietveld analysis showed that the *a* lattice parameter obeyed Vegard's Rule [42], but the *c* lattice parameter showed no trend with Al content.

The nucleation behavior of jarosite was as would be expected, showing a sharp increase in particle counts (correlating to the nucleation event) and then some growth or agglomeration mechanisms leading to sedimentation of the particles. In the case of Fe-containing alunite (jarosite I sample), the particle numbers did increase at approximately the same time as the pure jarosite but continued to steadily increase. In this case, a significant jump in the particle counts was not observed. This suggests that nucleation is not a single event but continues to occur throughout the measurement period.

5. Conclusions

Jarosite is a mineral of interest both from an environmental perspective (e.g., acid mine drainage systems) and more recently from a planetary perspective (the discovery of jarosite and alunite on Mars). Therefore, understanding how different formation temperatures and Al/Fe ratios impact infrared spectroscopy, XRD and thermal behaviour is important.

While alunite has the lower formation energy and, therefore, would be expected to be the thermodynamically stable product, high concentrations of Al are required in solution in order to form Al substituted jarosites. Molecular modelling results suggest that this could be due to the hydration energy of Al^{3+} compared to Fe^{3+}. The higher hydration energy of Al^{3+} cations means that this energy must be overcome in order to substitute Fe^{3+} for Al^{3+}.

The TGA results showed that the transition temperatures for structural water removal and de-sulphurization tend to higher temperatures as the Al content is increased in the structure. It was also found that the alunite structure will incorporate more "additional" water as well as structural water, and this was observed in the data as a broad endotherm below 300 °C. This again supports the modelling results that implicate Al^{3+} dehydration as a key intermediate step. Finally, DLS showed that the number of nuclei followed the expected trend based on surface energy considerations, but that, for alunite, nucleation did not appear to be a single event but rather a continuous one.

Supplementary Materials: The following are available online at www.mdpi.com/2075-163X/7/6/90/s1; Figure S1: XRD pattern and Rietveld fit for jarosite K (GOF 2.8); Figure S2: SEM image of particles formed in the synthesis of the jarosite J sample, Table S1: Fitting statistics for Rietveld refinement of samples; Figure S3: Vacancies and 'Additional water' found for samples versus iron content; Figure S4: Experimental XRD patterns for pure end members and their database match, Figure S5: Infrared spectra of pure jarosite (jarosite A) and pure alunite for wavenumbers 4000–1800 cm^{-1}; Table S2: Empirical potentials used in the modelling; Table S3: Comparison between simulated and experimental values (from supercell simulations).

Acknowledgments: I would like to thank Peter Chapman for the TGA analysis and acknowledge the Curtin Centre for Materials Research, Curtin University, WA, Australia for use of the SEM and XRD facilities.

Conflicts of Interest: The author declares no conflict of interest.

References

1. Bibi, I.; Singh, B.; Silvester, E. Akaganeite (beta-FeOOH) precipitation in inland acid sulfate soils of south-western New South Wales (NSW), Australia. *Geochim. Cosmochim. Acta* **2011**, *75*, 6429–6438. [CrossRef]

2. Das, S.; Hendry, M.J. Application of Raman spectroscopy to identify iron minerals commonly found in mine wastes. *Chem. Geol.* **2011**, *290*, 101–108. [CrossRef]

3. Long, D.T.; Fegan, N.E.; McKee, J.D.; Lyons, W.B.; Hines, M.E.; Macumber, P.G. Formation of alunite, jarosite and hydrous iron oxides in a hypersaline system: Lake Tyrell, Victoria. *Aust. Chem. Geol.* **1992**, *96*, 183–202. [CrossRef]

4. Dutrizac, J.E.; Jambor, J.L. Jarosites and their application in hydrometallurgy. *Rev. Miner. Geochem.* **2000**, *40*, 405–452. [CrossRef]

5. Klingelhofer, A.K.; Morris, R.V.; Bernhardt, B.; Schröder, C.; Rodinov, D.S.; de Souza Jnr, P.A.; Yen, A.; Gellert, R.; Evalanov, E.N.; Ming, D.W.; et al. Jarosite and hematite at Meridiani Planum from opportunity's Mossbauer spectrometer. *Science* **2004**, *306*, 1740–1745. [CrossRef] [PubMed]

6. Smith, R.T.; Comer, J.B.; Ennis, M.V.; Branam, T.B.; Butler, S.M.; Renton, P.M. *Toxic Metals Removal in Acid mine Drainage Tratment Wetlands*; Technical Report; Indiana Geological Survey: Bloomington, IN, USA, 2001.

7. Figueiredo, M.-O.; da Silva, T.P. The positive environmental contribution of jarosite by retaining lead in acid mine drainage areas. *Int. J. Environ. Res. Public Health* **2011**, *8*, 1575–1582. [CrossRef] [PubMed]

8. Smith, A.M.L.; Hudson-Edwards, K.A.; Dubbin, W.E.; Wright, K. Dissolution of Jarosite [KFe3(SO4)2(OH)6] at pH 2 and 8: Insights from Batch Experiments and Computational Modelling. *Geochim. Cosmochim. Acta* **2006**, *70*, 608–621. [CrossRef]

9. Murphy, P.J.; Smith, A.M.L.; Hudson-Edwards, K.A.; Dubbin, W.E.; Wright, K. Raman and IR spectroscopic studies of alunite-supergroup compounds containing Al^{3+}, Cr^{3+}, Fe^{3+} and V^{3+} at the B site. *Can. Mineral.* **2009**, *47*, 663–681. [CrossRef]

10. Smeaton, C.M.; Freyer, B.J.; Weisener, C.G. Intracellular precipitation of Pb by *Shewanella putrifaciens* CN32 during the reductive dissolution of Pb-jarosite. *Environ. Sci. Technol.* **2009**, *43*, 8086–8091. [CrossRef] [PubMed]

11. Leahy, M.J.; Schwarz, M.P. Modelling jarosite precipitation in isothermal chalcopyrite bioleaching columns. *Hydrometallurgy* **2009**, *98*, 181–191. [CrossRef]

12. Sasaki, K.; Takatsugi, K.; Hirajima, T. Effects of initial Fe^{2+} concentration and pulp density on the bioleaching of Cu from enargite by Acidianus brierleyi. *Hydrometallurgy* **2011**, *109*, 153–160. [CrossRef]

13. Drouet, C.; Navrotsky, A. Synthesis, characterization, and thermochemistry of K-Na-H_3O jarosites. *Geochim. Cosmochim. Acta* **2003**, *67*, 2063–2076. [CrossRef]

14. Drouet, C.; Pass, K.L.; Baron, D.; Drauker, S.; Navrotsky, A. Thermochemistry of jarosite-alunite and natrojarosite-natroalunite solid solutions. *Geochim. Cosmochim. Acta* **2004**, *68*, 2197–2205. [CrossRef]

15. Grube, E.; Nielsen, U.G. The stoichiometry of synthetic alunite as a function of hydrothermal aging investigated by solid-state NMR spectroscopy, powder X-ray diffraction and infrared spectrscopy. *Phys. Chem. Miner.* **2015**, *42*, 337–345. [CrossRef]

16. Stoffregen, R.E.; Alpers, C.N.; Jambor, J.L. Alunite-Jarosite crystallography, thermodynamics, and geochronology. *Rev. Mineral. Geochem.* **2012**, *9*, 454–479. [CrossRef]

17. Alpers, C.N.; Rye, R.O.; Nordstrom, D.K.; White, L.D.; King, B.-S. Chemical, crystallographic and stable isotopic properties of alunite and jarosite from acid-hypersaline Australian lakes. *Chem. Geol.* **1992**, *96*, 203–226. [CrossRef]

18. Ehlmann, B.L.; Swayze, G.A.; Milliken, R.E.; Mustard, J.F.; Clark, R.N.; Murchie, S.L.; Breit, G.N.; Wray, J.J.; Gondet, B.; Puolet, F.; et al. Discovery of alunite in Cross crater, Terra Sirenum, Mars: Evidence for acidic, sulforous waters. *Am. Mineral.* **2016**, *101*, 1527–1542. [CrossRef]

19. Becker, U.; Gasharova, B. AFM observations and simulations of jarosite growth at the molecular scale: Probing the basis for the incorporation of foreign ions into jarosite as a storage material. *Phys. Chem. Miner.* **2001**, *28*, 545–556. [CrossRef]

20. Desborough, G.A.; Smith, K.S.; Lowers, H.A.; Swayze, G.A.; Hammarstrom, J.M.; Diehl, S.F.; Leinz, R.W.; Driscoll, R.L. Mineralogical and chemical characteristics of some natural jarosites. *Geochim. Cosmochim. Acta* **2010**, *74*, 1041–1056. [CrossRef]

21. Scott, K.M. Solid solution in, and classification of, gossan-derived members of the alunite-jarosite family, northwest Queensland, Australia. *Am. Mineral.* **1987**, *72*, 178–187.

22. Scott, K.M. Origin of alunite and jarosite-group minerals in the Mt. Leyshon epithermal gold deposit, northeast Queensland, Australia. *Am. Mineral.* **1990**, *75*, 1176–1181.

23. Brophy, G.P.; Scott, E.S.; Snellgrove, R.A. Sulfate studies II. Solid solution between alunite and jarosite. *Am. Mineral.* **1962**, *47*, 112–126.

24. Basciano, L.C.; Peterson, R.C. Crystal chemistry of the natrojarosite-jarosite and natrojarosite-hydronium jarosite solid-solution series: A synthetic study with full Fe site occupancy. *Am. Mineral.* **2008**, *93*, 853–862. [CrossRef]

25. Nielsen, U.G.; Majzlan, J.; Grey, C.P. Determination and quantification of the local environments in stoichiometric and defect jarosite by solid-state H-2 NMR spectroscopy. *Chem. Mater.* **2008**, *20*, 2234–2241. [CrossRef]

26. Scarlett, N.V.Y.; Grey, I.E.; Brand, H.E.A. Ordering of iron vacancies in monoclinic jarosites. *Am. Mineral.* **2010**, *95*, 1590–1593. [CrossRef]

27. Maubec, N.; Lahfid, A.; Lerouge, C.; Wille, G.; Michel, K. Characterization of alunite supergroup minerals by Raman spectroscopy. *Spectrochim. Acta A Mol. Biomol. Spectrosc.* **2012**, *96*, 925–939. [CrossRef] [PubMed]

28. Spratt, H.J.; Rintoul, L.; Avdeev, M.; Martens, W.N. The crystal structure and vibrational spectroscopy of jarosite and alunite minerals. *Am. Mineral.* **2013**, *98*, 1633–1643. [CrossRef]

29. Rudolph, W.W.; Mason, R.; Schmidt, P. Synthetic alunites of the potassium-oxonium solid solution series and some other members of the group: synthesis, thermal and X-ray characterization. *Eur. J. Mineral.* **2003**, *15*, 913–924. [CrossRef]

30. Crabbe, H.; Fernandez, N.; Jones, F. Crystallization of jarosite in the presence of amino acids. *J. Cryst. Growth* **2015**, *416*, 28–33. [CrossRef]

31. Dutrizac, J.E.; Dinardo, O.; Kaiman, S. Factors affecting lead jarosite formation. *Hydrometallurgy* **1980**, *5*, 305–324. [CrossRef]

32. Mullin, J.W. Nucleation. In *Crystallization*, 3rd ed.; Butterworth-Heinemann: Oxford, UK, 1961; pp. 172–201.

33. Acero, P.; Hudson-Edwards, K.A.; Gale, J.D. Influence of pH and temperature on alunite dissolution: Rates, products and insights on mechanisms from atomistic simulations. *Chem. Geol.* **2015**, *419*, 1–9. [CrossRef]

34. Bunney, K.; Freeman, S.R.; Ogden, M.I.; Richmond, W.R.; Rohl, A.L.; Jones, F. Effect of La^{3+} on the crystal growth of barium sulfate. *Cryst. Growth Des.* **2014**, *14*, 1650–1658. [CrossRef]

35. Jones, F.; Richmond, W.R.; Rohl, A.L. Molecular modelling of phosphonate molecules onto barium sulfate terraced surfaces. *J. Phys. Chem. B.* **2006**, *110*, 7414–7424. [CrossRef] [PubMed]

36. Fleming, S.; Rohl, A.L. GDIS: A visualization program for molecular and periodic systems. *Zeitschrift für Kristallographie* **2005**, *220*, 580–584. [CrossRef]

37. Gale, J.D.; Rohl, A.L. The General Utility Lattice Program. *Mol. Simul.* **2003**, *29*, 291–341. [CrossRef]

38. Marcus, Y. Thermodynamics of solvation of ions Part 5. Gibbs Free energy of hydration at 298.15K. *J. Chem. Soc. Faraday Trans.* **1991**, *87*, 2995–2999. [CrossRef]

39. Sasaki, K. Raman study of the microbially mediated dissolution of pyrite by *Thiobacillus ferrooxidans.* *Can. Mineral.* **1997**, *35*, 999–1008.

40. Sasaki, K.; Konno, H. Morphology of jarosite-group compounds precipitated from biologically and chemically oxidized Fe ions. *Can. Mineral.* **2000**, *38*, 45–56. [CrossRef]

41. Majzlan, J.; Speziale, S.; Duffy, T.S.; Burns, P.C. Single-crystal elastic properties of alunite, $KAl_3(SO_4)_2(OH)_6$. *Phys. Chem. Miner.* **2006**, *33*, 567–573. [CrossRef]

42. Vegard, L. Die Konstitution der Mischkristalle und die Raumfüllung der Atome. *Zeitschrift für Physik* **1921**, *5*, 17–26. [CrossRef]

43. Kubisz, J. Studies on synthetic alkali-hydronium jarosites II: Thermal investigations. *Mineral. Pol.* **1971**, *2*, 51–59.

44. Kuçuk, F.; Yildiz, K. The decomposition kinetics of mechanically activated alunite ore in air atmosphere by thermogravimetry. *Thermochim. Acta* **2006**, *448*, 107–110. [CrossRef]

45. Piana, S.; Jones, F.; Gale, J.D. Assisted desolvation as a key kinetic step for crystal growth. *J. Am. Chem. Soc.* **2006**, *128*, 13668–13674. [CrossRef] [PubMed]

![minerals logo] *minerals*

MDPI

Article

Indications that Amorphous Calcium Carbonates Occur in Pathological Mineralisation—A Urinary Stone from a Guinea Pig

Denis Gebauer [1],*, Kjell Jansson [2], Mikael Oliveberg [3] and Niklas Hedin [2],*

[1]   Department of Chemistry, Physical Chemistry, University of Konstanz, Universitätsstraße 10,
    D-78464 Konstanz, Germany
[2]   Department of Materials and Environmental Chemistry, Arrhenius Laboratory, Stockholm University,
    SE-106 91 Stockholm, Sweden; kjell.jansson@mmk.su.se
[3]   Department of Biochemistry and Biophysics, Arrhenius Laboratories of Natural Sciences,
    Stockholm University, S-106 91 Stockholm, Sweden; mikael.oliveberg@dbb.su.se
*   Correspondence: denis.gebauer@uni-konstanz.de (D.G.); niklas.hedin@mmk.su.se (N.H.);
    Tel.: +49-7531-88-2169 (D.G.); +46-8-16-24-17 (N.H.)

Received: 29 January 2018; Accepted: 24 February 2018; Published: 27 February 2018

Abstract: Calcium carbonate is an abundant biomineral that is of great importance in industrial or geological contexts. In recent years, many studies of the precipitation of $CaCO_3$ have shown that amorphous precursors and intermediates are widespread in the biomineralization processes and can also be exploited in bio-inspired materials chemistry. In this work, the thorough investigation of a urinary stone of a guinea pig suggests that amorphous calcium carbonate (ACC) can play a role in pathological mineralization. Importantly, certain analytical techniques that are often applied in the corresponding analyses are sensitive only to crystalline $CaCO_3$ and can misleadingly exclude the relevance of calcium carbonate during the formation of urinary stones. Our analyses suggest that ACC is the major constituent of the particular stone studied, which possibly precipitated on struvite nuclei. Minor amounts of urea, other stable inorganics, and minor organic inclusions are observed as well.

Keywords: amorphous calcium carbonate; urinary stones; guinea pig; pathological mineralization; struvite

1. Introduction

Calcium carbonate is the most abundant biomineral and is the inorganic constituent of mussel shells, the carapace of crayfish and lobsters, the exoskeleton of corals, and an important biomineral in many more organisms [1]. Moreover, it is of major industrial (building industries), economic (scaling and incrustation), and ecological (climate) importance [2]. Studies of biomineralization serve as an inspiration for material scientists, as biogenic composites show outstanding performance in terms of material properties [3], which is rare in artificial counterparts [4–6]. The level of control over the crystallization processes in organisms is very high, but the underlying molecular mechanisms are still not understood in detail [7]. In general, it is accepted that organisms use amorphous intermediates during the biomineralization of calcium carbonates [8–10]. These intermediates appear to be crucial to the generation of complex hierarchical structures of many biomineralized exoskeletons [8,9].

Controlling crystallization is, however, not only linked to the generation of mineral-based hybrid materials but also to the inhibition of the unwanted precipitation of minerals. In organisms, pathological mineralization can lead to severe issues in soft tissues, the formation of mineral stones in bodily fluids, or the uncontrolled growth of hard tissues [11–14]. Organisms employ specific proteins to inhibit unwanted mineralization, which range from the blood-protein family of fetuins [11] to statherin (saliva) [15]. However, when it comes to pathological mineralization in mammals, the formation of less soluble calcium phosphates and oxalates is often more critical than uncontrolled precipitation of $CaCO_3$ [11,16,17]. An example of functionally formed calcium carbonate stones is the gastroliths of lobsters [8,10,18]. They serve as a calcium reservoir regularly needed upon molting, as the carapace cannot grow with the animal. Gastroliths are mostly constituted of amorphous calcium carbonate (ACC) [19], due to its higher solubility and better bioavailability as compared to crystalline calcium carbonates. One example of pathologically formed calcium carbonate is the urinary stones of guinea pigs. They typically contain large amounts of calcium carbonate, often up to 100%, and mostly constitute calcite, magnesian calcite, or monohydrocalcite [20,21]. The occurrence of ACC in such stones has not been reported so far, to the best of our knowledge.

In fact, the general relevance of ACC in the formation calcium carbonates is currently under debate [22,23]. From the point of view of classical nucleation theory, ACC is only accessible beyond a certain level of supersaturation, and is thus often assumed, arbitrarily, to not occur in many settings that are nominally undersaturated with respect to macroscopic phases of ACC [24]. However, there are indications from computer simulations of ACC being the thermodynamically stable form of $CaCO_3$ at small particle sizes (<~4 nm) [25]. Indeed, alternative views on the mechanism of phase separation, e.g., according to the so-called pre-nucleation cluster pathway, suggest that ACC would always form first, transiently [7], while recent in-situ electron microscopy studies suggest that crystalline calcium carbonate can form in solution via direct pathways [23]. The role of the electron beam in such experiments remains unclear, especially when it comes to the induction of crystallisation in amorphous intermediates [26].

In any case, the fundamental question regarding the occurrence of ACC is of ultimate relevance for pathological mineralisation as well. On the one hand, ACC can be overseen when techniques are employed that are only sensitive towards the detection of crystalline species, such as X-ray diffraction (XRD) [21]. This approach can lead to misinterpretations of stone compositions and bring about erroneous assessments of the consequences of dietary conditions. We do note that the occurrence of amorphous carbonated calcium phosphate (ACCP) in pathological mineralization is thoroughly established [27]; however, also, the role of ACC in pathological mineralisation could be a more widespread phenomenon. Strategies to inhibit nucleation and crystallisation, within the notions of alternative pathways to particle formation [7,28,29], may potentially allow for the development of new medical treatments, perhaps even beyond the calcium-carbonate-rich urinary stones in guinea pigs.

This study has been conducted on a particular stone of an individual guinea pig. It has a mixed inorganic composition of calcium carbonates and magnesium ammonium phosphate hexahydrate (struvite), indicating that the stone had formed upon urinary tract infection [30,31]. A global discussion of a combination of results indicates that a significant fraction of the calcium carbonate is amorphous. It appears as amorphous precursors, and intermediates could play a key role also in the pathological mineralisation of such stones.

2. Materials and Methods

Urinary stones. No animals were harmed in the course of this study; the urinary stone specimen was provided by Mia Winge of a non-profit organization for helping guinea pigs: http://www.marsvinshjalpen.se.

Scanning electron microscopy (SEM) and energy-dispersive X-ray spectroscopy (EDS). A urinary stone was cut through the centre utilising a blade saw, and gently polished under dry conditions, first using SiC paper (4000 mesh) and then diamond paper (1 micrometre). After polishing, the remaining loose abrasive material was blown away by a gentle air stream. This polished stone was imaged utilising the detector for back-scattered electrons in a SEM. To that end, a TM 3000 table scanning electron microscope (Hitachi, Tokio, Japan) was used with an accelerating voltage of 15 kV. EDS mapping was performed using a SDD detector (Quantax 70, Bruker, Billerica, MA, USA), and the spectra were analysed for three particular areas. This analysis is self-calibrating and relatively insensitive to local sample tilts and surface conditions.

Fourier transform infrared (FT-IR) spectroscopy. FT-IR spectra were recorded on a finely ground powder of a urinary stone with a 670-IR FT-IR spectrometer (Varian, Palo Alto, CA, USA) and its diamond-based attenuated total reflectance device (Golden Gate™, Specac, Orpington, UK) within a spectral range of 390–4000 cm^{-1} and a spectral resolution of 2 cm^{-1}. The room temperature detector of the spectrometer was used.

Thermal gravimetric analysis (TGA). TGA of a piece of the urinary stone was conducted on a TGA 7 analyser (Perkin Elmer, Waltham, MA, USA) by heating the sample from 25 to 900 °C at a rate of 2 °C/min under a flow of dry artificial air mixture.

X-ray powder diffraction (XRD). Diffractograms were recorded on a X'Pert Pro diffractometer (PANalytical, Almelo, The Netherlands) using Cu K$_{\alpha 1}$ (λ = 1.5406 Å) radiation and a PIXel detector. The diffractograms were acquired for 20° < 2θ < 60° using a scanning speed of 0.04° s^{-1} (acquisition time of ~17 minutes per diffractogram). Pieces of a urinary stone were finely ground, dispersed in isopropanol, and subsequently spread on silicon plates, uniformly.

3. Results and Discussion

The SEM images in Figure 1 of a gently polished stone exemplify successive regions of dense and porous areas that appear to have grown in rings. The dense rings alternate with loose and porous regions, similar to the concentric rings that Peng et al. observed by optical microscopy of a stone composed of calcium carbonate, magnesium ammonium phosphate hexahydrate (struvite), and calcium phosphate [32]. The infrared spectrum recorded on the stone displays distinct bands that can be assigned to the different vibrational modes of CO$_3$$^{2-}$ and PO$_4$$^{3-}$ ions (Figure 2). In addition, bands typical for CH$_3$, CH$_2$, and NH chemical groups are observed, as one would expect from the inclusion of proteins and other biomolecules. The assignments of bands in the fingerprint region (<1000 cm^{-1}) are consistent with the crystalline inorganic polymorphs as established by XRD data (see below) and can be assigned to phosphates and carbonates.

The band at 995 cm^{-1} is due to the anti-symmetric stretching mode (ν$_3$) of the phosphate ion, which is triply degenerate for free phosphates. It has earlier been reported to occur at 1006 cm^{-1} in magnesium ammonium phosphate hexahydrate (struvite) [33]. There is also a good match of the band frequency for the in-plane bending mode (ν$_4$) of the phosphate group (569 cm^{-1}), which has been reported to occur at 571 cm^{-1} in struvite [33]. The minor variances between the band positions and those in literature reports can be rationalized by the coupling with librational modes of water, which occur in this spectral region, as well as additional bands from calcium carbonate and organic compounds. The ν$_3$ band of the carbonate ion is located at 1402 cm^{-1}, whereas the band position and band width are rather inconclusive when it comes to discriminating in between ACC (split band 1392/1462 cm^{-1}) and calcite (1392 cm^{-1}) [34]. The broad band features a shoulder (~1469 cm^{-1}) that could relate to the split band that is normally observed for ACC and could include contributions from amide bands of proteins present in stone. The band for the out-of-plane bending mode (ν$_2$) at 872 cm^{-1} cannot be used directly to discriminate ACC from crystalline forms of CaCO$_3$ either, because its position varies for different types of ACC [35–40]. However, in this case, the band position agrees with literature reports on calcite [34]. Typically, the ν$_4$ bands of the carbonate ion are used to discriminate between ACC and different crystalline forms. These bands are sharp, although they are rather weak for calcite (713 cm^{-1}), vaterite (745 cm^{-1}), and aragonite (713/700 cm^{-1} split band) [41], and very weak and split for ACC (694/723 cm^{-1}) [34]. Actually, the ν$_4$ split bands of ACC are often not observable, since they are superimposed by a broad band that arises from the librational modes of (structural) water molecules [34]. In case of the urinary stone of the guinea pig, broad bands are located at 758/715 cm^{-1} with a minor shoulder at 702 cm^{-1}. The carbonate bands in the fingerprint region cannot directly be used to identify the very specific distribution of polymorphs of calcium carbonate present in the urinary stone of the guinea pig. Although some spectral features of calcite are observed, the vibrational modes of the carbonates are much broader than what is typical for calcite, which is indicative of substantial structural disorder that may arise from ACC and/or disordered Mg-calcite. Since the spectral region for the carbonate ν_2 and ν_4 bands is superimposed by contributions from the phosphate bands and bands that arise from organic constituents, the ν_2/ν_4-intensity ratio for the carbonate ions [42] cannot be unambiguously used to confirm the presence of ACC.

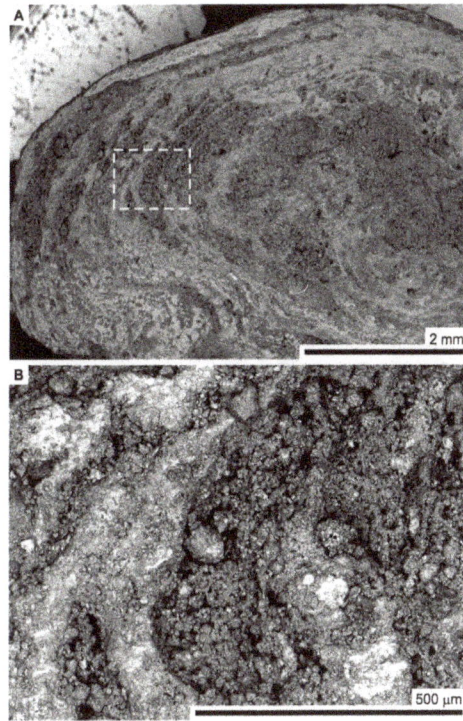

Figure 1. Scanning electron micrographs of a urinary stone from a guinea pig. The cross-section of the stone was polished, and images were recorded from backscattered electrons. (**A**) Overview and (**B**) magnified view of the dashed area indicated in (**A**).

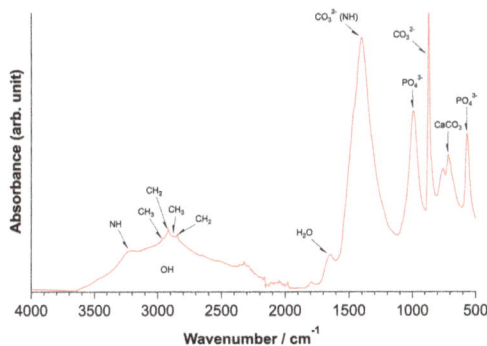

Figure 2. Fourier transform infrared spectrum of the powdered urinary stone.

XRD has often been used to identify and quantify the amounts of crystalline polymorphs in urinary stones of guinea pigs and other mammals [21]. The XRD diffractogram shown in Figure 3 indicates that struvite is present in the urinary stone studied here, alongside a poorly crystalline calcium carbonate phase, which is consistent with the IR analysis above; however, the XRD data cannot be unambiguously evaluated toward the amount of crystalline calcium carbonate owing to the unusual broad reflections and considerable noise level (see Figure 3, and further discussions below). Struvite has been observed in urinary stones of mammals in general [43]. The diffractogram (Figure 3) is a convolution of mainly two characteristic patterns of reflections, one set of reflections with narrow peak widths from struvite and one set with broad reflections that may be assigned to (Mg-)calcite.

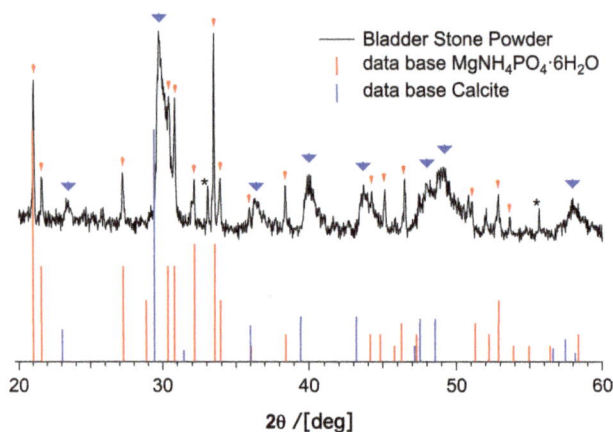

Figure 3. Top trace: X-ray diffractogram of the powdered urinary stone. Bottom lines: literature diffractograms for magnesium ammonium phosphate (struvite, $MgNH_4PO_4 \cdot 6H_2O$) (**red**) and calcite (**blue**). The asterisks (*) label reflections that cannot be unambiguously assigned.

It is important to note that the rather low signal-to-noise (S/N) ratio is not due to fast data acquisition, but rather due to a low amount of crystallinity in the sample. The reflections from struvite are marked with red arrows, and the broad reflections that may be assigned to calcite are marked with thick blue arrows (Figure 3). Interestingly, the pattern for the calcite phase is shifted by ~0.3° to larger angles as compared with the calcite database diffractogram. Given the internal struvite reference, the shift has to be considered real, although a constant shift in all reflections is surprising and usually due to a misalignment of the diffractometer. The shift of the reflections is not consistent with monohydrocalcite [44], which has a similar diffraction pattern as calcite, but with a shift to lower angles. Due to the low S/N ratio, despite a long acquisition time, and the broad reflections, we cannot unambiguously assign this reflection pattern to calcite. To our knowledge, such broad features have not been reported for calcite, which typically tends to exhibit narrow and intense reflections, but it may well be due to nano-calcite. Another possible explanation for this broadening is the accommodation of magnesium ions in the calcite lattice, which can cause significant broadening of the reflections [45,46]. Quantitatively, the observed shift in the (104) reflection of calcite without magnesium substitution ($2\theta = 29.429°$) to $2\theta \approx 29.8°$ (maximum of the reflection) gives a change in $d_{(104)}$ from 3.0326 Å (calcite) to ~3.0 Å, which is consistent with a Mg^{2+} content of ca. 10–20 mol % [45]. Owing to the low S/N ratio, it is difficult to extract the exact peak width in order to estimate the crystallite size via the Scherrer equation (assuming it is pure calcite) or, alternatively, to perform a Rietveld refinement for Mg-substituted calcite. In any case, the peak area of the broad reflections implies that this phase is more abundant than struvite.

Urinary stones of guinea pigs contain a small amount of organics in addition to the inorganic components [21]. We have used TGA in artificial air to determine the amount of organics, as well as the contents of struvite and calcium carbonate (Figure 4). The most pronounced mass loss can be assigned to the decomposition of the carbonate ions. Assuming that the observed loss of 30 % of relative weight, beginning at ca. 638 °C (Figure 4), is due to the calcination of calcium carbonate, yielding calcium oxide and carbon dioxide, we conclude that the urinary stone contains ~68 weight-% of $CaCO_3$ (see the Supplementary Materials). We note that in the light of the weak reflections in the XRD diffractogram (see above), the high carbonate content is actually surprising, and strongly suggests that a major part of the carbonates is in fact amorphous. Within this assumption, it follows that ca. 38 weight-% of the observed 49 weight-% remaining after reaching 820 °C is CaO (see the Supplementary Materials), indicating that ca. 10% are other (stable) inorganics. This level is consistent with the observed relative mass just before calcination (79%), which then is the sum of the relative masses of calcium carbonate and stable inorganics. The weight loss in between 112 and 162 °C (5%) may relate to decomposition/combustion of urea, which begins at 130 °C [47]. On the other hand, the decomposition of struvite, yielding magnesium hydrogenphosphate upon releasing ammonia and water, occurs in a single step between ~55 and 250 °C [48], and can hardly be separated from the decomposition/combustion of organics. Assuming that the ~10% of stable inorganics that remain after calcination of calcium carbonate, and that are not CaO, are magnesium hydrogenphosphate from struvite decomposition, the initial relative struvite content is ca. 20% (see the Supplementary Materials). This approach sums up to a total relative amount of inorganics of ca. 88%, suggesting that the stone contains ca. 12% organics, of which 5% may be urea. Based on this data, we cannot rule out the presence of amorphous magnesium carbonate or amorphous carbonated apatite, but since this TGA evaluation is consistent with quantitative EDS analyses (see below), these phases could constitute only minor fractions of the stone.

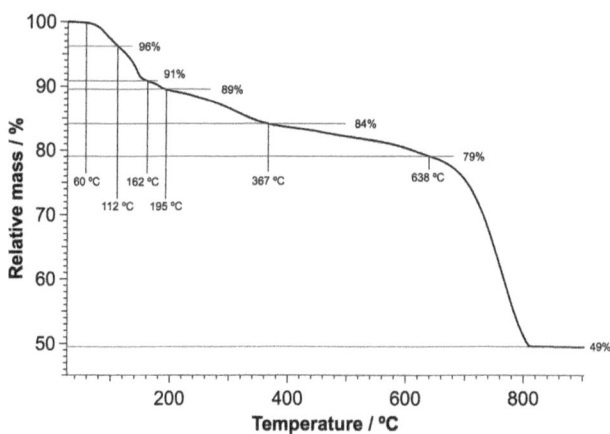

Figure 4. Thermal gravimetric analysis of the urinary stone of a guinea pig.

To study the compositional variation across the surface of the polished urinary stone, we recorded EDS maps. Figure 5 displays the EDS maps of the stone, in which clear and laminated ring patterns are observed not only for phosphorous but also for magnesium and oxygen. By comparison of the distribution of these elements with the SEM images, we conclude that struvite ($MgNH_4PO_4 \cdot 6H_2O$) is concentrated within the denser regions of the laminated stone, which precipitates from urine at a pH of 9 in the presence of relatively small amounts of Mg^{2+} ions [49]. The other elements are rather evenly distributed throughout the area of the gently polished urinary stone, although calcium displays a similar pattern as the elements of struvite. Hence, the majority component according to TGA (calcium carbonate) is present in both the dense and the loose regions and excluded from only some areas. Indeed, a close inspection of the EDS maps reveals that individual grains of struvite and calcium carbonate can be identified (white arrows in Figure 5). We speculate that the seed of the stone has been struvite (grain potentially located within the area indicated by a dashed yellow line in the SEM micrograph of Figure 5 considering the laminated pattern), which has served as a nucleus for further precipitation of calcium carbonate and inclusion or co-precipitation of other urinary constituents. The distribution of potassium, sodium, and chlorine is rather continuous, although some excluded areas are obvious, especially towards the outer perimeter. Lastly, silicon occurs randomly, likely being indicative of minor amounts of silica, taking the quantitative EDS data into account (see below, Table 1).

Figure 6 shows a magnified view on the porous region of the urinary stone with corresponding EDS mapping. Indeed, the calcium carbonate and struvite constituents of the stone are present as homogeneous and distinct phases in this part of the stone (white arrows mark micron-sized $CaCO_3$ particles in Figure 6). However, the EDS mapping also reveals regions that do not contain any of the traced elements (yellow arrow in Figure 6). We hypothesize that these regions are composed of urea. The compositional variations in the areas containing loosely packed inorganic particles appear to occur on relatively short length scales.

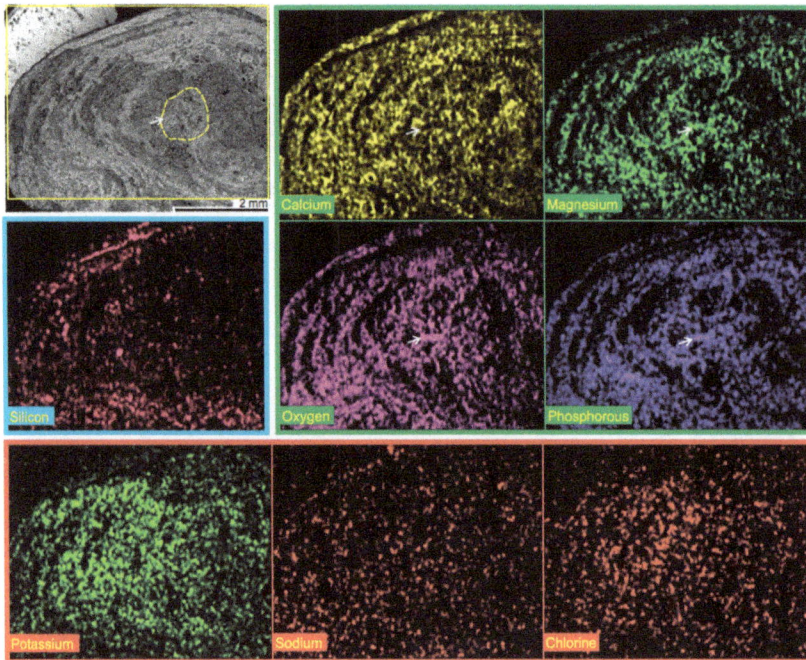

Figure 5. Energy-dispersive X-ray spectroscopy (EDS) maps of a cross-polished urinary stone from a guinea pig (overview, see Figure 6 for a magnified view). The top left image is the corresponding scanning electron micrograph, and the yellow frame indicates the area selected for EDS mapping. White arrows highlight an exemplified feature, where magnesium, oxygen, and phosphorous are abundant, but not calcium. This feature shows that although the contents of struvite and calcium carbonate are similarly distributed, larger individual grains can be identified. It may be speculated that struvite served as an initial seed within the yellow dashed area highlighted in the top left image. The EDS maps are grouped according to similar distributions within the urinary stone; calcium, magnesium, oxygen, and phosphorous (**green frame**); potassium, sodium, and chlorine (**red frame**); and silicon (**blue frame**).

The elemental composition of three spots of the urinary stone has been quantitatively determined by EDS, and is consistent with the composition determined from TGA (Table 1), which indicates ~68 weight-% of $CaCO_3$, ~10 weight-% of struvite and other stable inorganics, ~5 weight-% of urea, and ca. 17 weight-% of other organics (see above). The first entry in Table 1 represents a larger spot that covers a loose and a dense area, the second one a smaller porous area, and the last entry a dense region (corresponding SEM images with detailed EDS results can be found in the Supplementary Materials, Figures S1–S3). It can be seen that the weight percentages of oxygen, carbon, calcium, potassium, and chlorine do not vary significantly within the distinct areas, whereas magnesium and phosphorous, both indicative of struvite, are somewhat more abundant in the dense region of the stone. Silicon, which indicates the location of the minor silica content, is somewhat increased within the porous region of the stone. However, given the rather inhomogeneous distribution of Si (see Figures 5 and 6), this may be a coincidental finding rather than a general compositional feature of the stone. Altogether, this additional EDS mapping indicates that significant structural variations can only occur on length scales smaller than ca. 200 μm.

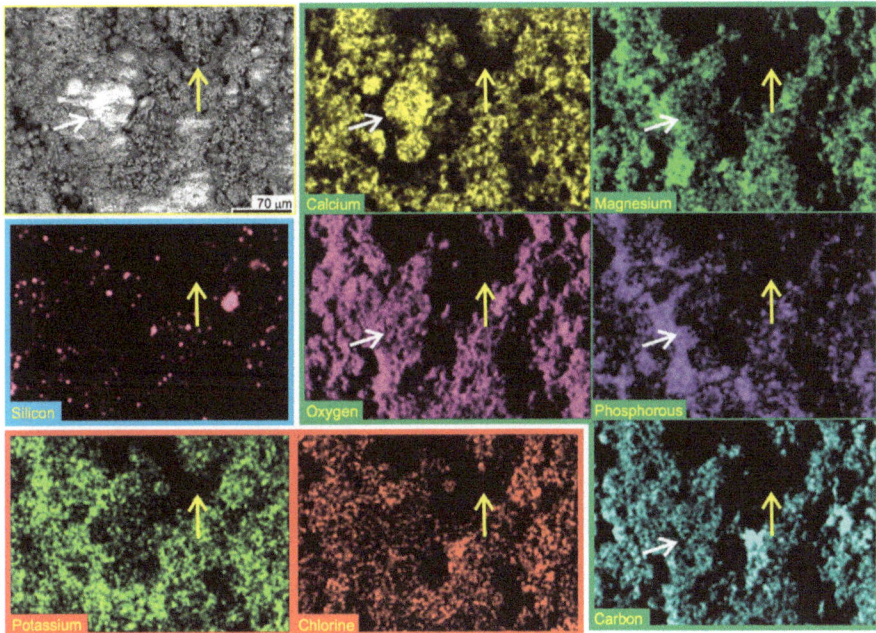

Figure 6. Energy-dispersive X-ray spectroscopy (EDS) maps of a cross-polished urinary stone of a guinea pig (magnified view of a porous region, see Figure 5 for an overview). The top left image is the corresponding scanning electron micrograph, and the yellow frame indicates the area selected for EDS mapping. White arrows highlight an exemplified feature in which calcium carbonate is abundant but magnesium and phosphorous is not. Yellow arrows mark a region where none of the investigated elements can be found to a significant extent. The EDS maps are grouped according to Figure 5; calcium, magnesium, oxygen, phosphorous, and carbon (**green frame**); potassium, and sodium (**red frame**); and silicon (**blue frame**).

Table 1. Elemental compositions estimated from electron-dispersive X-ray spectroscopy of three distinct regions in a cross-polished urinary stone of a guinea pig. The values represent weight percentages that have been normalized to 100%.

Element	Large Spot (weight-%)	Porous Region (weight-%)	Dense Region (weight-%)	Approx. Error (%)
Oxygen	48.5	49.5	50.2	6–8
Carbon	19.2	17.7	16.0	3–4
Calcium	20.7	18.6	16.9	0.5–0.7
Magnesium	5.1	6.4	7.2	0.3–0.4
Phosphorous	3.1	3.3	6.4	0.2–0.3
Potassium	2.2	2.6	2.1	0.1
Silicon	0.9	1.5	0.7	0.1
Chlorine	0.5	0.5	0.4	0

4. Conclusions

There is a variation in the components of the urinary stone of the guinea pig. The majority phase of the particular stone investigated here is calcium carbonate. The broad reflections for calcium (magnesium) carbonate and the fact that the stone contains 68 weight-% of calcium carbonate (based on TGA and EDS) indicate that a considerable fraction of the stone is ACC. ACC has to our knowledge not been documented for such urinary stones. The existence of ACC in the stone is consistent with the broad spectral features that can be assigned to carbonate vibrational modes in infrared spectroscopy (IR). It should be noted that ACC typically features very broad scattering in XRD. These are only evident for purely amorphous specimens and could not be observed for this stone that contains struvite and partially crystalline carbonates. The occurrence of ACC is generally also consistent with the presence of magnesium, which is known to kinetically stabilise ACC [40,50]. Transiently formed ACC [8] could be a precursor to monohydrocalcite (MHC) that can occur in such stones, whereas MHC might have transformed into Mg-calcite upon storage of the present stones [20]. The poor crystallinity of carbonates (Figure 3) might also indicate that they have formed via an amorphous precursor. While we cannot unambiguously prove the presence of ACC, further analyses based on TEM or polarized light microscopy would likely be inconclusive owing to the mixed composition of the stones. In any case, the present study shows that it could be important to further study if ACC is a common feature of urinary stones of guinea pigs. It should be emphasized that analyses [21] employing XRD cannot reveal the presence of ACC in samples with significant diffraction, and previously investigated stones may contain significant amounts of ACC. For example, the minute reflections due to calcium carbonate in the XRD analyses of the urinary stone of this study would have led to a dramatic underestimation of the calcium carbonate content, if the assessment would have been based on XRD alone. It is important to understand the occurrence of calcium carbonate, as it could potentially affect dietary recommendations. This can be a problem when the applied analytics is sensitive only for crystalline $CaCO_3$.

The stone displays successive laminated regions of dense and porous materials (growth rings), whereas crystalline struvite ($MgNH_4PO_4 \cdot 6H_2O$) appears to be somewhat concentrated in the dense regions of the stone when considering the relative amounts of magnesium and phosphorous determined by means of EDS in the different regions (Table 1). The calcium carbonate is more evenly distributed than the magnesium salt within the urinary stone, which could be indicative of the fact that the precipitation of ACC occurred on already nucleated struvite particles. Further studies of the nucleation of urinary stones could be relevant to supporting this hypothesis. In any case, our results suggest that amorphous intermediates likely play a role in pathological precipitation of calcium carbonate. The kinetic stabilisation of the ACC against crystallisation in this particular case may be based on the presence of magnesium ions and organic constituents, and/or their combination [51].

Supplementary Materials: The following are available online at http://www.mdpi.com/2075-163X/8/3/84/s1, Figure S1: SEM micrograph with EDS mapping of a large area of the urinary stone of the Guinea Pig, Figure S2: SEM micrograph with EDS mapping of a smaller porous region of the urinary stone of the Guinea Pig, Figure S3: SEM micrograph with EDS mapping of a smaller dense region of the urinary stone of the Guinea Pig.

Acknowledgments: D.G. is a Research Fellow of the Zukunftskolleg of the University of Konstanz and is supported by the Fonds der Chemischen Industrie. N.H. thanks the Institute Excellence Center CODIRECT for funds. We thank Mia Winge for providing the urinary stone. This research did not receive any specific grant from funding agencies in the public, commercial, or not-for-profit sectors.

Author Contributions: D.G., M.O., and N.H. conceived and designed the experiments; D.G. and K.J. performed the experiments; D.G., K.J., and N.H. analyzed the data; D.G., K.J., and N.H. wrote the paper.

Conflicts of Interest: The authors declare no conflict of interest.

References

1. Lowenstam, H.A. *On Biomineralization*; Oxford University Press: New York, NY, USA, 1989.
2. Geyssant, J.; Huwald, E.; Strauch, D. *Calcium Carbonate: From the Cretaceous Period into the 21st Century*, Tegethoff, F.W., Ed.; English Ed.; Birkhäuser Verlag: Basel, Switzerland; Boston, MA, USA, 2001; ISBN 3-7643-6425-4.
3. Fratzl, P. Biomimetic materials research: What can we really learn from nature's structural materials? *J. R. Soc. Interface* **2007**, *4*, 637–642. [CrossRef] [PubMed]
4. Farhadi-Khouzani, M.; Schütz, C.; Durak, G.M.; Fornell, J.; Sort, J.; Salazar-Alvarez, G.; Bergström, L.; Gebauer, D. A CaCO$_3$/nanocellulose-based bioinspired nacre-like material. *J. Mater. Chem. A* **2017**, *5*, 16128–16133. [CrossRef]
5. Gebauer, D. Bio-Inspired Materials Science at Its Best-Flexible Mesocrystals of Calcite. *Angew. Chem. Int. Ed.* **2013**, *52*, 8208–8209. [CrossRef] [PubMed]
6. Natalio, F.; Corrales, T.P.; Panthöfer, M.; Schollmeyer, D.; Lieberwirth, I.; Müller, W.E.G.; Kappl, M.; Butt, H.-J.; Tremel, W. Flexible Minerals: Self-Assembled Calcite Spicules with Extreme Bending Strength. *Science* **2013**, *339*, 1298–1302. [CrossRef] [PubMed]
7. Gebauer, D.; Kellermeier, M.; Gale, J.D.; Bergström, L.; Cölfen, H. Pre-nucleation clusters as solute precursors in crystallisation. *Chem. Soc. Rev.* **2014**, *43*, 2348–2371. [CrossRef] [PubMed]
8. Addadi, L.; Raz, S.; Weiner, S. Taking advantage of disorder: Amorphous calcium carbonate and its roles in biomineralization. *Adv. Mater.* **2003**, *15*, 959–970. [CrossRef]
9. Weiner, S.; Mahamid, J.; Politi, Y.; Ma, Y.; Addadi, L. Overview of the amorphous precursor phase strategy in biomineralization. *Front. Mater. Sci. China* **2009**, *3*, 104–108. [CrossRef]
10. Cartwright, J.H. E.; Checa, A.G.; Gale, J.D.; Gebauer, D.; Sainz-Díaz, C.I. Calcium carbonate polyamorphism and its role in biomineralization: How many amorphous calcium carbonates are there? *Angew. Chem. Int. Ed.* **2012**, *51*, 11960–11970. [CrossRef] [PubMed]
11. Jahnen-Dechent, W.; Heiss, A.; Schäfer, C.; Ketteler, M. Fetuin—A regulation of calcified matrix metabolism. *Circ. Res.* **2011**, *108*, 1494–1509. [CrossRef] [PubMed]
12. Wesson, J.A.; Ward, M.D. Pathological Biomineralization of Kidney Stones. *Elements* **2007**, *3*, 415–421. [CrossRef]
13. Solomonov, I.; Weygand, M.J.; Kjaer, K.; Rapaport, H.; Leiserowitz, L. Trapping Crystal Nucleation of Cholesterol Monohydrate: Relevance to Pathological Crystallization. *Biophys. J.* **2005**, *88*, 1809–1817. [CrossRef] [PubMed]
14. Walton, R.C.; Kavanagh, J.P.; Heywood, B.R.; Rao, P.N. Calcium oxalates grown in human urine under different batch conditions. *J. Cryst. Growth* **2005**, *284*, 517–529. [CrossRef]
15. Long, J.R.; Shaw, W.J.; Stayton, P.S.; Drobny, G.P. Structure and Dynamics of Hydrated Statherin on Hydroxyapatite as Determined by Solid-State NMR. *Biochemistry* **2001**, *40*, 15451–15455. [CrossRef] [PubMed]
16. Jahnen-Dechent, W.; Schäfer, C.; Ketteler, M.; McKee, M. Mineral chaperones: A role for fetuin—A and osteopontin in the inhibition and regression of pathologic calcification. *J. Mol. Med.* **2008**, *86*, 379–389. [CrossRef] [PubMed]
17. Qiu, S.R.; Wierzbicki, A.; Orme, C.A.; Cody, A.M.; Hoyer, J.R.; Nancollas, G.H.; Zepeda, S.; De Yoreo, J.J. Molecular modulation of calcium oxalate crystallization by osteopontin and citrate. *Proc. Natl. Acad. Sci.* **2004**, *101*, 1811–1815. [CrossRef] [PubMed]

18. Luquet, G.; Marin, F. Biomineralisations in crustaceans: Storage strategies. *C. R. Palevol.* **2004**, *3*, 515–534. [CrossRef]
19. Reeder, R.J.; Tang, Y.; Schmidt, M.P.; Kubista, L.M.; Cowan, D.F.; Phillips, B.L. Characterization of Structure in Biogenic Amorphous Calcium Carbonate: Pair Distribution Function and Nuclear Magnetic Resonance Studies of Lobster Gastrolith. *Cryst. Growth Des.* **2013**, *13*, 1905–1914. [CrossRef]
20. Skinner, H.C.W.; Osbaldiston, G.W.; Wilner, A.N. Monohydrocalcite in a Guinea Pig bladder stone, a novel occurrence. *Am. Mineral.* **1977**, *62*, 273–277.
21. Hawkins, M.G.; Ruby, A.L.; Drazenovich, T.L.; Westropp, J.L. Composition and characteristics of urinary calculi from guinea pigs. *J. Am. Vet. Med. Assoc.* **2009**, *234*, 214–220. [CrossRef] [PubMed]
22. Hu, Q.; Nielsen, M.H.; Freeman, C.L.; Hamm, L.M.; Tao, J.; Lee, J.R.I.; Han, T.Y.J.; Becker, U.; Harding, J.H.; Dove, P.M.; De Yoreo, J.J. The thermodynamics of calcite nucleation at organic interfaces: Classical vs. non-classical pathways. *Faraday Discuss.* **2012**, *159*, 509–523. [CrossRef]
23. Nielsen, M.H.; Aloni, S.; De Yoreo, J.J. In situ TEM imaging of CaCO₃ nucleation reveals coexistence of direct and indirect pathways. *Science* **2014**, *345*, 1158–1162. [CrossRef] [PubMed]
24. Andreassen, J.-P.; Beck, R.; Nergaard, M. Biomimetic type morphologies of calcium carbonate grown in absence of additives. *Faraday Discuss.* **2012**, *159*, 247–261. [CrossRef]
25. Raiteri, P.; Gale, J.D. Water is the key to nonclassical nucleation of amorphous calcium carbonate. *J. Am. Chem. Soc.* **2010**, *132*, 17623–17634. [CrossRef] [PubMed]
26. Nassif, N.; Pinna, N.; Gehrke, N.; Antonietti, M.; Jäger, C.; Cölfen, H. Amorphous layer around aragonite platelets in nacre. *Proc. Natl. Acad. Sci.* **2005**, *102*, 12653–12655. [CrossRef] [PubMed]
27. Bazin, D.; Daudon, M.; Combes, C.; Rey, C. Characterization and Some Physicochemical Aspects of Pathological Microcalcifications. *Chem. Rev.* **2012**, *112*, 5092–5120. [CrossRef] [PubMed]
28. Kellermeier, M.; Gebauer, D.; Melero-García, E.; Drechsler, M.; Talmon, Y.; Kienle, L.; Cölfen, H.; García-Ruiz, J.M.; Kunz, W. Colloidal stabilization of calcium carbonate prenucleation clusters with silica. *Adv. Funct. Mater.* **2012**, *22*, 4301–4311. [CrossRef]
29. Gebauer, D.; Cölfen, H. Prenucleation clusters and non-classical nucleation. *Nano. Today* **2011**, *6*, 564–584. [CrossRef]
30. Bichler, K.-H.; Eipper, E.; Naber, K.; Braun, V.; Zimmermann, R.; Lahme, S. Urinary infection stones. *Int. J. Antimicrob. Agents* **2002**, *19*, 488–498. [CrossRef]
31. McLean, R.J.C.; Nickel, J.C.; Cheng, K.-J.; Costerton, J.W.; Banwell, J.G. The Ecology and Pathogenicity of Urease-Producing Bacteria in the Urinary Tract. *CRC Crit. Rev. Microbiol.* **1988**, *16*, 37–79. [CrossRef] [PubMed]
32. Peng, X.; Griffith, J.W.; Lang, C.M. Cystitis, urolithiasis and cystic calculi in ageing guineapigs. *Lab. Anim.* **1990**, *24*, 159–163. [CrossRef] [PubMed]
33. Stefov, V.; Šoptrajanov, B.; Kuzmanovski, I.; Lutz, H.D.; Engelen, B. Infrared and Raman spectra of magnesium ammonium phosphate hexahydrate (struvite) and its isomorphous analogues. III. Spectra of protiated and partially deuterated magnesium ammonium phosphate hexahydrate. *J. Mol. Struct.* **2005**, *752*, 60–67. [CrossRef]
34. Gebauer, D.; Gunawidjaja, P.N.; Ko, J.Y.P.; Bacsik, Z.; Aziz, B.; Liu, L.J.; Hu, Y.F.; Bergström, L.; Tai, C.W.; Sham, T.K.; Edén, M.; Hedin, N. Proto-calcite and proto-vaterite in amorphous calcium carbonates. *Angew. Chem. Int. Ed.* **2010**, *49*, 8889–8891. [CrossRef] [PubMed]
35. Brečević, L.; Nielsen, A.E. Solubility of amorphous calcium carbonate. *J. Cryst. Growth* **1989**, *98*, 504–510. [CrossRef]
36. Aizenberg, J.; Lambert, G.; Addadi, L.; Weiner, S. Stabilization of amorphous calcium carbonate by specialized macromolecules in biological and synthetic precipitates. *Adv. Mater.* **1996**, *8*, 222–226. [CrossRef]
37. Günther, C.; Becker, A.; Wolf, G.; Epple, M. In vitro synthesis and structural characterization of amorphous calcium carbonate. *Z. Für Anorg. Allg. Chem.* **2005**, *631*, 2830–2835. [CrossRef]
38. Kojima, Y.; Kawanobe, A.; Yasue, T.; Arai, Y. Synthesis of amorphous calcium carbonate and its crystallization. *J. Ceram. Soc. Jpn.* **1993**, *101*, 1145–1152. [CrossRef]
39. Lam, R.S.K.; Charnock, J.M.; Lennie, A.; Meldrum, F.C. Synthesis-dependant structural variations in amorphous calcium carbonate. *CrystEngComm* **2007**, *9*, 1226–1236. [CrossRef]
40. Loste, E.; Wilson, R.M.; Seshadri, R.; Meldrum, F.C. The role of magnesium in stabilising amorphous calcium carbonate and controlling calcite morphologies. *J. Cryst. Growth* **2003**, *254*, 206–218. [CrossRef]

41. Vagenas, N.; Gatsouli, A.; Kontoyannis, C.G. Quantitative analysis of synthetic calcium carbonate polymorphs using FT-IR spectroscopy. *Talanta* **2003**, *59*, 831–836. [CrossRef]

42. Gueta, R.; Natan, A.; Addadi, L.; Weiner, S.; Refson, K.; Kronik, L. Local atomic order and infrared spectra of biogenic calcite. *Angew. Chem. Int. Ed.* **2007**, *46*, 291–294. [CrossRef] [PubMed]

43. Coe, F.L. Kidney stone disease. *J. Clin. Invest.* **2005**, *115*, 2598–2608. [CrossRef] [PubMed]

44. Effenberger, H. Kristallstruktur und Infrarot-Absorptionsspektrum von synthetischem Monohydrocalcit, $CaCO_3 \cdot H_2O$. *Monatshefte Für Chem.* **1981**, *112*, 899–909. [CrossRef]

45. Goldschmidt, J.R.; Graf, D.L. Relation between lattice constants and composition of the Ca-Mg carbonates. *Am. Mineral.* **1958**, *43*, 84–101.

46. Tompa, É.; Nyirő-Kósa, I.; Rostási, Á.; Cserny, T.; Pósfai, M. Distribution and composition of Mg-calcite and dolomite in the water and sediments of Lake Balaton. *Cent. Eur. Geol.* **2014**, *57*, 113–136. [CrossRef]

47. Schaber, P.M.; Colson, J.; Higgins, S.; Thielen, D.; Anspach, B.; Brauer, J. Thermal decomposition (pyrolysis) of urea in an open reaction vessel. *Thermochim. Acta* **2004**, *424*, 131–142. [CrossRef]

48. Bhuiyan, M.I.H.; Mavinic, D.S.; Koch, F.A. Thermal decomposition of struvite and its phase transition. *Chemosphere* **2008**, *70*, 1347–1356. [CrossRef] [PubMed]

49. Ronteltap, M.; Maurer, M.; Gujer, W. The behaviour of pharmaceuticals and heavy metals during struvite precipitation in urine. *Water Res.* **2007**, *41*, 1859–1868. [CrossRef] [PubMed]

50. Politi, Y.; Batchelor, D.R.; Zaslansky, P.; Chmelka, B.F.; Weaver, J.C.; Sagi, I.; Weiner, S.; Addadi, L. Role of magnesium ion in the stabilization of biogenic amorphous calcium carbonate: A structure–function investigation. *Chem. Mater.* **2010**, *22*, 161–166. [CrossRef]

51. Wolf, S.L.P.; Jähme, K.; Gebauer, D. Synergy of Mg^{2+} and poly(aspartic acid) in additive-controlled calcium carbonate precipitation. *CrystEngComm* **2015**, *17*, 6857–6862. [CrossRef]

minerals

MDPI

Article

Ammonium-Carbamate-Rich Organogels for the Preparation of Amorphous Calcium Carbonates

Zoltán Bacsik, Peng Zhang and Niklas Hedin *

Department of Materials and Environmental Chemistry, Stockholm University, SE 106 91 Stockholm, Sweden; zoltanb@mmk.su.se (Z.B.); peng.zhang@mmk.su.se (P.Z.)
* Correspondence: niklas.hedin@mmk.su.se; Tel.: +46-8-162417

Received: 31 May 2017; Accepted: 18 June 2017; Published: 27 June 2017

Abstract: Amine-CO_2 chemistry is important for a range of different chemical processes, including carbon dioxide capture. Here, we studied how aspects of this chemistry could be used to prepare calcium carbonates. Chemically crosslinked organogels were first prepared by reacting hyperbranched polyethylene imine (PEI) dissolved in DMSO with carbon dioxide. The crosslinks of the organogel consisted of ammonium-carbamate ion pairs as was shown by IR spectroscopy. These carbamate-rich organogels were subsequently subjected to aqueous solutions of calcium acetate, and amorphous calcium carbonate (ACC) precipitated. The ACC did not crystalize during the mixing for up to 20 h, as was shown by a combination of IR spectroscopy, X-ray diffraction, scanning electron microscopy, and thermal analysis. Some PEI had been included or adsorbed on the ACC particles. Traces of calcite were observed in one sample that had been subjected to water in a work-up procedure.

Keywords: amine; carbon dioxide; ammonium-carbamate ion pairs; organogel; amorphous calcium carbonate (ACC)

---

1. Introduction

The understanding of the details of precipitation and crystallization of calcium carbonates has changed. Now it commonly involves aspects of non-traditional mechanisms of crystallization where clusters and amorphous calcium carbonates (ACCs) are important objects [1]. An enhanced understanding of the formation of calcium carbonates is important as they contribute to the carbon balance globally, and there are, for example, indications that ocean acidification is already affecting the growth of coral reefs negatively [2]. In addition, calcium carbonates are important in various technical applications. For example, calcite can be added to a certain degree in Portland cement [3] and be used as a filler in papers [4]. Calcium carbonates are very important for many organisms in their biomineralized tissues, where the calcium carbonate contributes with a functional toughening [5,6]. ACCs occur in certain biomineralized tissues [5,7–9] but also in synthetic calcium carbonates [10–12] where they have been shown to display both structural and chemical differences [13–15].

Polymers have been shown to moderate the precipitation and crystallization of calcium carbonates by affecting the time scales of aggregation and crystallization, as well as the polymorphism and the morphologies of the particles [16–19]. Specifically, amines and polyamines have been shown to influence the precipitation and crystallization of calcium carbonates by various interactions with carbonate ions, affecting the buffering capacity, and their tendencies to adsorb to carbonate-rich interfaces [20–22].

Amines and polyamines have been studied to a large extent when it comes to their abilities to capture carbon dioxide from flue gas or air [23–26]. The CO_2-amine chemistry is surprisingly rich and involves coupled reaction networks with bicarbonate, carbamates, and carbamic acid moieties [27–29]. In many solvents, ammonium-carbamate ion pairs are the dominant species and form rapidly [30]. From a technological point of view, carbamates are typically preferred over bicarbonates in carbon-capture systems as they form and decompose relatively rapidly [27,31].

For polyamines in selected solvents, organogels have been shown to form when being contacted with $CO_2(g)$. For example, Carretti et al. have derived and studied such organogels that had been prepared from polyallylamine and $CO_2(g)$, and they related the gelation to ammonium-carbamate-based crosslinks [32]. Carretti et al. also extended these studies to include organogels of polyethyleneimine (PEI) induced by reactions with CO_2, and studied these ammonium-carbamate crosslinked gels in the context of art preservation [33,34].

We got inspired by the option to prepare ammonium-carbamate crosslinked organogels of PEI and CO_2 and wanted to investigate if such could be used to prepare calcium carbonates. The case of using such gels as sources of CO_2 was strengthened by the study of Prah et al. [35], who had shown that molecular ammonium carbamates could be used in the crystallization of calcium carbonates using solutions of calcium acetate.

The novelty of this study is that we use polymeric carbamates that were prepared from reactive CO_2 absorption in PEI solutions. It could be relevant to combined CO_2 capture and mineralization. The organogels were prepared from hyperbranched and low molecular weight PEI and CO_2, and we studied aspects of the precipitation of ACC when the organogels were mixed with aqueous solutions of calcium acetate.

2. Materials and Methods

A solution of hyperbranched PEI (Sigma-Aldrich, Saint Louis, MO, USA, CAS number: 25987-06-8, average Mw = 800) in dimethyl sulfoxide (DMSO) was prepared, by adding and mixing 0.3677 g (2.76 mmol) of PEI in 2 mL DMSO. The organogel was prepared by bubbling CO_2 through the PEI solution with a flow rate of 2.3 mL/min for 1 h. Subsequently, the gel was centrifuged to reduce the amount of DMSO. A solution of calcium acetate hydrate (Sigma-Aldrich, CAS number: 114460-21-8) was prepared in water; by adding and mixing 5 g of calcium acetate hydrate in 50 mL of deionized water. 4.6 mL of the calcium acetate solution was added at a flow rate of 0.75 mL/min to the CO_2-induced organogel of PEI using a syringe pump. Such reactive mixtures were stirred (500 rpm) for 1 h (samples SW and SE) or 20 h (samples LW and LE) at room temperature. The formed solids were filtered off and washed carefully with 5 mL of deionized water and 2 × 5 mL of ethanol (samples SW and LW) or with 3 × 5 mL of ethanol (samples SE and LE). The white solids were dried at room temperature. The sample nomenclature used is: S for short mixing and L for long mixing; W for washed with water and E for washed with ethanol.

The samples were studied with infrared (IR) spectroscopy using a Varian 670-IR spectrometer (Varian, Mulgrave, Australia) equipped with a single reflection ATR device (Specac, Orpington, UK) with a diamond ATR element. X-ray diffraction was used to study the crystallinity of the solids formed using a XPERT-PRO PANalytical powder diffractometer (PANalytical B.V., Almelo, The Netherlands) (reflection mode with an X'Celerator detector (Cu Kα1 radiation, λ = 1.5418 Å) from 5.0° to 85.0° (2θ). The step size was set to 0.0131°, and the step time was 0.0104 °/min, which resulted in a total accumulation time of 126 min. Scanning electron microscopy (SEM) was used to study the morphologies of the solids using a JEOL JSM-7401F scanning electron microscope (JEOL USA Inc., Peabody, MA, USA) that operated at a voltage of 1 kV. The amount of organics in the solids was estimated by thermal analysis using a TA Instruments gravimetric thermal analyzer. The samples were subjected to an air flow 20 mL/min, the temperature range was 25–900 °C, and the temperature rate was 5 °C/min.

3. Results and Discussion

Gels formed rapidly when CO_2 was bubbled through solutions of PEI in DMSO, and Figure 1 displays a corresponding image. For the gelation, we presumed that all the amine groups had reacted with CO_2 forming ammonium-carbamate ion pairs following the stoichiometry of amine: CO_2 = 2:1. In analogy to the study of Carretti et al. [32], we hypothesized that chemical crosslinks had formed among the PEI chains when the amino groups had reacted with CO_2. This hypothesis of chemical crosslinks with alkylammonium-alkylcarbamate ion pairs was supported by IR analyzes.

Figure 1. Photograph of the organogel formed by subjecting a solution of polyethylene imine in DMSO to CO_2.

Figure 2 shows the IR spectra of the organogel and the corresponding PEI. Alkylammonium-alkylcarbamate ion pairs were detected by analyzing the IR spectrum of the organogel. In this spectrum, presented in Figure 2a, the broad band at frequencies between 3350 and 2000 cm^{-1} and the bands at 1636 cm^{-1} and 1465 cm^{-1} are typical for ammonium groups, and the band at a frequency of 1565 cm^{-1} is a clear signature of the C=O group of carbamates [36]. Bands were also observed for DMSO, of which the one with the highest intensity appeared at a frequency of 1018 cm^{-1}.

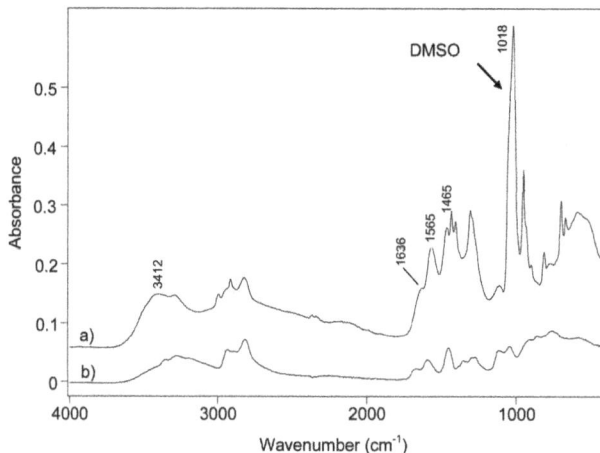

Figure 2. IR spectra of the (**a**) polyethylene imine (PEI)/ammonium-carbamate/DMSO organogel and (**b**) PEI.

The organogels were centrifuged to reduce the amount of DMSO. Subsequently, aqueous solutions of calcium acetate were added and white solid precipitates formed, see the reactions in Scheme 1. We added a stoichiometric amount of Ca^{2+} ions, corresponding to a 1:1 ratio of Ca^{2+}:carbamate. For this addition, we assumed that all of the amine groups of the PEI had formed ammonium-carbamates giving an N–carbamate ratio of 2:1. Two different groups of samples were prepared and subjected to short (samples SE and SW) and long (samples LE and LW) mixing after the calcium acetate solutions had been added.

In a control experiment when only water had been added to the organogel, the gel disappeared, which was expected by the instability of ammonium-carbamate ion pairs in aqueous solutions. Bicarbonate moieties tend to form on addition of water, and the PEI is expected to protonated and soluble in water.

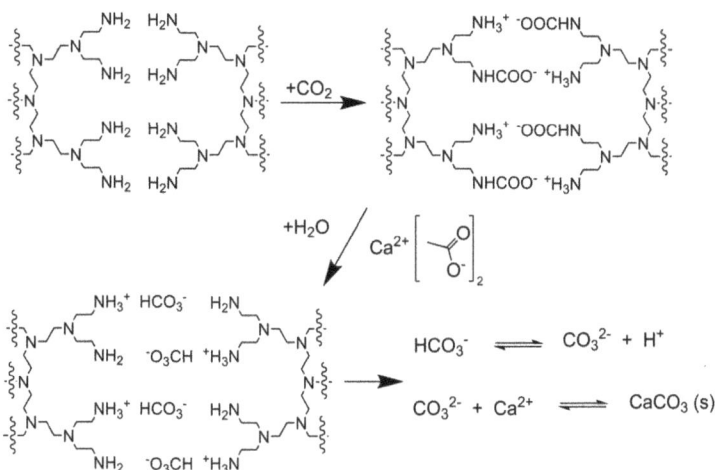

Scheme 1. Chemical reactions in the synthesis of amorphous calcium carbonate by bubbling CO_2 into solution of PEI and subjecting the formed organogel (PEI–ammonium-carbamate) to aqueous solutions of calcium acetate.

The solid precipitates were either washed in ethanol (samples SE and LE) or washed in water and ethanol (samples SW and LW). These experiments were performed to assess eventual contributions of water washing on the crystallization of calcium carbonate. The samples were studied with several methods. The XRD patterns revealed that the samples SE, SW, and LE were amorphous, and that the sample LW (20 h of mixing and water washing) contained a minor fraction of calcite with broad diffraction lines. Figure 3 shows the diffractograms of the samples, and the low signal-to-noise levels related to the amorphous nature of the samples.

The XRD diffraction patterns in Figure 3 indicated the absence of crystallinity but could not easily be used to detect ACCs, although the wide-angle scattering in the patterns is typical for ACC. IR spectroscopy, however, can be very useful for studying ACCs [14,37]. The IR spectra of samples SE, SW, LE, and LW are shown in Figure 4. The spectra clearly showed that the calcium carbonates in samples SE, SW, and LE consisted of fully of ACC, and mainly of ACC in sample LW. The bands observed at a frequency of 1380 cm^{-1}, with a shoulder at 1465 cm^{-1}, belong to the asymmetric stretching of carbonates. The band at a frequency of 1071 cm^{-1} corresponds to the symmetric stretching, and the band at 861 cm^{-1} to the out-of-plane bending modes. These frequency values were typical for ACCs. The band for the in-plane bending mode of the carbonate ions were observed at frequencies of around 700 cm^{-1}. This band is the most useful to distinguish ACCs and various polymorphs of crystalline calcium carbonates. For the ACCs of this study, a broad doublet was observed at a frequency of

~700 cm^{-1}. In the spectrum of sample LW, a sharp band also appeared at 711 cm^{-1}, which is typical for calcite and corroborated the findings from XRD of a small fraction of calcite in this sample. IR bands of the PEI were observed in the C-H stretching region at frequencies of 2800–3000 cm^{-1}.

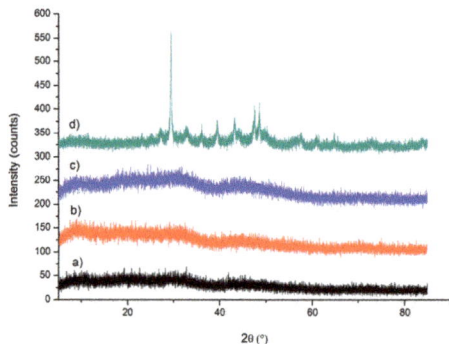

Figure 3. X-ray diffraction patterns of the amorphous calcium-carbonate-rich samples (**a**) SE, (**b**) SW, (**c**) LE, and (**d**) LW that had been precipitated when subjecting organogels (PEI–ammonium-carbamate) to aqueous solutions of calcium acetate. S and L stand for mixing for 1 h and 20 h; E and W stand for ethanol and water-and-ethanol washing.

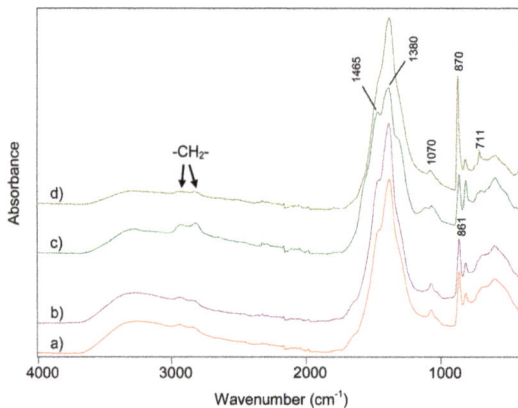

Figure 4. IR spectra of the amorphous-calcium-carbonate-rich samples (**a**) SE, (**b**) SW, (**c**) LE, and (**d**) LW. These samples had precipitated when subjecting the organogels (PEI–ammonium-carbamate) to aqueous solutions of calcium acetates. S and L stand for mixing for 1 h and 20 h; E and W stand for ethanol and water-and-ethanol washing.

The morphologies of the particles were studied by SEM, and the features observed in the images supported the amorphous nature of samples. Figure 5 shows SEM images of the ACC-rich samples of SE and LE and note that spherical particles are typically observed for ACC [14,38–40].

The amounts of amines in samples SE, SW, LE, and LW, were determined by thermogravimetric (TG) experiments. Figure 6 shows the TG curves, and they were surprisingly different. All curves had mass losses in three major temperature regions. The mass loss in the temperature range between room temperature and 200 °C was attributed to water loss. The second region of mass losses occurred at temperatures between 200 and 450 °C. These losses were attributed to the combustion of PEI. Sample SE (red), SW (black), and LW (green) had similar trends in this region, but sample LE (blue) had a significantly different tendency with a very significant step of mass loss related to combustion at a temperature of 370 °C. In the third regions, there were defined mass loss steps related to the decomposition of the carbonate groups (calcination) in all samples. From the mass losses observed in this region, the amounts of calcium carbonate were estimated. The amount of calcium carbonate in the samples SE, SW, and LE were similar and about 50 wt %, and sample LW consisted of ~70 wt % calcium carbonate. We speculate that the higher amount of $CaCO_3$ in sample LW can be explained by the calcite formation and consequently a more effective washing of the sample. Considering that the amount of water appeared to be around 5 wt % in each sample, the amount of polymer and solvent in the samples was between 25 and 45 wt %. In this context, note that the molecular formula for ACC is typically observed to be $CaCO_3 \cdot xH_2O$; however, the significant amount of organics in the samples made it very difficult to robustly establish the composition of the ACC in the studied samples.

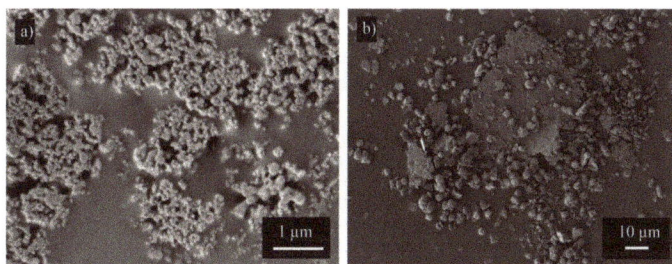

Figure 5. SEM image of the amorphous-calcium-carbonate-rich samples (**a**) SE and (**b**) LE. They had been precipitated when subjecting an organogel (PEI–ammonium-carbamate) to an aqueous solution of calcium acetate.

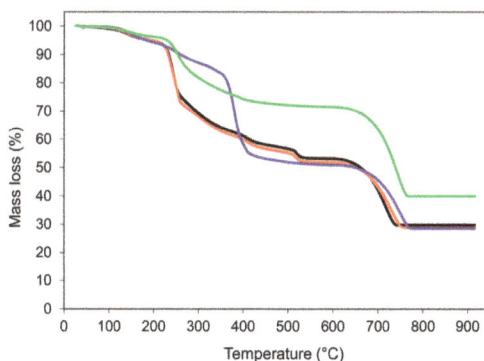

Figure 6. Thermogravimetric curves of the amorphous-calcium-carbonate-rich samples SE (red), SW (black), LE (blue), and LW (green). These samples had precipitated when subjecting organogels (PEI–ammonium-carbamate) to aqueous solutions of calcium acetates. S and L stand for mixing for 1 h and 20 h; E and W stand for ethanol and water-and-ethanol washing.

4. Conclusions

Ammonium-carbamate-rich organogels were successfully used to prepare ACC. It is our understanding that the ammonium-carbamate ion pairs are coupled chemically to bicarbonates in the aqueous calcium acetate solution used for the precipitation of ACC. The details of this chemistry were not studied. A natural extension of this study could be to study this chemistry in some detail. ACC was stabilized at the two mixing times used in this study (1 h and 20 h), the extent of this stabilization is relevant for further studies as well the interfacial details of the PEI–ACC interactions.

Acknowledgments: Swedish Research Council (2016-03568) and the Carl Trygger Foundation are thanked.

Author Contributions: Z.B. and N.H. conceived and designed the experiments, analyzed data, and wrote the paper; P.Z. performed most of the experiments and analyzed data. Z.B. performed the IR experiments.

Conflicts of Interest: The authors declare no conflict of interest.

References

1. Gebauer, D.; Kellermeier, M.; Gale, J.D.; Bergström, L.; Cölfen, H. Pre-nucleation clusters as solute precursors in crystallisation. *Chem. Soc. Rev.* **2014**, *43*, 2348–2371. [CrossRef] [PubMed]

2. Albright, R.; Caldeira, L.; Hosfelt, J.; Kwiatkowski, L.; Maclaren, J.K.; Mason, B.M.; Nebuchina, Y.; Ninokawa, A.; Pongratz, J.; Ricke, K.L.; et al. Reversal of ocean acidification enhances net coral reef calcification. *Nature* **2016**, *531*, 362–365. [CrossRef] [PubMed]

3. Matschei, T.; Lothenbach, B.; Glasser, F.P. The role of calcium carbonate in cement hydration. *Cem. Concr. Res.* **2007**, *37*, 551–558. [CrossRef]

4. Tegethoff, F.W. *Calcium Carbonate—From the Cretaceous Period into the 21st Century*, 1st ed.; Springer: Basel, Switzerland, 2001.

5. Politi, Y.; Arad, T.; Klein, E.; Weiner, S.; Addadi, L. Sea Urchin Spine Calcite Forms via a Transient Amorphous Calcium Carbonate Phase. *Science* **2004**, *306*, 1161–1164. [CrossRef] [PubMed]

6. Rodriguez-Navarro, C.; Burgos Cara, A.; Elert, K.; Putnis, C.V.; Ruiz-Agudo, E. Direct Nanoscale Imaging Reveals the Growth of Calcite Crystals via Amorphous Nanoparticles. *Cryst. Growth Des.* **2016**, *16*, 1850–1860. [CrossRef]

7. Gayathri, S.; Lakshminarayanan, R.; Weaver, J.C.; Morse, D.E.; Kini, R.M.; Valiyaveettil, S. In Vitro Study of Magnesium-Calcite Biomineralization in the Skeletal Materials of the Seastar Pisaster giganteus. *Chem. Eur. J.* **2007**, *13*, 3262–3268. [CrossRef] [PubMed]

8. Wilt, F.H. Developmental biology meets materials science: Morphogenesis of biomineralized structures. *Dev. Biol.* **2005**, *280*, 15–25. [CrossRef] [PubMed]

9. Lowenstam, H.A. Minerals formed by organisms. *Science* **1981**, *211*, 1126–1131. [CrossRef] [PubMed]

10. Brečević, L.; Nielsen, A.E. Solubility of amorphous calcium carbonate. *J. Cryst. Growth* **1989**, *98*, 504–510. [CrossRef]

11. Reddy, M.M.; Nancollas, G.H. The crystallization of calcium carbonate. *J. Cryst. Growth* **1976**, *35*, 33–38. [CrossRef]

12. Nakahara, Y.; Tazawa, T.; Miyata, K. Properties of Calcium Carbonate Prepared by Interfacial Reaction Method. *Nippon Kagaku Kaishi* **1976**, *5*, 732–736. [CrossRef]

13. Sun, S.; Chevrier, D.M.; Zhang, P.; Gebauer, D.; Cölfen, H. Distinct Short-Range Order Is Inherent to Small Amorphous Calcium Carbonate Clusters (<2 nm). *Angew. Chem. Int. Ed.* **2016**, *55*, 12206–12209. [CrossRef] [PubMed]

14. Gebauer, D.; Gunawidjaja, P.N.; Ko, J.Y.P.; Bacsik, Z.; Aziz, B.; Liu, L.; Hu, Y.; Bergström, L.; Tai, C.-W.; Sham, T.-K.; et al. Proto-Calcite and Proto-Vaterite in Amorphous Calcium Carbonates. *Angew. Chem. Int. Ed.* **2010**, *49*, 8889–8891. [CrossRef] [PubMed]

15. Farhadi-Khouzani, M.; Chevrier, D.M.; Zhang, P.; Hedin, N.; Gebauer, D. Water as the Key to Proto-Aragonite Amorphous $CaCO_3$. *Angew. Chem. Int. Ed.* **2016**, *55*, 8117–8120. [CrossRef] [PubMed]

16. Gower, L.B.; Odom, D.J. Deposition of calcium carbonate films by a polymer-induced liquid-precursor (PILP) process. *J. Cryst. Growth* **2000**, *210*, 719–734. [CrossRef]

17. Xu, A.-W.; Antonietti, M.; Yu, S.-H.; Cölfen, H. Polymer-Mediated Mineralization and Self-Similar Mesoscale-Organized Calcium Carbonate with Unusual Superstructures. *Adv. Mater.* **2008**, *20*, 1333–1338. [CrossRef]

18. Kim, Y.-Y.; Schenk, A.S.; Ihli, J.; Kulak, A.N.; Hetherington, N.B.J.; Tang, C.C.; Schmahl, W.W.; Griesshaber, E.; Hyett, G.; Meldrum, F.C. A critical analysis of calcium carbonate mesocrystals. *Nat. Commun.* **2014**, *5*, 4341. [CrossRef] [PubMed]

19. Abebe, M.; Hedin, N.; Bacsik, Z. Spherical and Porous Particles of Calcium Carbonate Synthesized with Food Friendly Polymer Additives. *Cryst. Growth Des.* **2015**, *15*, 3609–3616. [CrossRef]

20. Schenk, A.S.; Cantaert, B.; Kim, Y.-Y.; Li, Y.; Read, E.S.; Semsarilar, M.; Armes, S.P.; Meldrum, F.C. Systematic Study of the Effects of Polyamines on Calcium Carbonate Precipitation. *Chem. Mater.* **2014**, *26*, 2703–2711. [CrossRef]

21. Yasumoto, K.; Yasumoto-Hirose, M.; Yasumoto, J.; Murata, R.; Sato, S.; Baba, M.; Mori-Yasumoto, K.; Jimbo, M.; Oshima, Y.; Kusumi, T.; et al. Biogenic Polyamines Capture CO_2 and Accelerate Extracellular Bacterial $CaCO_3$ Formation. *Mar. Biotechnol.* **2014**, *16*, 465–474. [CrossRef] [PubMed]

22. Vinoba, M.; Bhagiyalakshmi, M.; Grace, A.N.; Chu, D.H.; Nam, S.C.; Yoon, Y.; Yoon, S.H.; Jeong, S.K. CO_2 Absorption and Sequestration as Various Polymorphs of $CaCO_3$ Using Sterically Hindered Amine. *Langmuir* **2013**, *29*, 15655–15663. [CrossRef] [PubMed]

23. Rochelle, G.T. Amine Scrubbing for CO_2 Capture. *Science* **2009**, *325*, 1652–1654. [CrossRef] [PubMed]

24. Xu, X.C.; Song, C.S.; Andresen, J.M.; Miller, B.G.; Scaroni, A.W. Novel polyethylenimine-modified mesoporous molecular sieve of MCM-41 type as high-capacity adsorbent for CO_2 capture. *Energy Fuels* **2002**, *16*, 1463–1469. [CrossRef]

25. Wang, X.; Ma, X.; Schwartz, V.; Clark, J.C.; Overbury, S.H.; Zhao, S.; Xu, X.; Song, C. A solid molecular basket sorbent for CO_2 capture from gas streams with low CO_2 concentration under ambient conditions. *Phys. Chem. Chem. Phys.* **2012**, *14*, 1485–1492. [CrossRef] [PubMed]

26. Bacsik, Z.; Atluri, R.; Garcia-Bennett, A.E.; Hedin, N. Temperature-Induced Uptake of CO_2 and Formation of Carbamates in Mesocaged Silica Modified with n-Propylamines. *Langmuir* **2010**, *26*, 10013–10024. [CrossRef] [PubMed]

27. Bacsik, Z.; Ahlsten, N.; Ziadi, A.; Zhao, G.; Garcia-Bennett, A.E.; Martín-Matute, B.; Hedin, N. Mechanisms and kinetics for sorption of CO_2 on bicontinuous mesoporous silica modified with n-propylamine. *Langmuir* **2011**, *27*, 11118–11128. [CrossRef] [PubMed]

28. Danon, A.; Stair, P.C.; Weitz, E. FTIR Study of CO_2 Adsorption on Amine-Grafted SBA-15: Elucidation of Adsorbed Species. *J. Phys. Chem. C* **2011**, *115*, 11540–11549. [CrossRef]

29. Mafra, L.; Čendak, T.; Schneider, S.; Wiper, P.V.; Pires, J.; Gomes, J.R.B.; Pinto, M.L. Structure of Chemisorbed CO_2 Species in Amine-Functionalized Mesoporous Silicas Studied by Solid-State NMR and Computer Modeling. *J. Am. Chem. Soc.* **2017**, *139*, 389–408. [CrossRef] [PubMed]

30. Masuda, K.; Ito, Y.; Horiguchi, M.; Fujita, H. Studies on the solvent dependence of the carbamic acid formation from ω-(1-naphthyl)alkylamines and carbon dioxide. *Tetrahedron* **2005**, *61*, 213–229. [CrossRef]

31. Choi, S.; Drese, J.H.; Jones, C.W. Adsorbent Materials for Carbon Dioxide Capture from Large Anthropogenic Point Sources. *ChemSusChem* **2009**, *2*, 796–854. [CrossRef] [PubMed]

32. Carretti, E.; Dei, L.; Baglioni, P.; Weiss, R.G. Synthesis and Characterization of Gels from Polyallylamine and Carbon Dioxide as Gellant. *J. Am. Chem. Soc.* **2003**, *125*, 5121–5129. [CrossRef] [PubMed]

33. Carretti, E.; Dei, L.; Weiss, R.G.; Baglioni, P. A new class of gels for the conservation of painted surfaces. *J. Cult. Herit.* **2008**, *9*, 386–393. [CrossRef]

34. Carretti, E.; Bonini, M.; Dei, L.; Berrie, B.H.; Angelova, L.V.; Baglioni, P.; Weiss, R.G. New Frontiers in Materials Science for Art Conservation: Responsive Gels and Beyond. *Acc. Chem. Res.* **2010**, *43*, 751–760. [CrossRef] [PubMed]

35. Prah, J.; Maček, J.; Dražič, G. Precipitation of calcium carbonate from a calcium acetate and ammonium carbamate batch system. *J. Cryst. Growth* **2011**, *324*, 229–234. [CrossRef]

36. Bacsik, Z.; Hedin, N. Effects of carbon dioxide captured from ambient air on the infrared spectra of supported amines. *Vib. Spectrosc.* **2016**, *87*, 215–221. [CrossRef]

37. Andersen, F.A.; Brečević, L. Infrared Spectra of Amorphous and Crystalline Calcium Carbonate. *Acta Chem. Scand.* **1991**, *45*, 1018–1024. [CrossRef]

38. Zou, Z.; Bertinetti, L.; Politi, Y.; Jensen, A.C.S.; Weiner, S.; Addadi, L.; Fratzl, P.; Habraken, W.J.E.M. Opposite Particle Size Effect on Amorphous Calcium Carbonate Crystallization in Water and during Heating in Air. *Chem. Mater.* **2015**, *27*, 4237–4246. [CrossRef]

39. Gorna, K.; Hund, M.; Vučak, M.; Gröhn, F.; Wegner, G. Amorphous calcium carbonate in form of spherical nanosized particles and its application as fillers for polymers. *Mater. Sci. Eng. A* **2008**, *477*, 217–225. [CrossRef]

40. Foran, E.; Weiner, S.; Fine, M. Biogenic Fish-gut Calcium Carbonate is a Stable Amorphous Phase in the Gilt-head Seabream, Sparus aurata. *Sci. Rep.* **2013**, *3*, 1700. [CrossRef] [PubMed]

MDPI

St. Alban-Anlage 66

4052 Basel

Switzerland

Tel. +41 61 683 77 34

Fax +41 61 302 89 18

www.mdpi.com

Minerals Editorial Office

E-mail: minerals@mdpi.com

www.mdpi.com/journal/minerals